KUNST UND ZETTEL IM MESSER

Kunst und Zettel im Messer

Bavarian State Library
Cgm 582

Hans Lecküchner

Robert Brunner • Daniel Burger • Casper J. van Dijk
Olivier Dupuis • Jessica Finley • Adam Franti • Falko Fritz
Dierk Hagedorn • Johann Heim • Alexander Kiermayer
Carsten Lorbeer • Julia Lorbeer • Oskar ter Mors

Edited by Michael Chidester

HEMA Bookshelf

This book is published as a companion to the facsimile of the Bayerische Staatsbibliothek's manuscript Cgm 582 produced by HEMA Bookshelf.

The transcription in this book first appeared in 2006 on the website *Gesellschaft für Pragmatische Schriftlichkeit.* http://www.pragmatische-schriftlichkeit.de/cgm582.html

Published by HEMA Bookshelf, LLC.
411a Highland Ave #141
Somerville, MA, 02144
www.hemabookshelf.com

HEMA Bookshelf endeavors to respect the copyright in a manner consistent with its educational mission. If you believe any material has been included in this publication improperly, please contact us.

Version 1.0, 2021

ISBN 978-1-953683-17-5

Typeset in Libertinus Serif Display and Libertinus Sans, which are used under the Open Font License http://libertine-fonts.org/
German text typeset in Humboldt Fraktur, used under the 1001Fonts Free For Commercial Use License https://www.1001fonts.com/licenses/ffc.html

Printed by Lightning Source.

Contents

Manuscript Abbreviations

This is not a comprehensive list of manuscript fencing treatises, just those referenced in this book.

A *Codex Amberger*
Towson, Sammlung Amberger

AD *Albrecht Dürer*
Wien, Albertina (Graphische Sammlung), Hs. 26-232

ADL *Albrecht Dürer-London*
London, British Library, Sloane MS No. 5229

AR *Anton Rast* (AR)
Augsburg, Stadtarchiv, Reichsstadt Schätze Nr. 82

B *Baumann's fight book*
Augsburg, Universitätsbibliothek, Cod.I.6.4°.2

F1 *Tower/Walpurgis Fechtbuch*
Leeds, Royal Armouries, FECHT 1 or MS I.33

FB *Fechtbüchlein*
Wolfenbüttel, Herzog August-Bibliothek, Cod. Guelf. 1074 Novi

G *Goliath*
Kraków, Biblioteka Jagiellonska, Berol Ms. Germ. Qu. 2020

GE *Gregor Erhart*
Glasgow, R. L. Scott Collection, E.1939.65.354

GK *Gladiatoria-Kraków*
Kraków, Biblioteka Jagiellonska, Berol Ms. Germ. Qu. 16

GN *Gladiatoria-New Haven*
New Haven, Yale Center for British Art, U860 .F46 1450

HF *Hans Folz*
Weimar, Q 566

HM *Hans Medel* or *Sigmund Schining*
Augsburg, Universitätsbibliothek, Cod.I.6.2°.5

HS *Hans von Speyer*
Salzburg, Universitätsbibliothek, M.I.29

HTB *Hans Talhoffer-Berlin*
Berlin, Kuperstichkabinett der Stiftung Preußischer Kulturbesitz, 78 A 15

HTG *Hans Talhoffer-Gotha*
Gotha, Forschungsbibliothek Schloss Friedenstein, Chart. A.558

HTK *Hans Talhoffer-København*
København, Det Kongelige Bibliotek, Thott 0290 2°

HTKa *Hans Talhoffer-Kassel*
Kassel, Landesbibliothek, 2° Ms. iurid. 29

HTM *Hans Talhoffer-München*
München, Bayerische Staatsbibliothek, Cod. icon. 394a

HTM2 *Hans Talhoffer-München #2*
München, Bayerische Staatsbibliothek, Cod. icon. 394

HW *Hugo Wittenwiler*
München, Bayerische Staatsbibliothek, Cgm 558

JML *Joachim Meyer-Lund*
Lund, Universitetsbibliotek, MS A.4°.2

JMM *Joachim Meyer-München*
München, Bayerisches Nationalmuseum, MS Bibl. 2465

JWA1 *Jörg Wilhalm-Augsburg #1* or *Lienhart Sollinger-Augsburg*
Augsburg, Universitätsbibliothek, Cod.I.6.2°.2

JWA2 *Jörg Wilhalm-Augsburg #2*
Augsburg, Universitätsbibliothek, Cod.I.6.2°.3

JWA3 *Jörg Wilhalm-Augsburg #3*
Augsburg, Universitätsbibliothek, Cod.I.6.4°.5

JWM1 *Jörg Wilhalm-München #1*
München, Bayerische Staatsbibliothek, Cgm 3711

JWM2 *Jörg Wilhalm-München #2* or *Lienhart Sollinger-München*
München, Bayerische Staatsbibliothek, Cgm 3712

JWW *Jörg Wilhalm-Wolfenbüttel* or *Lienhart Sollinger-Wolfenbüttel*
Wolfenbüttel, Herzog August-Bibliothek, Cod. Guelf. 38.21 Aug. 2°

L *Lew fight book*
Augsburg, Universitätsbibliothek, Cod.I.6.4°.3

LH *Hans Lecküchner-Heidelberg*
Heidelberg, Universitätsbibliothek, Cod. Pal. germ. 430

LM *Hans Lecküchner-München*
München, Bayerische Staatsbibliothek, Cgm 582

LQ *Liber Quodlibetarius*
Erlangen, Universitätsbibliothek, Ms. B 200

K *Kölner Fechtregeln*
Köln, Historisches Archiv, Best. 7020

N *Nürnberger Hausbuch* or *Nicolas Pol fight book*
Nürnberg, Germanisches Nationalmuseum, Cod. Hs. 3227a

P *Paris fight book*
Paris, Musée National du Moyen Age, CL23842

PD	*Peter von Danzig fight book*	Rome, Cod. 44.A.8
PF	*Peter Falkner*	Wien, Kunsthistorisches Museum, KK 5012
PKB	*Paul Kal-Bologna*	Bologna, Universitätsbibliothek, Ms. 1825
PKG	*Paul Kal-Gotha*	Gotha, Forschungsbibliothek Schloss Friedenstein, Chart. B.1021
PKM	*Paul Kal-München*	München, Bayerische Staatsbibliothek, Cgm 1507
PKS	*Solothurner Fechtbuch*	Solothurn, S554
PKW	*Paul Kal-Wien*	Wien, Kunsthistorisches Museum, KK 5126
PMA	*Paul Hektor Mair-Augsburg*	Augsburg, Universitätsbibliothek, Cod.I.6.2°.4
PMD	*Paul Hektor Mair-Dresden*	Dresden, Sächsische Landesbibliothek, Mscr. Dresd. C.93/94
PMM	*Paul Hektor Mair-München*	München, Bayerische Staatsbibliothek, Cod. icon. 393
PMW	*Paul Hektor Mair-Wien*	Wien, Österreichische Nationalbibliothek, Cod. 10825/6
SE	*Sigmund Emring* or *Glasgow Fechtbuch*	Glasgow, R. L. Scott Collection, E.1939.65.341
SK	*Shermkunst*	Chicago, Newberry Library, MS folio U 423 .792
SR	*Sigmund Ainringck fight book*	Dresden, Sächsische Landesbibliothek, Mscr. Dresd. C.487
W01	*Wolfenbüttel fight book #1*	Wolfenbüttel, Herzog August-Bibliothek, Cod. Guelf. 83.4 Aug. 8°

This table is modified from a manuscript cataloging system developed by Dierk HAGEDORN. For more information about some of these, see his article in this book, "Many magnificent Messer manuscripts (plus plentiful picturesque prints)".

Figures and Tables

A Comparative Analysis of the Nomenclature of Johannes Lecküchner and Johannes Liechtenauer

Many Magnificent Messer Manuscripts

Fighting with Long Knife for Leisure or Self-Defense

The Martial Arts Tradition of a Fifteenth Century Bavarian Priest

Art and Symbolism in the Genre of *Fechtbücher*

Acknowledgements

I would like to thank Dr. Wolfgang-Valentin Ikas and the Bavarian State Library for agreeing to license the manuscript scans to us; Tracy Seyfert and the artisans at Grimm Book Bindery who produced the beautiful facsimile itself; and the contributing authors whose work is found in this companion volume. Martin Enzi helped correct digital artifacts in the scans that were beyond my abilities, and Carrie Patrick provided an invaluable outside editor's eye. Most importantly, I thank the many supporters who took a leap of faith and contributed money to this project with no guarantee that it would produce anything worthwhile.

Facsimile Project Supporters

Christian Albrecht • AGEA Editora • Scott Aldinger • Nicholas Allen • Michael Allenson • José António Alves da Cunha Coutinho • Gabriele Ammann • Kelly Anderton • Grant Arnold • Gabriel Asatryan • Athena School of Arms • Cameron Atkinson • Saskatoon Bagua • Ross Bailey • Flora Bajolet • Jack Baker • Bernat Bardagil Mas • Volker Baringer • Jonathan Bartlett • John Bauschatz • Michael Bazar • Tim Beerens • Thomas Belloma • Muli Ben-Yehuda • Simon Berg • Joe Berry • Christopher Bertell • Thomas Biebauw • Hildo Biersma • Florian Binder • Erwan Bineau • Charel Bodé • Leah Bonser • Doug Bostic • Tomas Brosnan • William Brown • Shane Brown • Kendra Brown • Daniel Burger • Christoph Busche • Charles Buschmann • Chris Castle • J. Ceirante • Pierre-Alexandre Chaize • Stephen P. Cheney • Michael Chidester • Laura Chidester • D. G. Church • Samuel Ciortea • Nick Clark • Dustin M. Collier • Benjamin Conan • Peter Concannon • Christy R. Conley • Josué Coufleau • Conner Craig • Myles Cupp • David D'Antonio • James Darling • Dean Davidson • De Taille et d'Estoc • James T. DeVito, Jr. • I. D. van Dijk-Struik • Florian Docter • Bas Doeksen • Mark Driggs • Eric Dussel • Matthieu Dutheil • Caius Ehmke • Martin Enzi • Anna Ermakova • Heikki Europaeus • Martin Fabian • Jessica Finley • Max Fishman • Dean Flagg • Peter Frank • Philip L. Frank • Chris J. Franklin • Elizabeth Free • Falko Fritz • Kevin Frost • Anthony Garnier • Nathan Gede • Rolf Geissbühler • Paul Geyer • Glenn M. Gibeson • David Girard • Joseph Giuliano • Koen van Gorp • Nathan Grepares • Kyle Griswold • Dion Groot • Daniel Gutiérrez Martínez • Richard Harvey • Chad Healey • Ethan Heilman • G. Helleman • Charlotte Herbert • Alexander Hereford Lewis • Aldo Hernandez • Stijn van Hijfte • Aaron Himmler • Roger Hobden • Mark Holgate • Annie Holmes • Steven Hradsky • Mairi Hunter • J. T. Hutchison • Antti Ijäs • Allison Jacobsen • Jason James • Oliver Janseps • Emily Jennings • Alex Jonischkies • Ronneberger Julian • Peter Kahle • Wolfgang Kainbacher-Ott • Alexander Kalywas • Christoph

Karpe • Chris Kerr • Leelund Kim • Don Kindsvatter • Holger Klasen • Anthony Klon • Hugh T. Knight • Dylan Knowles • Arne Koets • Ryan Kohler • Matthys Kool • Alex Kotarakos • José Juan Laguarda Pradas • Brent Lambell • L. Lee • Craig Ligon • Joseph Lilly • Charles Lin • Joakim Linde • London Longsword • Chalin Lukas • Magnus Lundborg • Julian Maddox • Jan-Willem Maessen • Tim Magnuson • Łukasz Majewski • Matti Mäki-Petäys • Sophie Marshall • Carey Martell • Gilles Martinez • Eric Mauer • Clemens Mayer • Vero McMillan • R. Meister • Louis Melancon • Emanuel Meyer • Jerrod Miller • Russ Mitchel • Andrew Mohn • Matthew Mole • Eliot Mook • D. A. Morineau • Oskar ter Mors • P. Muir • Samuel Munilla • Kevin Murakoshi • Derek Nash • Keith Nelson • Walter Neubauer • Noble Science Academy • Joel Norman • Jacob Norwood • Maxim Nossevitch • J. P. Olivian • Benjamin J. Olson • Scarlett Ord • Enric Ortuno • Christian Ouellet • Palm Beach Sword School • Julius Pedersen • Catarina Encarnação Pereira • Patrick Perreault • Jennifer Perry • Arjan Peters • Stefan Peterson • Julia Plöderl • Gregory Poulos • Bryce S. Powell • Riitta Pulkkinen • Dean Pye • Nancy D. Reimers • Ashton Ricks • Paul Rimell • Anthony Rischard • Thomas Riviere • Frédéric Robert • Daniel Rogers • Eetu Röpelinen • John Rothe • Nathan Rowe • James Samuel • Berry Schmaal • Marik Schoenke • Barbara Schönfeld • Thomas Schratwieser • Matthias Schrott • Till Schultz • A. J. Sedlmair • Henry Sharum • Adam W. Sheppard • Trevor Sinz • Edward Sleight, Jr. • Joe Smart • Mark Smead • Jason Smith • Marcin Smolka • Richard Spellman • L. R. Spivack • S. L. Stocki • Leopold Stoffels • Ottawa Swordplay • Thomas Sznigir • Joeli Takala • J. W. Tamplin • Tim Teino • Craig Templeman • Ian Terry • Christian Tobler • Elizabeth Townshend • Dan True • Charlie Underwood • Kirby Urbanek • Burak Urgancioglu • Christopher Valli • Ashley Vogt • Collin Vredenburg • Bartlomiej Walczak • Christopher Walsh • Roland Warzecha • Ken Weber • Jesse Whitfield • Daniel Wickstrom • Paul-Michael Wiedow • Michael Richard Wilcox • Sebastian Wilhelmsson • Guy Windsor • Christopher Wolfla • Hartmut Writh • D. E. Wyatt • Conrad Yu • Jakub Zalesak

Preface

There's something magical about holding a Medieval book in your hands. It's a tangible piece of history produced by craftsmen of another age—parchmenters or papermakers, scribes or printers, rubricators, illuminators, tanners, smiths, metalworkers, and binders, each applying their arts to contribute to something that might soon be discarded or might last for a thousand years (sometimes more).

The content of such a book starts with what they wrote and drew in it, which tells us what the person who designed or ordered the book thought was important, but it goes far beyond that—we can learn about how it was intended to be used based on its size and shape and the materials used to create it, and we can learn how it was actually used based on the damage it suffered and the marks left by readers. Notes and doodles on the pages, pages pasted in or torn out, even the crayon scribbles of children—these tell us about the readers, and their relationship to the book.

In this facsimile, I've tried to capture some of this magic and create something that is a bit more than just a fancy book.

High-resolution scans of the manuscript were created by the Bavarian State Library and then carefully cleaned and corrected by HEMA Bookshelf. These were then sent on to Grimm Book Bindery, which printed them on sheets of rag paper similar to the original 15th century paper stock and then collated and sewed them following the current construction of the manuscript (this is why you'll find tiny page stubs in a few places in the book: that's where single leaves were inserted into the binding). Once sewn, the pages were sent on to Liberty Book & Bible to be gilded before returning to Grimm to be bound.

The manuscript received its current binding in the mid-20th century, but I decided to try to simulate the original binding it might have had in the 15th. After a lot of research, I chose to base the cover design on the Folger Shakespeare Library's ms. INC P700, a manuscript made in Nürnberg in 1481 and prominently featuring both the Imperial Eagle and the Eagle of St. John, both of which seem particularly fitting for a book produced by a priest and dedicated to a Wittelsbach prince.

The facsimile was therefore bound in beech boards (designed by Capital Woodworking) and brown pigskin. Artist Elliot RAZO extracted the tooling design from this manuscript, which was made into a letterpress stamp to apply to the leather.

Clearly this facsimile is a thoroughly modern creation, made by 21st century artisans using 21st century techniques, but it was built based on something old and worn and full of stories.

The purpose of this companion volume is to further illuminate some of those stories. In its pages, you will find articles from some of the leading

Fig. 1: The current covers to LM (top) and INC p700 (bottom)

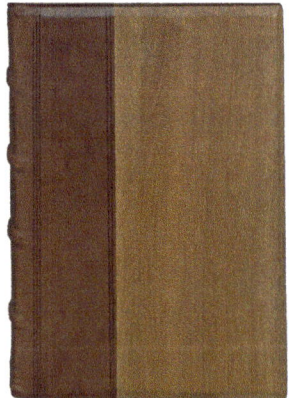

scholars in the field of historical European martial arts studies that high-light various aspects of the manuscript and the world it came from.

The first section contains a detailed description of the manuscript and its history, as well as a full transcription by Carsten Lorbeer, et al.

The second section relates to Lecküchner himself. Daniel Burger describes the life and times of Hans Lecküchner in great detail. Falko Fritz discusses the physical properties of Messers and how they relate to Lecküchner's teachings, and Jessica Finley delves into the unusual terminology used by Lecküchner (and Liechtenauer).

The third section looks at the wider landscape of Messer fencing that Lecküchner existed in. Dierk Hagedorn offers the first catalog of all surviving Messer and Dussack treatises. Olivier Dupuis discusses the concept of fencing in jest and in earnest and how it's reflected in fencing treatises. Casper J. van Dijk and Oskar ter Mors evaluate the Messer teachings that predate Lecküchner and look for possible influences on his teachings. Finally, Adam Franti looks considered the place of Messers and Dussacks in art and society throughout the early modern period.

I have elected not to include a translation; instead, it's my hope that this transcription will be read alongside Jeffrey Forgeng's 2015 translation, or on its own for German readers. (Grzegorz Żabiński, et al.'s 2012 effort, of course, already includes their own transcription alongside the translation.) While neither of these works is in need of an update, I hope that this volume will serve as a worthwhile complement to them.

MICHAEL CHIDESTER

Codicological Description

To His Serene Highness, the Noble Prince and Lord Philipp Ludwig, Count Palatine of Rhein, Duke in Bavaria, Count of Veldenz and Sponheim, my gracious Prince and Lord.

Sein Durchlauchten hochgebornen Fürstenn vnnd Herrin Phillips Ludwigenn Pfaltzgrauenn bey Rhein Hertzegenn in Bayern, Grauenn zue Veldentz vnndt Svonhaim Meinen gnedigen Fürstenn vnndt Herren

~ Johan Tettelbach, 1579

 hile our manuscript has received considerable attention from scholars and credible codicological descriptions have previously been published in German,[1] to date there has been no comprehensive treatment in English.[2] I aim to remedy this with the following description and discussion of the physical characteristics, history, and contents of the manuscript.

München, Bayerische Staatsbibliothek • paper, I + 215 + I • 1482 • Nürnberg

Cgm 582[3] is a paper manuscript[4] with 217 leaves in folio format, measuring 300 mm × 207 mm. The manuscript shows signs of wear and small tears in many pages; larger tears on the bottom of folia 100 and 140 seem to have been glued, while a tear on the edge of 112 was repaired with tape. On folio 115v, a paper patch has been glued over the original illustration, containing a new illustration by the same artist.

The edges of the page block are gilded and gauffered, a process by which a heated tool is rolled across the gilding to impress a pattern into it (fig. 1); this process also resulted in the wavy pattern of ridges visible along the page edges. It currently has a mid-20th century half-pig binding with beech boards; the leather is blind tooled with a simple pattern of one double- and one triple-line.

---

[1] RASPE 1905 and DÖRNHÖFFER 1910 discuss the manuscript from an art historical perspective, whereas WIERSCHIN 1965 discusses the manuscript briefly and HILS 1985A at more length as a fencing treatise. SCHNEIDER 1978 makes a fair effort at a codicological description but is hampered by the brevity required by her project. MÜLLER 1992 and 1994 provide extensive discussion of Lecküchner's works, but little from a codicological perspective. Most recently LENG 2008 offers another description focused on the manuscript's artwork.

[2] The only substantial work in English is FORGENG 2018, which contains ample discussion of the manuscript but doesn't seek to present the information systematically or comprehensively.

[3] Or *Hans Lecküchner-München* (LM).

[4] FORGENG 2018, p xviii, describes it as a parchment manuscript and sees significance in that choice, but this is obviously incorrect given the clearly visible watermarks and paper lines, and it contradicts every other description of the manuscript (including his own on p xvii, which is particularly puzzling).

Dating and origin

Folio 216v of the manuscript is dated St. Sebastian's Eve (19th January) in 1482.[5] This date is supported by watermark analysis; there are three different 'bull's head' watermarks visible in the paper, all of which first appear in the third quarter of the 15th century (fig. 2).[6] Watermarks also show that the paper comes from Northern Italy, possibly in the Brescia area, but the language and artwork (see below) support an origin in Northern Bavaria, probably the city of Nürnberg.

History and provenance

LM was created or ordered by Hans Lecküchner († 1482) in Nürnberg in 1482 based on an earlier manuscript from 1478.[7] This is typically assumed to be *Hans Lecküchner-Heidelberg* (LH), but that's a poor candidate because it's clearly a presentation copy, not a draft or working copy. Furthermore, the posthumous copies of the treatise, of which *Hans von Speyer* (HS) is the earliest and the only complete one, follow the general pattern of LH but include plays unique to LM. This suggests either a lost intermediary copy between LH and LM,[8] or more likely, a lost original copy from 1478 that was the source for LH, LM, and subsequent copies.

Both LH and LM are dedicated to Philipp 'the Upright' of Wittelsbach (1448–1508), Elector Palatine of the Rhine. HILS states that this was in thanks for Lecküchner's appointment to the parish of Herzogenaurach,[9] but there's no evidence to support speculations of his motives.

---

[5] *Composita Est materia illa per dominum Johannem Leckuchner tunc tempore plebanus Jn hertzogaurach Anno domini M° cccc° septuagesimo octauo sed iste liber scriptum est et completus Anno 8° secundo Jn vigilia sancti Sebastiani etc* ("This material has been composed by Reverend Hans Lecküchner, at the time parish priest in Herzogenaurach, in the year of our Lord 1478, but this book has been written and completed in the 82nd year on St. Sebastian's Eve, etc.")

[6] SCHNEIDER 1978, p 177.

[7] See note 5.

[8] FORGENG 2018, p xix.

[9] HILS 1985A, p 91.

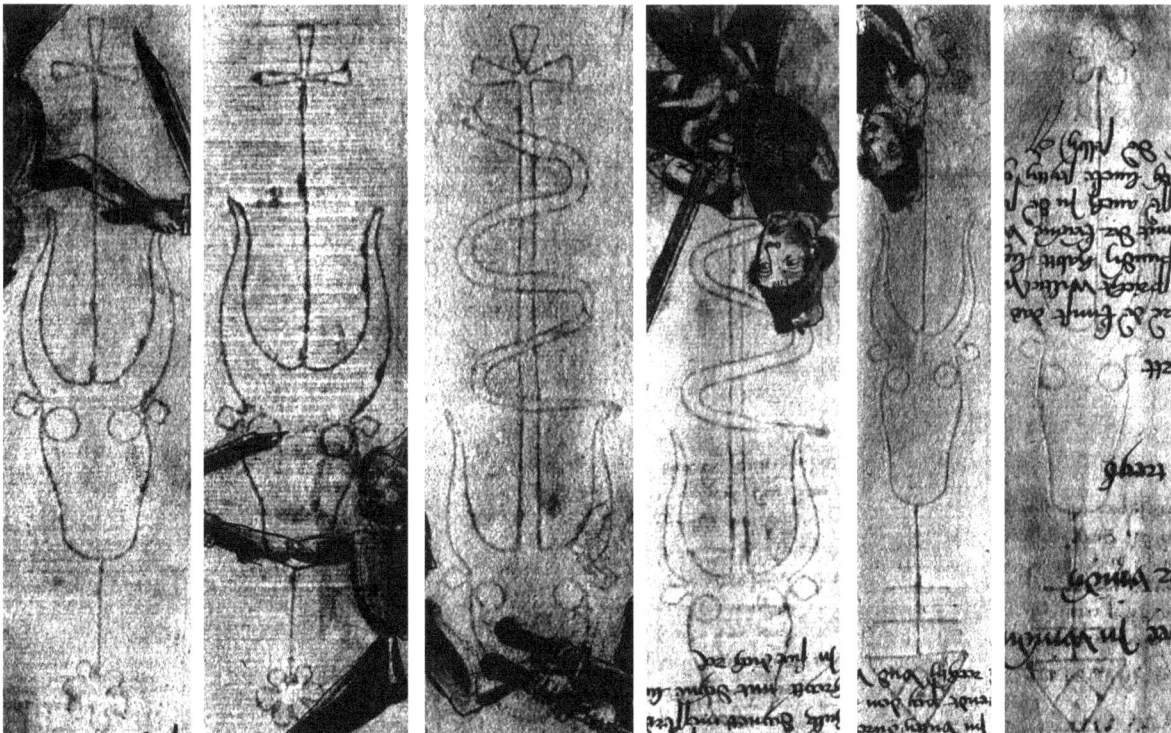

Perhaps owing to Lecküchner's death at the end of 1482, LM was never delivered to the Palatine library. If it were still in Lecküchner's possession at that time, it most likely remained with the church, either in his parish library or transferred to some higher church official.[10]

The remaining provenance of LM is recorded in the covers of the manuscript itself.

Its first known owner was the priest Johan Tettelbach (1517–1598). In 1579, he was ecclesiastical superintendent in Burglengenfeld, a small city about 50 miles southeast of Nürnberg. On 24th August, he gifted the manuscript to Philipp Ludwig of Wittelsbach (1547–1614), Duke of Palatinate-Neuburg, and his dedication is recorded inside the front cover (fig. 3).

When Philipp Ludwig's son and successor, Wolfgang Wilhelm (1578–1653), moved his residence from Neuburg to Düsseldorf in 1636, he brought his library with him. In 1664, the first known catalog reference places the manuscript in this library (as noted in pencil inside the front cover).[11]

Wolfgang Wilhelm's son, Philipp Wilhelm (1615–1690), became Elector Palatine in 1685 and the manuscript finally joined the library of the Palatinate as Lecküchner intended. A large copperplate bookplate was placed inside the front cover with the label 'Bibliotheca Patatina'[12] (fig. 3). It

---

[10] FORGENG 2018, p xviii. For more on Lecküchner's life and times, see Daniel BURGER's chapter, "Hans Lecküchner of Nuremberg and his fencing treatise in the Long Knife", in the present work.

[11] *Ehemals (1664) in Düsseldorf vgl. Cbm C. 555* ("Formerly (1664) in Düsseldorf, see Cbm C. 555"). This is a reference to the BSB ms. Cbm Cat. 555, p 130, which records the cataloging and can be consulted online here: http://mdz-nbn-resolving.de/details:bsb00133824 (accessed 23.08.2021).

[12] An apparent misspelling of 'Palatina'.

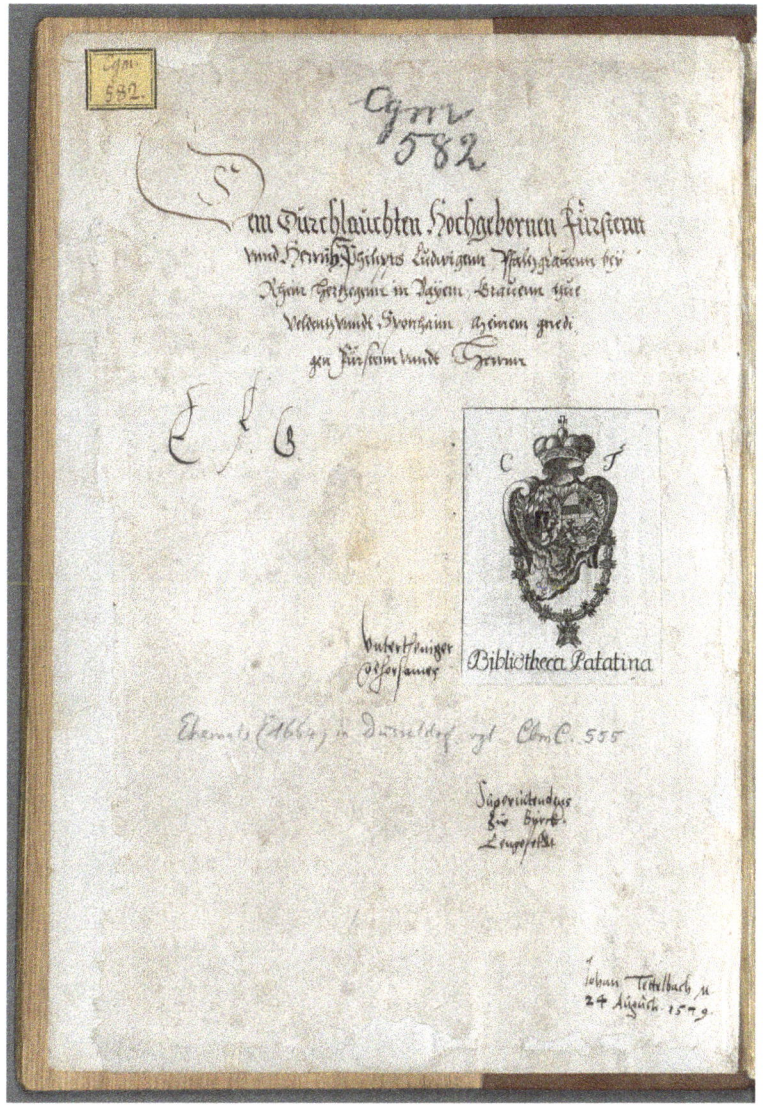

subsequently moved to Mannheim when his son, Karl III Philipp (1661–1742), transferred the library there in 1731. Here it received the signature *J. 167*, which is recorded in pencil on folio Iv.[13]

The Wittelsbach moved the entire Palatine library when Mannheim was ceded to the Duchy of Baden in 1802, and in 1803, consolidated it into the Bavarian ducal library in München, which was renamed the *Bibliotheca Regia Monacensis* ("Royal Library of München") when the Kingdom of Bavaria was established in 1805 (this appears frequently in library stamps throughout the manuscript). In München, LM was cataloged under its current signature: *Cgm 582*,[14] which is found twice inside the front cover (on a sticker and written in pencil; fig. 3), once each on folia Ir and 109r (this last along with the date 1831).

In 1919, after the Kingdom of Bavaria was overthrown by the German Revolution and ultimately replaced by the Free State of Bavaria, the library was renamed 𝕭𝖆𝖞𝖊𝖗𝖎𝖘𝖈𝖍𝖊 𝕾𝖙𝖆𝖆𝖙𝖘𝖇𝖎𝖇𝖑𝖎𝖔𝖙𝖍𝖊𝖐 ("Bavarian State Library"). It owns the manuscript to this day.

A final note on the history of the manuscript is that it was rebound in September 1961[15] as part of conservation activities (and a sticker was affixed inside the back cover). The old half-leather cover was discarded, and no information remains to say whether it included any part of the original 15th century binding or was entirely the product of an intermediate rebinding (see below).[16]

---

[13] SCHNEIDER 1978, p 178. LENG 2008, p 75, states that it received this signature earlier, during the Düsseldorf years, but related manuscript PKM never resided in Düsseldorf and was cataloged as *J. 156* in Mannheim.

[14] 'Cgm' is an abbreviation for *Codices germanici monacenses*, the German-language books of München.

[15] SCHNEIDER 1978, p 178.

[16] The library's practice in this time was to replicate the materials of the binding being replaced so it may be assumed that the previous cover was also wood and blind-tooled pigskin. I'd like to thank Dr. Irmhild CEYNOWA, for this information.

Content

The manuscript bears no title but is referred to in the incipit as ķŭnſt unð ʒeðel ym meſſer ("Art and Recital in the Messer").[17] The treatise is organized into a series of brief mnemonic verses ('the text') which are followed by prose exegesis ('the gloss') that lays out the series of plays and martial teachings. Beginning on folio 3v, nearly every page also contains an illustration.

The entire manuscript appears to be the work of Hans Lecküchner, though the manuscript frequently ascribes the text to an unnamed master (possibly an acknowledgement of the otherwise absent grand master Johannes Liechtenauer, upon whose verses Lecküchner based his own). He organizes his teachings into twenty-three "main sections" (ḥawbtſtucke), of which the firſt six are labeled "hidden ſtrokes" (verporgen ḥew).[18] These are liſted on the table.[19]

Collation and foliation

Fig. 4 presents the collation of the manuscript, which is composed of 215 leaves in 18 quires:[20]

$$(VII - 2)^{12} + 2\ VI^{36} + (VI + 1)^{49} + 3\ VI^{85} + VII^{99} + (VI + 1)^{112} + 4\ VI^{161} +$$
$$(VI + 1)^{174} + VI^{186} + (VI + 1)^{199} + V^{209} + (IV - 1)^{216}$$

There are single guard leaves on either end (I + 215 + I) which were added during a rebinding in the 16th or 17th century;[21] these were attached to the firſt and laſt quires to replace the original firſt and laſt folio, which were in turn paſted to the covers as endpapers.[22]

---

[17] LM f 1r. The manuscript is also sometimes called *Kunst des Messerfechtens* ("Art of Messer Fencing") in the literature, but this seems to result from confusion with Egenolff's 1530s abridgement.

[18] For more on Lecküchner's main sections, including a comparison to those in the teachings of Liechtenauer, see Jessica FINLEY's chapter, "'What's in a Name?'", in the present work.

[19] Spelling reflects that of the manuscript.

[20] SCHNEIDER 1978, p 177.

[21] The new guard page was already present when the manuscript was cataloged in Mannheim no earlier than 1731. Without watermarks, the age of this paper can't be narrowed down further.

[22] FORGENG 2018, p xvii.

I
_I_   1   2   3   4   5   6

II
13   14   15   16   17   18

III
25   26   27   28   29   30

IV
37   38   39   40   41   42

V
50   51   52   53   54   55

VI
62   63   64   65   66   67

VII
74   75   76   77   78   79

VIII
86   87   88   89   90   91   92

IX
100   101   102   103   104   105

X
113   114   115   116   117   118

XI
125   126   127   128   129   130

XII
137   138   139   140   141   142

XIII
150   151   152   153   154   155

XIV
162   163   164   165   166   167

XV
175   176   177   178   179   180

XVI
187   188   189   190   191   192   193

XVII
Lost   200   201   202   203   204

XVIII
210   211   212   213

7	8	9	10	Lost	11	12	I
19	20	21	22	23	24		II
31	32	33	34	35	36		III
43	44	45	46	47	48	49	IV
56	57	58	59	60	61		V
68	69	70	71	72	73		VI
80	81	82	83	84	85		VII
93	94	95	96	97	98	99	VIII
106	107	108	109	110	111	112	IX
119	120	121	122	123	124		X
131	132	133	134	135	136		XI
143	144	145 - 146	147	148	149		XII
156	157	158	159	160	161		XIII
168	169	170	171	172	173	174	XIV
181	182	183	184	185	186		XV
194	195	196	197	198	199		XVI
205	206	207	208	209	Lost		XVII
214	215	216	217				XVIII

There seems to be a folio missing between 10 and 11, as well as a complete missing bifolium on the outside of quire XVII (between 199 and 200, and 209 and 210). The latter can be confirmed by comparison with LH, which has content in both of those positions that's missing from LM; this would mean that XVII was also initially a sexternion.[23] Further material lacunae are suggested by the singletons 49, 112, 174, and 187, but there are no gaps in the text or gloss compared to LH so it's impossible to say what the contents of the hypothetical missing folia could have been.

Folia 174 and 187 might alternatively have formed the original outer layer of quire XV, cut or torn in half and then attached to XIV and XVI by mistake during rebinding.[24] This would make XV a third septernion.[25]

Based on the continuous content, single hand (see below), and consistent quire pattern of one septernion followed by six sexternions, the manuscript appears to be a single production unit.

Folio numbers were added in the 15th century (1–216).[26] The foliation jumps from 144 to 146, skipping 145; a later hand added '145 –' in pencil to 146 and also numbered the final guard leaf 217. The foliation is otherwise continuous, so any material losses had already occurred by that time.

Writing and decoration

The manuscript is written in a bastarda in a single hand using a northern Bavarian dialect;[27] the incipit on 1r and explicit on 216v might be read as indicating an autograph by Lecküchner, but that can't be

---

[23] A quire of six bifolia/twelve folia

[24] They share the rubrication that is otherwise only found in quire XV and on ff 1r, 2r, and 193r.

[25] A quire consisting of seven bifolia/fourteen folia

[26] Ibid.

[27] SCHNEIDER 1978, p 178.

proven.[28] Folia 1r–3r have a writing frame of 215–230 × 133 mm (only ruled on the sides) with 30–33 lines.[29] Beginning on folio 3v, the format changes to include half-page illustrations on most folia, limiting the writing to 4–23 lines and omitting any visible frame. The text is generally written in a larger and bolder style than the gloss, making them easy to differentiate.

A second 15th century hand, perhaps the same as the foliator, added minor annotations on folia 148r, 149v, 154r, 181r, and 216r.

There are no decorative borders or Lombards, though an 'O' seems to have been intended on 1r but never executed (fig. 5). 1r and 2r have minor rubrication in the form of red slashes (faded to a pale pink) highlighting the first letter of each line of text and paragraph of gloss; this also appears on folia 174–187 (fig. 6) and 193r, but is absent from the rest of the manuscript.

Illustrations

There are 414 watercolor pen and ink drawings by "a master very close to Michel Wolgemut"[30] from a Nürnberg school.[31] The same master or workshop may be responsible for the similar illustrations in two 1490s treatises[32] which may be attributed to the Marxbrüder captain Peter Falkner.[33]

LENG suggests that the initials B and M (on 28v and 45v respectively) drawn on the chests of fighters might be the artist's monogram,[34] but

[28] MÜLLER 1992, p 253 (note 12). Another possibility is that the dedication in LH, ff 115r–116r, is the only part written personally by Lecküchner.

[29] SCHNEIDER 1978, p 177.

[30] ...dem Michel Wolgemut sehr nahestehenden Meisters besitzt. DÖRNHÖFFER 1910, p XVII.

[31] Ibid. Also RASPE 1905, p 42.

[32] Peter Falkner (PF) and Paris fight book (P). Suggested by LENG 2008 (p 73) among others.

[33] PF also quotes Lecküchner's verses extensively.

[34] LENG 2008, p 75.

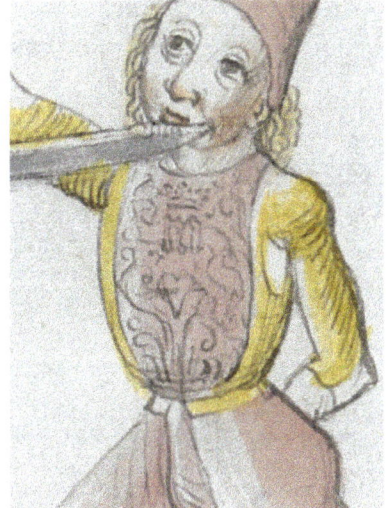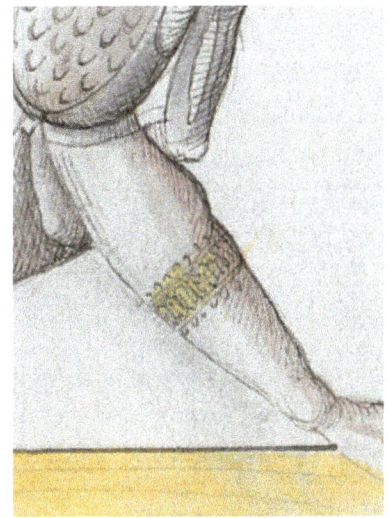

FORGENG proposes that they're an abbreviation for *Beata Maria* ("Blessed Mary") and compares them to the *Ave Maria* ("Hail Mary") garter on 202r (fig. 7).[35]

The illustrations are 120–200 mm tall and span the width of the page with no frame. Only 1r–3r, 26r, 34v, 63r, 72v, 91r, 147r, 153v, 170v, 198v, 211v, and 216v lack illustrations, whereas 49rv, 107v, 180rv, and 181rv have illustrations but no text or gloss. Paints include ocher, rose, red, yellow, brown, gray, and blue.[36]

The illustrations were drawn before the text was added to each page, as indicated by the caption on 97r which complains that the illustration doesn't correspond to any part of the treatise.[37] On 115v, a new illustration was glued onto the page, perhaps covering an older one (raising the question of why the same procedure wasn't used to correct 97r).

The illustrations generally depict two figures in attitudes of fencing or wrestling on a flat yellow floor,[38] occasionally with landscape embellishments.[39] The fencers carry Messers and wear tight hosen and doublets, shoes (and perhaps also gloves), and sometimes hats, all with varying levels of detail. There are no signs or tokens indicating which of the two figures is performing the technique described in the gloss.

Other noteworthy artistic embellishments include: a manicule pointing to a fencer,[40] a dog watching the fencers,[41] a fencer playing backgammon while sitting next to his defeated and trapped opponent,[42] a fencer being cast into a sack by three others,[43] and a second pair of fencers on a hill in the background with birds associating them to the text.[44]

---

[35] FORGENG 2018, p xviii.

[36] LENG 2008, p 76.

[37] *Item das vntten gemalt stett Ist nichtz wedeuttden etc* ("Also, this that is painted below means nothing, etc.")

[38] Or red on f 7v.

[39] LM ff 95r, 96v, 97r, 99v, 112v, 132r, 160r, 161r, 174r, and 182v. 50r features storm clouds.

[40] LM f 33r.

[41] LM f 90r.

[42] LM f 91v.

[43] LM f 92r.

[44] LM f 161r.

CARSTEN LORBEER, JULIA LORBEER, JOHANN HEIM, ROBERT BRUNNER, and ALEX KIERMAYER[45]

Scientific Transcription

Editor's note: This transcription was prepared between 2001 and 2005 by a team from the Gesellschaft für Pragmatische Schriftlichkeit *and* Schwertkampf Ochs. *It was released online in 2006 but has never been published in print. I translated the transcription notes and footnotes from German, and any errors in them are my own.*

ur transcription of *Hans Lecküchner-München* (LM) was made with the help of the microfilm from the Bayerische Staatsbibliothek in München. In disputed cases, Lecküchner's concept manuscript *Hans Lecküchner-Heidelberg* (LH) was used, the content of which largely corresponds to LM. Such cross-referencing is noted in the footnote.

Furthermore, the following guidelines were used in the preparation of the transcription:

Formatting

- The page formatting of the manuscript was retained in the transcription, i.e. the line dimension of the verses and glosses corresponds to that in LM. Only the foliation was moved to the margin.
- Verses are transcribed in bold; the scribe has chosen a more careful, larger font there.
- Resolutions of abbreviations are given in underline, e.g. It<u>em</u>.
- The division of the chapters, as indicated in the header, is based on the division of the manuscript by the scribe into chapters or main parts.

Abbreviations

- A nasal stroke is resolved with *-e*, *-n*, or *-en* (e.g., lern<u>en</u> 1r, line 14). A nasal line over a double-n or -m is replaced by an appended *-e*, e.g. arm<u>e</u>, armm<u>e</u>, Wenn<u>e</u>, dann<u>e</u>.
- In a few places in the text, the writer uses an L-shaped abbreviation which stands for an appended *-em* or *-em*.

[45] Special thanks to Andreas MEIER and Marita WIEDNER for their corrections to the first draft.

- The frequently occurring r abbreviation is supplemented accordingly as *-r* or *-er*, e.g. mess<u>er</u>, her<u>r</u>.
- At the end of a text passage or to indicate a logical caesura, the writer uses the 'et cetera' abbreviation, which is resolved as etc or <u>etc</u>. A nasal line above the etc is ignored for the transcription.

Use of capital letters

- Capital letters are only used in the transcription if they are clearly recognizable as such. Exceptions are verses and the beginning of a new paragraph. A letter is not considered to be capital if it is only set high or written too long. Frequently found capital letters are I, E, S, G, T.
- The first letter of 'Item' or 'In' is transcribed as I.

Diphthong and umlaut spelling

- Umlauts and diphthongs are reproduced as they were used as far as possible. Letters that are simply overdotted are not transcribed as umlauts, as it is usually unclear whether the point was made deliberately or was caused by accidental removal of the pen.
- A double line is transcribed as a normal umlaut mark.

Sound shifts

- The transcription does not reflect any sound shifts. This means that a written u is not reproduced as a New High German v or f (zuuechten, zwiuach) and a b is not reproduced as v or w (verborffen, bayssen).

Transcription of certain letters

- In handwriting, a and o are sometimes very difficult to distinguish from one another. Therefore, both vowels are transcribed as a or o if they are unambiguous, and accordingly in cases of doubt. In a few cases in which the New High German word meaning deviates from the spelling, this is indicated in a footnote.

Word corrections and additions

- If the writer has obviously forgotten individual letters, i.e. if there is a spelling mistake, the missing letter is added in square brackets [...].
- Crossed out letters, words, or sentences are also reproduced as such in the transcription. Word supplements are inserted at the corresponding text passage and the insertion is noted as a footnote.

The transcription

(Museum sticker)   **Inside Cover**
Cgm. 582.

(Pencil)
Cgm 582

(Neo-Gothic sixteenth century script)
Sein Durchlauchten hochgebornen Fürstenn
vnnd Herrin Phillips Ludwigenn Pfaltzgrauenn beÿ
Rhein Hertzegenn in Baÿern, Grauenn zue
Veldentz vnndt Svonhaim Meinen gnedi=
gen Fürstenn vnndt Herren

.E.K.G.

(Sticker of museum crest)
Bibliotheca Palatina

(Pencil)
Ehemals (1664) in Düsseldorf vgl. Cbm C. 555

(Neo-Gothic sixteenth century script)
Vntertheniger
Gehorsamer

Sūperintendens
Zue Būrck=
Lengefeldt

Johan Tettelback u.
24 Augusti 1579.

Cgm. 582   **Ir**

J. i67   **Iv**

|Das ist herr hannsen Lecküchner[46] von Nurenberg künst vnd zedel   **1r**
ym messer dy er selbs gemacht vnd geticht hatt Den Text
vnd dy auslegung dar über Dem hochgeporen fursten vnd
herren hertzogen philippen phaltzgraffen Bey reyn Ertzdruckseß
vnd kurfürst vnd hertzog yn Bayern etc

---

[46] There are both an *-er* abbreviation and a small mark
which may be an attempt to insert the "er".

1r (cont.)　　　|Dy vor rede

|[O]|B dw wilt achten
|Messer vechten betrachten
|So leren ding das dich zirtt
|Zu schimpff ze ernnst hofirt
|Do mit du erschreckest
|Vnd dy meyster künstenlich erbeckest

|Hye hebt sich an dy vor rede ym messer vnd sagt wer woll
vechten ym messer Das er schül lernen[47] rechte kunst vnd sich
geben auff dyse nachgeschriben artt vnd kunst So mag er woll
westan mit rechter kunst vor fursten vnd vor herren ym
sol auch pillich seyner kunst paß gelönet werden den anderen
meysteren des dings nicht weysen vnd sy nicht dar eyn kunnen
schicken wenn es seyn vill meyster des Swertz dy nicht wissen
von der art des messers noch recht auß synnen mügen wer sich
yn dyse ding vnd art schicken kan der syndt vill ernnstlicher
stück do mit er dy meyster woll mag auff becken vnd kunstenlich
weschlissen Das sy sich an danck schlagen stossen furne werffen
halten müß lassen;

|Wer newer versetzet
|Mit allen kunsten wirt er geletzet

|So nw der meyster geendet hatt dy vor rede nw gibt er dir ein
gutte ler dy verste alzo Wenn dw mit dem zwuechten ze dem
man kumbst so soltu nicht stil sten mit dem messer vnd seynen
hewen nach wortten wenn dy vechter dy newer wellen versetzen
dy werden ser geschlagen wann dy weil sy versetzen mügen sy
1v　　kain rechte kunst treyben vnd so sy anders nicht thun wollen
dan versetzen vnd auff ander lewt schleg vnd erbeyt wollen
sehen So werden sy geschlagen vnd geswecht bey allen iren
kunsten.

Yndes / vor / vnd /nach / dy wortt
Seyn aller kunst eyn hortt
Swech / vnd / sterck prüff weyslich
So dw wildt vechten künstenlich

Hye sagt der meyster vnd melt den grundt des messers vechtens
vnd spricht yndes vor vnd nach etc Das soltu alzo verstan
das dw vor allen dingen recht vernemen solt dy zway ding das
ist das vor vnd nach. vnd dy swech vnd dy sterck vnd das wort

---

[47] After LH: *söll lerenen*

yndes wann dar auß kumbt aller grundt aller der kunst des
vechtens vnd wenn dw dy ding recht vernymbst vnd verstest
vnd zwvoran[48] des wortz yndes nicht vergist yn allen stucken
dy dw treyben pist So magstu woll eyn gutter meyster des
messers sein vnd magst fursten vnd herren leren das sy mit
rechter kunst wol bestann yn schimpff vnd yn ernnst

Item das vor ist. So dw ee kumbst mit dem haw dann er das er
dir versetzen muß So erbayt yndes behendiklich mit
dem messer fur dich oder anderen stücken vnd laß yn dar nach
zw keyner erbeit kummen alzo hastu ym an gebunen das vor
vnd wenn er ee kumbt mit der erbeyt oder mit dem haw
denn dw vnd ym das versetzen müst So erbeyt yndes behen=
dicklich mit dem messer oder sunst mit stucken So nymbstu
ym das vor mit dem nach vnd das heyst das vor vnd nach

Item dw solt auch vor allen dingen wissen dy swech vnd dy sterck
des messers nw merck von dem gehültz piß an dy mitten des
messers das ist dy sterck vnd von der mitten vntz an den ortt
das ist dy swech vnd wye dw dar auß vechten scholt das vindest
dw her nach geschriben vnd was das wort yndes ist

Sechs hew lere
Auß eyner handt wyder dy were
Dy linck handt leg auff den rucken
Auff dy prust wiltu were zucken

Hye lert der meyster wye man sich halten sol yn dem vechten
des messer vnd wye man sich dar ein sol schicken vnd zum
ersten So soltu mit eyner hendt vechten mit dem messer vnd dy
ander soltu auff dem ruck haben ob dw aber wildt mit lerer handt
vechten alz mit messer nemen vber greffyen arme beschlissen
So soltu dy handt von dem ruck auff dy prust wenden waß
dw ym wildt ynbendig vber den arme treyben

Item ze dem anderen so werden yn dem text sechs verporgen hew
gemelt vnd werutt[49] auß dem kummen vil gutter stuck wer sy recht
treyben kann vnd sych yn dy weyß kunstenlich zw schicken
vnd wye dw dy treyben solt mit dreyen stucken das wirstu her
nach vnterricht.

---

[48] From LH, f 3v: *zw vor an*

[49] LH, f 4v: *peruertt* comes from *beruren, berüeren*
(BAUFELD 1996, p 29); *ruoren = rüeren* (LEXER 1885, p 204)

2r (cont.)

Zoren |haw |vecker
Entruſt: hat zwinger / gefer mit |vincker

Hye nendt der meyſter dy sechs verporgen hew wye yetzlicher
mit namen heyſt auff das das dw dy ſtück deſter paß vernemen
mügſt vnd spricht das der erſt heyſt der zoren haw Der ander
der wecker der dritt der entruſthaw der vyerdt der zwinger
Der funft der gefer haw vnd der sechs hayſt der wincker
Das seyn sechs ſtuck vnd seyn der hawbt ſtuck alz dw her
nach hōren wirſt;

Dy paſteyn |versetz
Nachrayß |vberlauff |vnd absetz
Den wechsel |durch / zuck
Lauff durch |dy abschneydt |druck
Ablauff |benym
Durchge |pogen |dy wer nymb
Heng dy winden |gen plōssen

2v Schlag dy ſtraych: ler verdrossen

Hye zelt der meyſter dy anderen hawbtſtuck Der seyn sibenzehen
Das erſt seyn dy vyer leger dy werden werürt So der text
spricht dy paſteyen
Item das ander ſtuck heysset das versetzen das wird beruert So
der text spricht versetz
Item Das dritt seyn dy nachreysen das wirt beruert So gespro=
chen wirtt ~~vberlauff dy~~ nachraysen
Item das viert sein dy vberlauffen das wirt berütt So gesprochen
wirtt vber lauff dy
Item Das funft seyn dy absetzen das wirtt berütt so gesprochen
wertt absetz
Item das sechſt sein dy durch vechselen das berütt der meyſter
so er spricht vechsel durch
Item das sybent heyſt das zucken das wirtt berurtt So
er spricht zuck
Item Das acht sein dy durchlauffen dy werden werütt so ge=
sprochen wirtt lauff durch
Item das newndt sein dy abschnidt oder dy vier schnidt das
wirtt beruret so gesprochen wertt dy abschnidt
Item das zehendt heyſt dy hendt drücken das wirt beruret
so er spricht druck
Item das eylfft sein dy ablauffen das wirdt berütt So er spricht
ablauff
Item das zwelfft seyn dy pnemen das wirdt perütt so er spricht
pnym

It[em] das dreyzehent das sein dy durchgen vnd das wirtt   **2v (cont.)**
berut so er spricht durchge
Item das virzehent heyst der pogen das wirt beruert so er
spricht pogen
Item das funftzehent heyst das messer nemen das wirt beruett so er
so gesprochen wirtt wer nymb
Item das sechzehndt sein dy hengen dy werden berüert so gespro   **3r**
chen wirt heng
Item das sibzehendt sein dy winden yn dem messer das
beruert vns der meyster so er spricht wind gen den plössen Also
hastu der hauptstuck sein dreyundzaynzig[50] etc

Was auf dich wirt gericht
Zorenhaw ortt das gar pricht
Wiltu yn beschemen
Am messer ler abnemen

So der meister das vechten des messers yn dy stuck geteylt hat vnd
eyn ytlichs mit namen genent nw hebt er an ze sagen von dem
ersten glid der tailung als von dem zorenhaw vnd ist zw wissen das
der zorenhaw mit dem ortt bricht all ōberhaw vnd ist doch eyn
schlechter pawren schlag

Item treyb den alzo wenn er von der rechten seitten oben zu dem   **3v**
kopff schlecht so haw von deiner rechten seitten So haw von
deiner rechten seitten auch von oben mit ym zōrniklich gleich
an all versatzung oben ein Indes laß den ortt fuer ein schiessen
ym zu dem gesicht oder prust vnd wendt Indes dein messer
gegen dem seinen das dy lang schneid oben stee vnd dy kurtz
vnden wirtt er des ortz gewar So windt wider auff sein lincke
seiten den ortt zu dem gesicht das Ewchling dy lang schneid
oben stee wirtt er des ortz gewar So reiß am messer auff oben
an des messers klingen wider von seinem messer vnd haw Im
zu der anderen seytten zu dem kopff das heisset ab genommenn
am messer;

Zorenhaw ortt auff dich gericht   **4r**
Haw das recept[51] senck lanck ortt prüst sticht

Hie lertt der meyster dy pruch wider den zorenortt vnd spricht
der text zoren haw ortt etc Das soltu alzo verstan Macht er
den zorenhaw auff dich mit dem ortt so haw Im nach seiner
handt In das gelenck Inbendigs auff sein Ewerliche handt etc

---

[50] From LH, f 6v: *drey vndzwayntzyg*   [51] *recept = Rezept* (HENNIG, p 263)

4v   Item macht er auff dich den zorenhaw mit dem ortt vnd will
dir zu dem gesicht ſtechen So senck dein messer nider mit ge=
ſtracktem arme vnd senck Im den ortt auff sein pruſt vnd scheub
In woll ze ruck vnd setz das linck pain woll zu rück etc

5r
Den ortt soltu durch sencken
Das haubt lanck mit zwirch krencken

Hie lertt der meiſter ein ſtuck so dw den zorenhaw gemacht
haſt mit dem ortt so laß den ortt nider sencken vnd var zwischen
dich vnd In fur dein leib vnd schlag In mit der zwirch vnd
mit der langen schneiden zu seinen rechten oren etc

5v
Will er zwirch schlagen zu den oren
Senck ortt zu pruſt wiltu In betoren

Hie lertt der meiſter ein bruch wider das Ee gemelt ſtuck vnd
sprich will er zwirchen etc Das soltu alzo verſtan So er dy
zwirch zu deinen oren iſt schlagen vnd iſt hoch mit den arme so
senck ym den hangenden ortt in sein pruſt vnd scheub In mit
dem messer alzo von dir

6r
Auß zorenhaw ler gesenckt auff recht treyben
Schlach kürtz oder ler durch das mawl schneiden

Hie lertt der meiſter aber ein ſtuck das dw auß dem zorenhaw
treiben solt vnd sprich auß zorenhaw lere etc Das soltu alzo
verſtan so dw den zorenhaw auff In gemacht haſt So soltu den
ortt aber sencken vnd schalt mit gesencktem messer hoch mit
dem arme zwischen dich vnd In faren das das gehultz obersich
ſtee Indes verbendt dein handt vnd schlag Im dy kurtz schneid
In sein gesicht oder schneid In durch sein antlitz mit der kurtzen
schneiden etc

6v
Kurtz mit schneiden iſt er pinden
Windt ortt recht so wirſt In vinden

Hie lertt der meiſter ein bruch widr das ee gemelt ſtuck vnd
spricht kurtz will mit schneiden etc Das soltu alzo verſtan
will er dich mit er kurtzen schneiden dich In dein antlitz schne[i]den
oder schlagen wie vor gesagt Iſt So windt Im Indes den ortt
auff gegen seinem messer vnd ſtich Im zu dem gesicht seiner
rechten seitten etc

Gerad ab nemen lere
Lanck mit der schneiden ze haubt schlag sere

Hie lertt der meister abnemen mit dem langen messer vnd spricht
alzo Gerad ab nemen etc So dw den zorenhaw hawest so windt
dein messer gegen dem seinen das dy lang schneid übersich stee Indes
windt wider auff dein rechte seitten gegen seiner lincken das dy lang
schneid vnden stee vnd dy stumpff oben vnd stich Im gerad zu
dem gesicht Indes schlag behentlich ab am messer zu seiner rechten
seitten mit der langen schneiden zu dem kopff etc

Am messer geradt ist er abnemen
weseit tretten soltu dich ~~treten~~ remen
Beleib / schlag / schneid / ader / stich
Dein flech auff sein wendt / alzo dy abnemen prich

Hie lertt der meister dy pruch widr dy abnemen vnd sprich am
messer gerad etc Das soltu alzo verstan So er abnybt auff dich
mit geradem messer wie vor gesagt ist Indes trit beseit auff dein
rechte seitten woll auß seinem schlag vnd beleib alzo sten vnd
schlag In mit der langen schneiden deines messers auff sein kopff
oder wendt dein messer gegen dem seinen das dy lang schneid ob stee
vnd stich Im zu seinem gesicht oder schlag Im dy kurtz schneid In
das antlitz.

Item oder far Im mit der flech auff sein messer Im abnemen vnd schlag
In zu seinem haubt mit der langen schneiden vnd merck So du mit
der flech auff sein messer vellest das du sewberlich dar auff stossest
Indes far auff mit der langen schneiden vnd schlag In zu dem kopff
wie vnden gemalt stet etc

Will flech abnemen prechen
Wind auff recht wiltu dich rechen

Hie lertt der meister ein pruch wider den ee gemelten pruch
vnd spricht alzo will flech etc Das soltu alzo verstan So
du abnemen pist Im messer stet er alzo still mit dem messer vnd
lest dich ab nemen indes felt er mit der flech seines messers
auff das dein vnd will dich zu dem kopff schlagen indes windt
auff gegen seinem messer auff sein rechte seitten So ruerstu In an
das haubt wie vnden gemalt stet etc Oder harr piß er auff
gett vnd zu dein kopff schlagen will so gee mit auff vnd
windt vber sein messer auff sein lincke seitten das dy kurtz schneid

8r (cont.) ob stee vnd stich In dem gesicht der selbigen seitten vnd das ist
ser ein gutter bruch sequitur textus[52]

8v [53]**Mit abnemen will er dich letzen**
Ortt auff recht zu gesicht ler setzen

Hie lertt der meister aber ein pruch wider dy abnemen vnd sprich
alzo mit abnemen will etc Das soltu alzo verstan Ist er abne=
men Im messer mit der langen schneiden so wend dein flech an
dy sein vnd beleib alzo sten leg den dawmen auff dy flech dey=
nes messers vnd stich Im zu dem gesicht seiner rechten seitten etc

9r **Setzt er zugesicht auff recht ortt**
Ortt senck var gerad auff prust ist der hortt

Hie lertt der meister ein pruch widr den ee gemelten pruch
vnd spricht Setzt er zu gesicht etc Soltu das alzo versten
so du wild abnemen Ist er dir das prechen mit dem ortt wie ee
gemelt ist so senck deinen ortt lanck auß gestracktem arme vnd
setz den lincken fuß woll hinden hin auß vnd stich In geradt
auff sein gurgel ader prust etc

9v **Hew stich merck**
Im pandt weich oder hērtt
Vor vnd nach Indes hab acht
Dy leüf des krigs recht betracht

So nw der meister von dem ersten stuck gesagtt hat als von
dem zorenhaw nu sagtt er ein gutte ler das ist wenne einer
mit dir ein haut oder sticht so soltu gar eben mercken wenne ein
messer an das ander klitz ob er Im pandt waich oder hertt sey vnd
als pald dw des empfindest so arbeit Indes mit dem krig nach
der weich vnd nach der hertt zu der negsten plöß: Vnd waß
das vor vnd nach ist das pistu vor unterrichtt

10r **Den krig auff löß**
Oben / nyden / wartt der plöß

Hie lertt der meister was der krig sey vnd was man darauß treyben
sol zu den vier zinnen Das ist zu den vier plössen Nu ist zu
wissen Das der krieg sein dy winden Im messer vnd dy arbeit
dar auß mit dem ortt zu den vier zinnen Das heist der krieg
Nu den krig treib alzo wenne du mit Im ein hawest den zorenhaw
vnd als pald er den versetz So var auff mit dem arme windt

---

[52] Latin: *Sequitur Textus* = "The text follows"

[53] There is an unknown character on the left side of the sheet, or the scribe has started too far to the left.

Im den ortt an sein messer oben ein zu dem gesich seiner lincken seitten ***10r (cont.)***
setzt er denn den stich ab So var auff dy anderen seitten auff dy andren
zinnen unter seinem rechten arme auff dy prust volgt er denn aber dem
stich nach So arbeit zu der negsten plöß dy du vinden magst etc

Vier sein der zinnen ***10v***
Dy du solt gewinnen
Der wach nymb war
Wo sy sey mit gefar

Hie lertt der meister Das dw Im zuuechten So dw zu dem man
kumbst soltu nicht dem man nach dem messer hawen Sunder dw
solt dich fleyssig remen der vier zinnen die erst das ist dy recht
seitten oberhalb der gurtel Dy ander dy linck seytten auch ober=
halb der gurtel dy anderen zwō sein vnderhalb[54] der gurtel des
mans nymb dir der zinnen eine fur vnd acht nicht was er
gegen dir vicht haw dar zu kunniklich[55] versetzt er dir dan So haw
schnel zu der anderen zinnen der negsten ploß die dw gehaben
magst etc

Hastu das vernummen ***11r***
Dy zinnen sein gewunnen
Wiltu dich rechen
Dy vier zinnen kunstenlich prechen
Oben duplir
Do nyden gantz mutir

Hie sagt der meister ob dir einer ernnstlich ein hawett wiltu dich den
rechen an Im vnd dy zinnen kunstenlich prechen So treib dy dupliren
gegen der sterck des messers vnd das mutiren gegen der swech Ich sag
dir furbar das er sich hartt vor schlegen schützen mag vnd hart zu
schlegen kummen kan

Item hawstu Im von deiner lincken seitten vnd pindest Im an sein messer
mit der langen schneiden so far pald auff mit dem arme vnd pleib
alzo sten an dem messer vnd schlag Im hintter seines messers klingen
mit der stumpffen schneyden etc

Item das dupliren treyb alzo wenne er dir oben zu hawet von ***11v***
seiner rechten achsel So haw auch von deiner rechten mit Im gleich
starck oben zu dem kopff versetz er denn den haw mit sterck So
wendt das gehultz vnder dein rechten arme vnd far pald auff

---

[54] From LH, f 10r: *nyderhalb*   [55] From LH, f 10r: *kunelich*

11v (cont.)
mit dem arme vnd schlag In mit der langen schneyden hinter seines
messers klingen auff den kopff

Item nü gesagt Ist von den dupliren Nu wirtt gesagt von den mutiren
dy soltu treyben als vnden gemalt stat vnd das mutiren treyb alzo
wenne dw Im von deiner rechten achsel oben starck ein hawest versetz
er vnd ist waich Im messer So windt Im auff dein lincke seytten auff
sein messer vnd far da mit woll auff mit dem arme vnd heng
Im den ortt oben vber sein messer vnd far da mit dem arme auff
vber sein messer vnd stich Im zu der andren zinnen etc

12r
Item hawestu Im von deiner lincken seytten mit der langen schneyden
an sein messer gepunden so far auff mit dem arme vnd pleyb mit
der selben schneiden an seinem messer vnd windt Im auff sein lincke
seitten vber sein messer vnd stich Im zu der andren zinnen seiner
lincken seitten alzo magstu dy zway stuck treyben auß allen hewen nach
dem du empfindest ob er waich oder hertt Im messer ist etc

12v
Wil er mütiren
Zuck ortt das wirtt dich ziren

Hie lertt der meister Ein pruch wider dy mutiren vnd sprich will er etc
Das soltu alzo verstan will er dy mutiren machen Dy weill er vber dein
messer will varen mit dem seinen So windt mit deinem messer auff
dein lincke seitten In den hangunden ortt vnd stich Im schnel zu
seinem gesicht etc

13r
Will er mütiren auff recht machen
Wind ym hangünden ortt solt wachen

Hye lertt der meyster aber wie man die mutiren prechen soll
vnd spricht will er etc Das soltu alzo verstan will er dy
mutiren machen auff dein rechte seytten so wind gegen ym schnell
den hangenden ortt Das dy recht schneyd ob stee etc

13v
Hangenden ortt auff dy linck seytten
windt: wiltu dy mütiren ab leytten

Hye lertt der meyster wie man dy mutiren prechen soll
auff der lincken seytten vnd spricht hengenden ortt etc Das
sol dw alzo versten Macht er dy mutiren auff deiner
lincken seytten So wind gegen Im den hangenden ortt vnd
stich Im nach der negsten plöß etc

14r
Dy mütiren ym hangenden ortt pisz ab leytten
Geradt zu gesicht magstu reytten

Hie lertt der meyster aber ein pruch wider dy mutiren vnd spricht

14r (cont.)

alzo Dy mutiren ym etc Das soltu alzo versten So dw
nü gewunden hast wider dy mutiren vnd ligst In dem
hang[e]nden ortt Indes far Im zu dem gesicht mit dem langen
ortt vnd stich Im Inn das gesicht etc

Weck auff behendt

14v

Ortt zu gesicht wendt
Wer mit schritten woll weckett
Vill hew er ser streckett

So nw der meyster das erst capitel auß gelegt hatt vnd erzelt
mit seiner eygenschafft Nw hebtt er an das ander capitel vnd
spricht weck auff behendt etc Nü merck Das der wecker
ist der vier versetzen eyns wider die vier leger wann da mitt
gewindt man den stiren vnd den eber vnd dy vnterhew etc

Item Den wecker mach alzo wenn dw mit dem zuuechten zu
dem man kumbst stet er denne gegen dir vnd helt sein messer
für den kopff In der hutt des stiren auff seiner lincken seitten
so setz den lincken füß fur vnd hab dein messer auff deiner rechten
achsel oder In der schranck wey[56] dem rechten pein vnd spring
mit dem rechten fuß woll auff dein rechte seitten vnd schlag
In mit der rechten schneyden wol auff sein messer Indes windt Im
den ort In das gesicht etc

Ortt zu gesicht wecker ist wenden

15r

Vberwindt ortt ist das enden

Hie lertt der meyster ein pruch wider den zoren ortt vnd spricht
ortt zu gesicht etc Das soltu alzo versten macht er dir den
wecker vnd will dir zu dem antlitz stechen auff deiner lincken
seitten Indes windt gegen seinem messer auff sein rechte seitten
vnd stich Im zu dem gesicht seiner rechten seitten etc

Recht lere wecken

15v

Kurtz scholt[57] das auß ecken

Hie lertt der meister aber ein stuck auß dem wecker vnd
spricht Recht lere etc Das soltu alzo versten Machstu den
wecker von deiner lincken seitten Im auff sein rechte seitten Indes
so dw den wecker gemacht hast schlag In mit der kurtzen
schneyden auff sein kopff wie vnden gemalt stett

---

[56] From LH, f 12r: *pey*   [57] *schalt* = *Schwung* (HENNIG 2001, p 279)

16r   Recht mit ortt weck

Heng den oder gerad ſtreck

Hie lertt der meyſter wie dw den ortt ausß dem wecker prauchen
sollt vnd spricht Recht mit ortt etc Vnd das soltu alzo verſtan
wenn du den wecker auff sein rechte ^{seitten} gemacht haſt auff von
deiner lincken seitten Indes windt gegen seiner messer In den han=
genden ortt Das dy lang schneid ob ſtee vnd ſtich Im zu dem
gesicht seiner rechten seitten etc

16v   Weck gesicht ortt leſt varen

Wider windt linck ortt nit solt sparen

Hie sagt der meiſter ein pruch wider das eegemelt ſtuck vnd
spricht weck gesicht etc Das soltu alzo verſtan vertt er
mit dem hangenden ortt In dein gesicht Indes laß dein messer
sincken vnter sein arme vnd far auff mit dem langen ortt
vnd ſtich Im zu seinem gesicht etc

17r   Item wiß das dy schranckhuet iſt gutt vnd vill ſtuck mach
man dar ausß treyben Dw scholt den wecker auch dar auß
treyben von payden seitten vnd schick dich In dy schranckhutt
alzo wenne dw mit dem zuuechten zu dem man kumbſt
so seꜩ deinen rechten fuß vor vnd hallt dein messer mit dem
ortt auff der erden auff deiner lincken seitten das dy ſtumpff
schneid gegen dem man ſtee vnd gib dich alzo ploß mit der
rechten seitten haut er dir dan oben zu der ploß ein So
spring auß dem haw mit deinem lincken füß woll auff sein
rechte seitten vnd wind das gehulꜩ deines messers vnder
dein arme vnd schlag In mit der langen schneyden auff
sein messer vnd wendt Im dy kurꜩ schneyd auff sein haubt
vnd erbeit Indes als du woll weiſt etc

17v   Item alzo schick dich In dy schranckhuet deiner rechten seitten
wenn du mit dem zuuechten zu dem man kumbſt So seꜩ
deinen lincken fuß für vnd halt dein messer mit dem
ortt auff deiner rechten seitten das der ortt auff der erden ſtee
vnd dy kurꜩ schneid gegen dem man ſtee vnd gib dich mit
der ~~rechten~~ ^{lincken} seitten ploß vnd schreyt mit deinem rechten füß
woll auß dem haw auff dein rechte seitten vnd schlag In
mit der kurꜩen schneiden auff sein messer nohendt bey der
handt vnd wendt Im dy lang oder kurꜩ schneyd auff
sein haubt oder ſtich Im zu dem gesicht als vor auch
magſtu den wecker treyben auff den ſtir wie vnden
gemalt iſt etc

Weck linck nicht kurtz haw
Den wechsel recht paw

Hie sagt der meyſter ein ſtuck wider dy hutt auß dem lüginslandt
vnd wider den Stiren vnd treib das allzo wenne du mit dem
zuuechten zu dem man geſt ſtet er denn In diſen obgemelten
hutten oder legeren vnd du ſteſt Inder schranckhut so thu Im
gleich als du Im den wecker wolleſt hawen zu seiner lincken
seitten vnd verpring den haw nicht Sünder wechsel Indes
durch auff dy andren negſten zinnen vnd ſtich Im zu dem
gesicht auff sein rechte seitten oder In dem wechsel schlag
Im dy kurtz schneid vber sein messer auff seiner rechten seitten
vnd senck den ortt woll vber sein messer auff seiner lincken
seitten vnd haw Im dy recht schneid durch sein kopff amen

Durchwechsel linck iſt er wecken
Pleib gerad iſt er ſtrecken

Hie lertt der meiſter ein pruch wider das durchwesseln Im wecker vnd
spricht durchwechsel linck etc Das soltu alzo verstan wenn er durchwech-
sen will Im wecker von seiner rechten seitten auff dein rechte vnd thut
Sam er wöll hawen zu deiner lincken seitten Indes wechselt er durch vnd
ſtich dir zu dem gesicht deiner rechten seitten So pleib alzo ſteen mit dem
messer vnd ſtich Im mit dem langen geraden ortt geſtrackt zu dem gesicht
vnd den hin derſten fuß setz weid hin auß auff das das du deſter weitter
raichen mugſt etc

Den wecker linck haw
Durchwechsel da mit schaw

Hie sagt der meiſter aber ein gut ſtuck So einer ſtet in der hut
lunginslandt oder In dem Stiren oder so er auff dich schlecht ausß
dem luginslandt So lig in der schranckhut auff deiner lincken seitten
vnd haw Im dy ſtumpff schneid auff sein messer Indes far hoch
auff mit dem arme vnd wechsel Im durch vnd ſtich Im zu dem
gesicht oder pruſt auff seiner lincken seitten.

Item wiltu In paß versüchen so laß Indes dein messer mit dem
ortt auff dein lincke seitten vnd haw Im wider ein auff sein
rechte seitten mit der scharpffen oder kurtzen schneid etc

Weckt linck ortt will er helen
Senck ortt geradt ſtoß dy kelen

Hie lertt der meiſter ein pruch wider das durchwechslen auß
dem wecker So er den wecker machen wil von seiner lincken
seitten auff dein rechte seitten vnd spricht weckt linck etc

19v (cont.) Das soltu alzo verstan haut er von seiner lincken seitten den
wecker auff dein rechte seitten vnd will Indes durchwechslen
auff dein lincke seitten so peleib geradt mit deinem esser auff
seiner lincken seitten vnd stich Im mit geradem ortt oder
messer zu der kelen oder antlitz wie vnden gemalt stett etc

20r Weck dy flechen
Den meistern wiltu sy swechen
So es pucht oben
So stand ab das will ich loben

Hie sagtt der meister aber ein gutt stuck das soltu alzo verstan
wenne du mit dem zuuechten zu dem man kumbst So leg
dein messer In dy schranckhut oder halt das auff deiner rechten
achsel haut er dir danne oben zu der plöß So haw auch
starck gegen seinen haw mit der recht[e]n schneiden vnd als pald
dy messer zusammen klitzen So windt Indes den ortt zu dem
gesicht vnd stich In zu der öbren zinnen oder mach das stuck
das ist auch dy maynung des textz als pald dy messer zusammen
klitzen So haw Indes mit der kurtzen schneiden zu dem kopff
seiner lincken seitten Indes haw wider behentlich von einer zinnen
zu der anderen als du weist etc

20v Der wecket vmb süst[58]
Henget ortt er trifft dy prust

Hie sagtt der meister ein gemein pruch wider den wecker vnd
spricht Der wecket etc Das soltu alzo verstan Macht einer
auff dich den wecker so merck dar auff wenne ein messer an
das ander klitzen will Indes windt dein messer gegen deiner
lincken seitten vnd heng Im den ortt auff sein prust vnd sch=
eub In woll von dir etc

21r Mit wecker will er placken
Lanck kopff soltu hacken

Hie lertt der meister ein anderen pruch wider den wecker vnd
spricht mit wecker etc Das soltu alzo verstan Ist er den wecker
auff dich machen so merck dar auff So er gegen deinem messer ist
hawen Indes ~~ku~~ tzuck dein messer gegen dir vnd senck den ortt
gegen deiner lincken seitten vnd haw Im vber sein messer geradt
auff sein kopff wie vnden gemalt stet etc

---

[58] *vmb süst = umsonst; sus = sunst* (BAUFELD 1996,
p 231)

Will er den wecker prauchen
Zuck durchge auff recht sölt In gauchen

21v

Hie lertt der meister aber ein pruch wider den wecker vnd spricht
will er den wecker etc Das soltu alzo verstan So er macht
den wecker auff dich wie vor Indes gee durch auff sein rechte
seitten vber sein messer vnd stich Im zu dem kopff oder gesicht
wie vnden gemalt stet etc

Den wecker ist er hawen
Entrüst tzuck ortes soltu dich fräwen

22r

Hie lertt der meister aber ein pruch wider den wecker vnd
spricht Den wecker ist etc Das soltu alzo verstan hawt
er von seiner rechten seitten den wecker auff dein lincke seitten
Indes windt dein messer auff dein lincke seitten vnd far mit
dem hangenden ortt auff vnd schlag In mit dem entrusthaw
zu seinem kopff oder setz Im den ortt an als du woll
waist etc

Weck dy flech grob mit doben
Stand ab: oder pleib oben

22v

Hie sagtt der meister ein guttes stuck wider dy die nicht
vechten kunnen oder wider dy pawrenstraich vnd mach
das alzo haut einer auff dich mit einem grossen pawrenschlag
vnd will dich ernnstlich durch den kopff hawen So
halt dein messer in der schranckhutt oder pey der rechten
achsel oder neben dem knye das dy spitz seines messers gegen
Seinem[59] antlitz stee vnd wie du denn In eynem standt stest
so magstu das stuck prauchen vnd thu Im alzo hautt
er auff dich so haw auch gleich mit Im eyn zwischen
messer vnd man mit der langen schneiden vnd als pald
dy messer zu sammen klitzen So far Indes gegen deiner
lincken seitten vmb den kopff auff dein rechte seitten mitt
deynem messer vnd haw Im zu dem hals will du Im nit
zu dem hals oder kopff hawen so schlag In auff den lincken
arme oder In den pauch seiner lincken seitten grob vnd pew=
erisch das er seyn woll empfindt etc

Mitt pogen ler abetzen
Waß grob ist letzen

23r

---

[59] Corrected from *Deinem*.

23r (cont.) Hie lertt der meiſter ein pruch widr das ee gemelt ſtuck
vnd spricht mit pogen etc Das soltu alzo verſten So er vmb
sein haubt will hawen dich an deinen halß schlagen will
so pewg auff vnder seyn messer vnd verſetz mit dem pogen
ob du wildt Indes windt vber sein messer auff seiner rechten
seitten vnd trit Indes hintter In vnd schlag In zu seynem kopff etc

23v ### Grob will er wecken vnd gaffen
Starck ſtoß kelen das gibt den affen

Hie lertt der meiſter aber ein pruch widr das ee gemelt ſtuck
vnd spricht Grob will er etc Das soltu alzo verſtan will
er den groben schlag In deyn pauch hawen od[e]r an dein kopff
So merck als pald er vmb seyn haubt faren will vnd will
dich schlagen Indes far behentlich Im zu seinem hals oder
angesicht vnd ſtoß In ſtarck vnd scheub In da mit von
dir wie vnden gemalt ſtet etc

24r ### Weck höflich mit wincken
Auff recht von der lincken

Hie lertt der meiſter ein hübsch ſtuck auß dem wecker vnd
spricht weck höflich etc Das soltu alzo verſtan Mach
den wecker von deiner rechten seitten auff sein lincke seitten
Indes als pald ein messer auff das ander klitz So wendt
dein handt Ewchling gegen seinem messer Das dy kurtz schneid
In ein wenig rür auff seiner lincken seitten Indes senck den
ortt gegen der erden vnd far zwischen dich vnd In auff gegen
deiner lincken seitten vnd schlag In auff seiner rechten seitten
dy kurtz oder lang schneiden auff sein kopff oder ſtich schneid
vnd thu was du wild etc

24v ### Der swech nymb war
Den halß vach vngevar

Hie sagt der meiſter ein ſtuck wider die swach In der versatzung
sein vnd treib das alzo So dw Im mit dem wecker ſtarck
auff sein messer pindeſt iſt er denne swach Im pandt So windt
Indes auff sein rechte seitten deyn messer vnd nymb In bey
dem hals hebt er den mit der ſterck auff sein messer So
haw Im wider schnel vnder sein messer zu seinem haubt
auff seyner rechten seitten iſt er aber als ſtarck das du das
ſtuck nicht treyben kanſt So erbeit schnel mit der swech
zu der negſten plöß eyner andren zynnen

Item Du solt wissen das du nit lang magſt eynen halten mit
dem messer bey dem hals mit eyner handt dar vmb merck

als pald du Im nach dem hals arbeittest als pald ſtich Im
zu dem gesicht etc

Der entrüſt ny[m]bt
waß on oben dar kümpt

Hier heb der meiſter an zu sagen von der dritten ſtuck vnd capit[e]l
Im messer vnd erlewttert das mit seiner eygenschafft Nu merk
der entruſthaw pricht dy hut vom luginslandt vnd alle
hew dy von oben nyder gehawt werden vnd treibt den alzo wenne
du mit dem zuuechten zu dem geſt vertt er denn gegen dir vnd
helt sein messer mit ausgeracktem arme vber das haupt In der
hut luginslandt vnd wartt auff dich So merk wenn du nohent
zu Im kumbſt So setz deinen linken fuß fur vnd halt dein
messer auff der erden das dy flech gegen deinem leibe ſtee an deiner
rechten seitten tritt er denn zu dir vnd drött dir zu schlagen so
soltu vor kummen vnd spring mit dem fuß woll auf dein rechte
seitten vnd Im sprung wendt dein messer mit dem gehultz fur
dem haubt das der dawm vnden ſtee vnd schlag Im zu mit
der ſtumpffen schneyden zu seiner lincken seytten zu der obren zinnen
kumpt er aber Ee mit dem haw den du so spring mit dem rechten fuß
vnd mit der vor geſtriben[60] versatzung auß dem haw
wol auff dein rechte seytten vnd schlag In mit
dem entruſthaw lanck auß dem arme zu seinem
haubt etc

Ruſt zu der ſterck
Dein erbeyt da mit merck

Es iſt zu wissen wenne dw Im mit dem ~~et~~ entruſthaw zu
haweſt das soltu thun mit ſterck deines messers versetzt er
denn so far auff mit der ſterck deynes messers nach der
swech seines messers So du pegreuffeſt mit der ſterck seyn
swech so soltu mutiren Im vber sein messer zu der andren
zinnen oben zu dem hals so du aber nicht dar zu kummen
magſt So erbeyt auß dem dupliren hinter seynes messers klingen
vnd schlag Im zu dem kopff Iſt er aber mit der versatzung zu
ſtarck das du zu den ſtücken nicht kummen magſt so erbeyt mit
dem kreutz oder mit dem wernagel vnd ſtoß Im sein messer
da mit hyn weck vnd schlag Im mit dem entruſthaw zu
der andren zynnen etc

---

[60] *geschrieben*

26r **Rüſt zu den zynnen**
Kumbſt dar eyn dir wirtt gelingen

Hie wirſt gelertt das du dich fleyssig solt remen der vyer zynnen
So magſtu den man deſter gewiser haben vnd dem kunelich[61] zu
hawen vnd nicht haw zu dem messer als du den vor vnterricht
piſt wenne einer mit dem zuuechten zu dir kumbt ſtet er danne
gegen dir yn der hut luginslandt So spring mit dem rechten fuß
gegen Im woll auff deyn rechte seytten vnd schlag In mit dem
entruſthaw zu der obren zynnen verſetzt er den so schlag aber pald
zu der vndren zynnen seyner rechten seytten alzo haw den haw
von eyner zynnen zu der andren vnd wiltu das dir geling so
soltu mit ytlichem entruſthaw zu der obren zynnen schlagen
verſetzt er denn So schlag aber pald zu der obren zynnen seyner
rechten seytten alzo haw den haw von eyner seytten zu der andren
vnd wiltu das dir geling so soltu mit ytlichen Entruſthaw
weyt auß springen auff eyn seytten vnd mit dem arme gib dich
woll hyn vmb So haſtu den man deſter gewyſter vnd magſtu
Im an seyn haubt woll eyn gutte Rür geben Indes soltu dich
pewaren das du dein schilt das iſt dy versatzung woll fur dein
haubt halteſt etc

26v **Feler verfuret**
Dy zinnen nach wunsch er rüret

Hie lertt der meyſter wie man mit dem entruſthaẅ felen
sollt wenne mit den feleren werden dy meyſter ser geplewet
dye dy newer versetzen wollen vnd hawen nach dem messer des
manß Nu den fclcr trcyb alzo Thu als sam du yn wolleſt
hawen mit dem entruſthaw zu den obren zynnen seyner linken
seytten vnd ym haw So du nohent bey dem messer piſt So
haw ihm zu der andren zynnen ~~seytten~~ seyner rechten seytten zu
dem haubt etc

27r **Ob du feleſt zwiuach**
Dy schnydt schneydt mit mach

Item du solt das alzo verſtan halt deyn messer auff deyner
rechten seytten das der ortt auff der erden sey vnd der dawm vnden
an der flech des messers vnd so er denn hoch ligt Im lunginslandt
vnd haſt gemerckt das er sich leichtiklich verfuren leſt vnd
gerne der versatzung nach gett so var auff fur deyn leyb mit dem
messer vnd laß den ort hangen gegen der erden vnd var mit dem

---

[61] LH, f 17r: *künstenlichen*

messer vber deyn linke achselen vntz vber dy mit ~~sey~~ deynes   27r
haubtz das der ortt vber ruck hyn auß gee vnd das gehultz
gegen dem man ſte vnd thu sam du In wolleſt schlagen zu seyner
lincken seyten yndes schlag In zu dem rechten oren mit dem ent=
rüſthaw etc

Item wiltu den feler recht machen so thu Im alzo ſtee   27v
mit deynem lincken fuß fur als vor So du nw das
messer fur deynen leib fureſt So schreyt mit dem rechten
fuß zu deynen lincken vnd dro ym als du wolleſt zu
seyner lincken seytten schlagen vnd wenck mit dem
obren teyl deynes leybs gegen seyner lincken seytten dy
weyll yndes schreit mit deynem lincken fuß auff seyn rechte
seytten vnd schlag In mit dem entruſthaw zu seynen
rechten oren yndes far auff mit dem messer vnd thu aber
samm du In wolleſt schlahen auff seyn lincke seytten vnd wenck
eyn wenig mit dem leyb auff seyn lincke seytten Indes
schreyt aber mit deyner rechten fuß hinter seynen rechten vnd
schlag In wider zu der rechten seytten eyn mit dem dupliren
vnd schlag Im dy kurtz schneid hinter seynes messers
klingen In das angesicht Indes schreit nochpaß vmb
In vnd far aber auff mit dem messer als vor vnd schreyt
mit dem lincken fuß woll hinter seynen rechten vnd schlag
In In seyn genick Indes var aber auff vnd schreyt
aber hinter yn vnd schlag Im in den pauch das heyſt
der treffer Indes vmbschreyt In gar vnd haw Im
wider zu dem nacken etc ¶

¶Item wiltu disß ſtuck Eegemelt nicht treyben So schneyd In mit   28r
dem dupliren durch das gesicht mit dem dupliren oder will er yndes
gen auß dem pandt so schneyd In ſtarck yn dy negſten ploß etc

Will er zu den zynnen Rüſten vnd messen
Sprechuenſter soltu nicht vergessen

Hie lertt der meyſter eyn pruch wider dy veler vnd entruſthew
vnd spricht will er zu etc Das soltu alzo verſtan will
er mit den ruſthewen feler machen So soltu das prechen mit
dem sprechuenſter das iſt mit dem langen ortt vnd dy plöß
da mit meyſterlich süchen als du den woll weyſt etc

Der zwinger eyn pricht   28v
Waß püffel haüt oder ſticht
Wer wechsel ~~träet~~ ᵖⁱˢᵗ traüen
Der zwinger will In berauben

28v (cont.)

Hie sagt der Meyster von dem virden stuck vnd capitel das
da heysset der zwinger den treyb alzo wenn du mit dem
zuuechten zu dem man kumbst So setz deynen lincken fuß für
vnd halt deyn messer auff der erden mit dem ortt das der dawm
vnden sey auff dem messer haut er denn von oben zu dem
haubtt so verbendt deyn messer vnd haw gegen seynen haw
eyn mit der stumpffen schneyden lanck außgestrackten arme oben
vber seyn messer Im zu dem gesicht Wer er aber als listig
vnd felet mit dem vnd wechslet vntten durch so pleyb mit
dem langen ortt lanck vor seym gesicht so mag er nichtz
schaffen vnd kan vnden nicht durch kummen etc

29r

Der gefer mit seyner artt
Des antlitz vnd der prust wart

Hie sagtt der meyster von dem gefer von dem fünften stuck vnd
capitel den soltu alzo machen ligt eyner In der hut pasteyn
vnd so du mit dem zuuechten zu Im kumbst so setz deynen
lincken fuß für vnd halt deyn messer mit aufgerecktem arme
hoch vbersich vber deyn haubtt In der hut luginslandt vnd
spring zu Im mit dem rechten fuß vnd haw mit der langen
schneyden oben eyn vnd pleyb mit dem arme hoch vnd senck
Im den ortt vntersich zu seynem gesicht oder prust etc

29v

Gefer haw
Durchwechsel yndes schaw

Item das stuck soltu alzo treyben ligtt er gegn dir Im lug=
inslandt So du zu Im kumbst So haw gegen seinem messer
Ist er hoch vnd will den haw ab setzen Indes laß deyn
ortt faren nyder sincken vnd wechsel durch vnd stich ym
auff seyn rechte seytten zu dem gesicht mit gestracktem arme
Indes mach waß dir eben ist etc

30r

Den wincker soltu erstrecken
vnd dy meyster do mit erbecken
zwiuach ler wincken
zu der rechten von der lincken

Hie sagt der meyster von dem sechsten stuck das do heisset der
wincker vnd ist der hauptstuck eyns ym messer vnd geret woll auff
dy freyfechter dy frey versetzen vnd ist eyn newer haw vnd stuck ym
messer vnd ist seltzam vnd gutt wiltu das stuck machen So thu
Im alzo stee mit deynem lincken fuß fur vnd halt deyn messer auff
deyner rechten achsl stet er dan Im luginslandt So haw von deyner
rechten achsel mit stracktem arme zu Im lanck eyn mit der langen

schneyden vnd Im haw so windt gegen seynem messer vnd schlag
In auff das haubt mit der kurzen schneyden yndes laß deyn messer
sincken fur dein leib vnd ge durch Indes auff deyn lincke seytten mit
gesenckten ortt vnd schlag In auff den kopff auff seyner rechten seytten
mit der kurtzen schneyden etc

30r (cont.)

Item leg deyn messer auff deyn lincke seytten das der ortt auff der
erden ſtee vnd der daum oben auff dem messer lig oder ſtee In
der schranckhutt das der recht fuß vor ſtee ligtt er den In der
hutt luginslandt So haw Ewchling eyn zu seyner rechten
seytten dy kurtz schneid yndes gee durch mit dem messer vnd
senck den ortt vnd far mit dem messer auff seyn lincke seytten
alzo haſtu das ſtuck zwiuach von payden seytten etc

30v

Item wiltu aber nicht durchgen das er dich zu letz dunck
durch zugen von seyner rechten seytten So pleyb alzo ligen mit
dem messer vnd windt gegen dyner lincken seytten das dy
kurtz schneyd vnten ſtee vnd ſtich Im zu dem gesicht etc

Windt linck ym treffen
Haw recht: lanck wiltu yn effen

31r

Das soltu alzo verſtan ſtet er gegen dir Im luginslandt So haw
von deyner rechten achsel gerad oben eyn zu Im eynen oberhaw mit
langem oder geſtracktem arme zu seyner lincken seytten Nu merck ym
haw dy weill der haw nach dem kopff gett yndes windt gegen
seynen messer das dy kurtz schneyd vnden ſtee vnd verendt den schlag
nicht Sunder yn dem schlag laß den ortt sincken vnd ge vntten
durch mit dem messer auff deyn lincke seytten vnd schlag In mit
der kurtzen schneyden auff den kopff yndes wendt deyn messer
gegen dir vnd schlag In auff der selbigen seitten auff den kopff
mit der langen schneyden Indes mach waß dir eben iſt Das ſtuck
gett auch zu zue peyden (sic) seytten seytten etc

Item so du haweſt mit der langen schneyden auff seyner rechten seytten
nach seynem haubt Indes windt gegen dir dein messer
vnd var auff mit dem arme vnd ſtich Im zu dem ges
icht seyner rechten seytten Indes windt wider gegen Im
auff sein rechte seytten vnd ſtich auff seyner rechten seytten
wider zu dem gesicht etc

31v

Wiltu dy meiſter plewen
Des winckers soltu dich frewen
Was kumpt krumpt oder schlecht
Das iſt dem wincker als gerecht

32r

32r (cont.) Hie sagt der meyſter voɳ dem ſtuck das da heyſt der winckeꝛ
vnd spricht wiltu dy meyſtꝛ etc Das soltu alzo verſtan
das daß ſtuck recht zu get auff eynem yꜩlichen er verſeꜩ
krump odeꝛ schlecht So hatt das ſtuck seinen furganck Thu
Im alzo versetzt er krump so haw voɳ deyneꝛ rechten achsel
oben auff sein messeꝛ eynen freyen oberhaw vnd Im haw So
verbendt dein messeꝛ Ee den das eyns an das andeꝛ kliꜩet
vnd var mit dem arme hoch auff vnd schlag Im dy
kurꜩ schneyd auff sein kopff etc

32v Item haſtu In nicht getroffen als yeꜩund gesagt iſt mit
dem haw So far mit dem ortt gegen deineꝛ lincken
seytten vnd ſtich Im zu der kelen odeꝛ pruſt odeꝛ nymb
ab auff dein messeꝛ als du woll weiſt etc

33r **Vier leger soltu mynnen**
Im messeꝛ vechten wild gewynnen
Paſtey vnd luginslandt
Stir vnd eber sy dir bekandt

Saend[62] So nw̄ der meyſteꝛ gesagt hat vnd außgericht dy
Sechs hew odeꝛ capitel nw̄ sagt er hie voɳ dem sibenden capitel
vnd haubtſtuck vnd zu dem erſten von den vier legern im
messeꝛ Da voɳ man halten mag vnd sich mit recht kunſt daꝛ
auß weren

Item Das erſt legeꝛ oder hüt heysset dy paſtey Schick dich in dy
alzo Seꜩ deinen rechten füß vor vnd halt dein messeꝛ mit
geſtracktem arme vor dir mit dem ortt auff der erden das dy
kurꜩ schneyd oben ſtee etc

33v Item dy ander hut oder legeꝛ iſt der luginslandt schick dich
dar ein alzo seꜩ deynen rechten fuß für vnd halt dein messeꝛ
mit aufgerecktem arme hoch fur dein haubt vnd dy
lang schneid odeꝛ scharpff schneid gegen dem man ſtet
als oben gemalt ſtett da du ein handt gemalt sichſt

Item In den Ebereꝛ Schick dich alzo seꜩ deinen linck[e]n füß
fur vnd halt dein messeꝛ bey dem rechten payn auff der
rechten seytten mit dem gehulꜩ neben der hüff das dy
ſtumpff schneid oben ſtee vnd der ortt fursich auff ſtee
dem man zu dem gesicht

---

[62] *Sa* from *sar* = "immediately afterward", "then immediately" (either alone or reinforced by related expressions) (LEXER 1885, p 205), *end* = *ehe*, "before" (LEXER 1885, p 43)

Item du solt den Eber treyben auch von der lincken seytten
mit dem messer vnd schick dich In dy hutt alzo halt
dein messer neben deiner lincken seytten neben bey der hüff
das dy lang schneyd oben gewendt sey vnd das gehultz
sey eyn benigß gesenckt gegen der erden vnd der ort
vbersich ſtee dem man zu dem gesicht wie vnden gemalt
ſtet etc

33v (cont.)

Die vierdt hut heysset der Stir Schick dich In den alzo
ſtee mit deynem lincken fuß fur vnd halt deyn messer zu
der lincken seytten mit dem gehultz fur das haubt das dy
kurtz schneyd gegen dir ſtee vnd der ortt zu dem gesicht etc
Item In den Stiren auff der andren seytten Schick dich wie
vnden gemalt ſtet etc

34r

Vier versetzen soltu ab synnen
Wiltu dy leger zwingen
Vor versetzen dich hüet
Versetzen dick den man müet

34v

Hie sagt der meyſter von dem achten capitel oder ſtuck
Das ~~seyn dye vier versetzen dy selben prechen dy vier~~
~~hüt Nu soltu wissen das dy vier versetzen synd die vier~~
~~hütt~~ Das sein dy vier versetzen dy selben prechen dy vier hutt
Nu soltu wissen bas dy vier versetzen sindt vnd wiß das
keyn versetzen wider dy vier hutt gehortt den dy vier hew
dy prechen dy vier hutt
Item der erſt haw iſt der wecker Der pricht dy hut auß dem
Stiren
Der ander iſt der entrußthaw der pricht dy hut vom luginslandt
Item der dritt heysset der zwinger der pricht dy hut auß dem
eber
Item der vierd iſt der geferhaw der pricht dy hut dy da heyſt
dy paſtey wie du dye ſtuck treyben solt das piſtu zu gutter
maß woll vnterricht Da vorne da dy haw erzelt werden
vnd wye man dy vier hutt prechen soll
Item Es spricht auch der text vor versetzen hutt dich Das iſt
als vill geredt das du nicht vill versetzen sollt willtu
anders nicht geschendt vnd geschlagen werden als du den
vor vnterricht piſt etc

Vermeynt sich Eber ſtir zu weren
Langen geraden ortt soltu im keren

35r

Hie lertt der meyſter pruch wider den eber vnd den ſtiren
vnd spricht vermeynt sich etc Das soltu alzo verſtan ſtet

35r (cont.)   er Im eber auff seiner lincken seitten vnd will dir ſtechen
auff dein rechtn seytten So far mit langen geſtracktem ortt
Im Indes zu seym gesicht vnd vall mit dem hindreſten
fuß woll hinauß So trifſtu In in das antlitz vnd du solt
wissen das die pruch mit dem langen ortt pesser seyn denne
dy vor gemelten auß den hewen worvmb dy hew kunnen
sich selbſt vor dem langen ortt nich bebaren sy werden da
mit geprochen etc

35v   Item Stet er In dem Eber auff seyner rechten seytten vnd
will dir ſtechen auß der hut zu deyner lincken seytten So
leg dich mit deynem messer auff dy erden das der dauwm oben
ſtee auff der flech vnd dy kurtz schneyd gegen dem man vnd
far Im mit geradem langen ortt zu dem gesicht wie vnden
gemalt ſtet etc

36r   Item ſtet er in dem Stiren auff seyner rechten seytten vnd
will dir zu ſtechen auß der selbigen hut Indes var aber
gerad auff vnd ſtich Im zu dem gesicht mit dem langen ort
wie vnden gemalt ſtet etc

36v   Item Stet Er In dem Stiren auff seyner lincken seytten vnd
will dir alzo zu ſtechen wie vor So far auff mit geſtra=
cktem ortt vnd ſtich Im zu seym gesicht oder kelen oder
pruſt wie vnden gemalt ſtet vnd mit disen pruchen
prichſtu dy vier hut am aller paſten etc

37r   **Dw solt an setzen**
vnd an vyer enden In letzen
Er kumb oben oder vnden
Ortt gerad iſt yn verbunden

Hie sagtt der meyſter von den an setzen vnd das ansetzen iſt ein
gutz ernnſtlichs ſtuck vnd gett nohent zu den vier zynnen
vnd gehortt zu treyben als pald du endt will geben mit dem
messer vnd treyb dy ansetzen alzo leg dich In dy hut paſtey
will er den oben ein hawen oder ſtechen von seyner rechten seytten
(sic) Sokum vor vnd scheuß Im den langen ortt zu der negſten
ploß seyner lincken oder rechten seytten etc

37v   Item haut er von oben eyn seyner lincken seytten Sokum vor
mit der versatzung vnd scheuß Im den ortt lanck eyn zu
seyner rechten seytten zu der negſten plöß etc

38r   Item haut er dir denn von vnden auff seyn rechte seytten so scheuß
Im den ortt linck eyn zu der negſten ploß seyner lincken seytten etc

Item oder hawt er dir zu von vnden auff von seyner lincken
seytten So scheuß Im den ortt lanck ein zu der vndren ploß
seyner rechten seytten wirtt er deß schuß gewar vnd
versetzt So pleyb mit dem messer auff seynem vnd erbeyt
behentlich zu der negsten ploß etc

Item du solt auch wissen als pald Ir zu sammen kumbt
Im zuuechten vnd als pald er seyn messer newer hebt
vnd vmbschlagen will alzo pald soltu In den ortt fallen
vnd stechen nach der negsten ploß will er aber mit dem
messer nicht gan so soltu selbst mit deynem messer gen
vnd als offt du eyn schlag endest als offt vall Indes In den
ortt kanstu dy ansetzen recht treyben so kann er an schaden
hartt von dir kummen sequir textus;

Ler nachraysen
Zu wer dy schnydt thu weysen
Zwiuach ler dy machen
mit künsten dich solt besachen

Hie sagtt der meyster von dem newden capitel vnd stuck das ist
das nachrayßen der ist vill vnd manigerlay dy gehören zu treyben
mit grosser fürsichtikeyt gegen den freyuechteren dy auß langen
freyen hewen vechten vnd sunst von rechter kunst des messers
nicht halten vnd die sich verhawen vnd schlagen nach den
peuchen vnd nach den seytten vnd dy sy vor dem man abzyhen
vnd mach das nachraysen alzo wenne du mit dem zuuechten
zu dem man kumbst so setz deynen lincken fuß für vnd stee In der
hutt luginslandt vnd sich eben was er gegen dir vicht hawt
er dir den von oben lanck eyn von seyner rechten achsel so wartt
das er dich mit dem haw nicht erlang vnd merck dy weil er
eyn messer vntersich mit dem haw lest gen gegen der erden so
spring mit dem rechten fuß zü vnd haw Im oben zu der ploß
seyner rechten seytten Ee wenn Er mit dem messer wider
auff kumpt etc

Item ein anders stuck wenne er sich verhawet vnd dw Im
nach hauest fertt er denne pald auff mit dem messer vnd
versetzt So pleyb mit der rechten schneyden auff seynem
messer starck hebtt er denne mit dem messer vbersich So
spring mit deynem linken fuß woll hinter seynen rechten
vnd schlag Im zu dem kopf seyner rechten seyten Indes
mach waß du wild vnd merck albey ob er weich
oder hertt im pandt sey Dar vmb soltu wissen dy enpfinden
des messers etc

40r   Item aber ein stuck wenne er sich vor dir verhawet vnd du
nach Im hawest pindest denn auf seyn messer gegen seyner
lincken seytten Schlecht er denne auß der versatzung mit dem
entrusthaw pald vmb dich Zu deiner rechten seytten So kum
yndes mit dem entrusthaw vorne vnder seyn messer gegen
seyner lincken seytten an seyn hals oder treyb dy schnyt oder
var auf mit dem gehultz vnd senk den ortt In sein antlitz
ader prust vnd scheub In alzo zu ruck Indes erbeyt nach den
zynnen etc

40v   Item aber ein stuck leg dich In dy hut pastey stee mit deynem
lincken fuß fur will er dir den von seyner rechten seytten hawen
auff das messer Indes weich auß dem haw mit dem messer
auff seyn rechte seytten vnd stich Im mit gestracktem arme
vber seyn messer vnd schreyt mit dem rechten peyn hyn
nach Ist er als behend vnd versetzt den stich Indes far
auff behendlich mit gestracktem arme windt auff seyn
messer Im den ort starck zu dem gesicht etc

41r   Item du magst die nachraysen zu payden seytten machen vnd
dy schnyt auch dar eyn pringen vnd thu Im alzo wenne er
sich vor dir verhawet Es sey von der rechten oder linken seytten
So haw Im frŏhlich nach der ploß fertt er dann auff vnd
pindt dir vnden an das messer So merck als pald eyn messer
an das ander klitzt so fal Im Indes mit der langen schneyden
In seyn arme vnd druck mit dem schnyt vntersich oder schneyd
Im In das angesicht etc

41v   **Umb das hawbt will er gagen**
            Zu halz linck soltu schlagen

    Hie lertt der meyster aber eyn nachraysen vnd sprich vmb das
haubtt etc Das soltu alzo verstan ligt er In der zwirch
auff seyner linken seytten gegen deyner rechten vnd will vmb
seynen kopff schlagen zu deyner linckn seytten Indes pleyb
mit deynem messer alzo ligen vnd var gerad nyder vnd
haw In durch seyn hals wie vnden gemalt stett etc

42r   **Verhawen sy sich dy Iüflappen**
            Nachrayß gib In der alten rappen

    Dasoltu alzo verstan will dich ainer In deynen pauch
schlagen vnd will da mit hin weck springen so merk als
pald er mit dem haw nyder gett So schlag Im nach seynem
haubt wie vnden gemalt stet etc

Pogen mit schilt recht dich solt bebaren　　　　　42v
Nachraysen will er den nach faren

Hier lertt der meyster wie man dy nachraysen versetzen
sol vnd spricht pogen mit etc Daß soltu alzo versten hastu
dich verhauet auff dein rechte seytten so soltu mit dem pogen
faren vnder seyn messer daß der dawm vnten stee an deynem
messer vnd das gehultz gegen deyner lincken seytten will er
alzo nach raysen so vach den schlag auff dy zwirch
deines messers das dy kurtz schneid gegen Im stet etc

Pogen vberwindt wang ler stechen　　　　　43r
vnd mit ortt den pogen prechen

Das soltu alzo versten will er auff faren mit dem pogen vnd
versetzen dy nachraysen Indes windt vber sein messer auff sein
rechte seytten vnd stich Im mit dem ortt nach dem rechten
wang seyner rechten seytten etc

Vberbintten mit ortt will er possen　　　　　43v
vberfar mit cloß: schlag zu gesicht thu stossen

Hie lertt der meyster eyn pruch wider den widerpruch vnd
solt das alzo versten windt er dir zu deynen rechten
oren vnd will dich zu dem wang stechen wie vor Indes
far mit dem gehultz vber seyn swech des messers vnd
schlag oder stich Im zu seym angesicht etc

Auff linck solstu auff pygen　　　　　44r
Will nachraysen dich betrigen

Das soltu alzo verstan hastu dich verhawen auff dein lincke
seytten vnd er will dir nachraysen so far auff mit dem
pogen vnter seyn messer das dy lang schneyd gegen Im stee
wie vnden gemalt stett etc

Windt ortt linck auff pogen　　　　　44v
Mach den krig er wirt betrogen

Das soltu alzo versten gett er gegen dir mit dem pogen von scyner
lincken seytten So erbeyt Im mit dem krig zu seynen oren des
lincken wangs vnd windt auff sein lincke seytten das
der dawmen Innen stee gegen dir vnd dy lang schneyd vnden vnd
stich Im Zu seynem antlitz oder wo du hyn wild etc

Windt er vnd lest ortt schawen　　　　　45r
Far vbersich ortt stich oder magst hawen

45r (cont.)   Hie lertt der meyster eyn pruch wyder den Eegemelten
pruch windt er dir zu dem angesicht wye for So haw
Indes vbersich mit gesenktem ortt mit der langen schneyden
In seyn gesicht Oder stich In zu der kelen etc

45v   **Will er als vor hawen**
Ortt gerad macht Im grawen

Das soltu alzo versten hawt er dir zu dem gesich als Eegemelt
ist oder will dich dar eyn stechen so pleyb alzo sten vnd stich In
mit dem langen ortt mit gestracktem arme In das angesicht
So kann er dich mit seynem haw nicht erlangen etc

46r   **Vberlauff dy vntren ram**
Sterck vnd In bescham
Ist dy versatzung dar pracht
kurtz ortt wind piß bedacht

Hie sagtt der meyster von dem zehnden capitel als von
dem vberlauffen vnd spricht vberlauff etc Das verstee
alzo wenne du mit dem zuuechten zu den man kumbst
hautt er dir danne vntten zu das versetz nit Sunder merck
wenn seyn vnterhew oder halphew gegen dir gen So haw
Im von deyner rechten achsel von oben lanck eyn vnd scheuß
Im den ortt lanck zu dem gesicht vnd setz Im an so kann
er dich vnden nicht erraichen fertt er denn von vntten
auf vnd versetzt so pleyb mit der langen schneyden starck
auff seynem messer vnd Indes far auff mit dem arme wind
vnd heng Im den ortt Zu dem gesich Amen

46v   **Entrüst in vberlauff**
wind vnd erhoch den knauff

Hie sagtt der meyster wye man Im messer vberlauffen sol
vnd spricht Entrust In etc Das verste alzo ligtt er In
der hut luginslandt So hab dein messer auff der rechten
seytten stee mit dem lincken fuß fur vnd gee mit dem Ent
rusthaw zu seyner lincken seytten vast an In Indes far
vnden vber seyn messer mit dem gehultz auf seyn rechte
seytten vnd zeuch auff seynem messer dein messer zu ruck
vnd far auff mit dem gehultz vnd vall mit der lincken
hand In dy mit deynes messers vnd stich Im In das
antlitz mit gewappender hand oder[63] etc

---

[63] Appears to be cut off compared to LH.

Vberlauff mit schlagendem ortt
Zu kopff Zu pauch: hals schlag den mort

Das soltu alzo verstan so Du vber faren pist seyn messer
als vor geschriben stet So du deyn messer alzo an dich zogen
hast so vall In dy mit deynes messers mit deyner lincken
handt vnd wendt dy recht an deynem messer vmb das der
dawm vbersich stee vnd schlag In mit dem gehultz auff
sein kopff vnd schreyt Indes mit dem rechten fuß woll
hinter seynen rechten vnd schreyt mit deynem lincken fuß
auch woll hinter In vnd schlag In an alle versatzung
mit dem gehultz In seynen pauch Indes hintertrit In noch
paß schlag In den mit dem gehultz In seynen nacken etc

Wiltu dich vberlauffens mo°ssen
Vach ^hals^ vber peyn ler stossen

Hie sagtt der meyster wie du solt vberlauffen den man vnd
pey dem hals vahen vnd werffen vnd solt das alzo verstan
Ge zu dem man als vor von deyner rechten seytten zu seyner
lincken Indes windt stark aber mit dem gehultz vber
seyn messer vnd kum der rechten handt zu hilff vnd
far mit der lincken hand In dy swech deyneß messers
klingen vnd var Im an den hals seyner lincken seytten
vnd druck In mit der handt oder arme starck vnter seynen
hals vnd schreyt mit deynem linken peyn hinter seyn
rechteß peyn vnd stoß In alzo dar vber etc

Deyn schneyd wendt
Schneyd: mit ortt piß behendt
Will er ringen vnd drucken
| Stoß ring wer ler zucken

Hie sagtt der meyster ein pruch wider dy Eegemelten verß
vnd yre stuck vnd zum ersten wider dy ersten zwen
verß so er sprich Entrust In etc

Item den ersten pruch mach alzo fertt er dir mit dem
gehultz vber dein messer auff zu ruck wie vor gemelt
ist vnd will dich mit geboppender hand mit dem kurzen
ortt In das gesicht stossen Indes dy weill er mit dem gehultz
auff fertt auff deynem messer Indes wendt dein schneydt
vbersich gegen Im vnd schneyd Im mit der langen schneyden
vber seyn hand vnd druck vast vndersich auff seyn lincke
seytten etc

48v Item ein ander pruch fertt er dir mit dem gehultz als vor
vnd will dich mit ~~gep~~ gebappendem ortt In das gesicht
stechen so windt deyn messer gegen deyner lincken seytten
das dy lang schneyd ob ste Indes senck deyn ortt auff
seyn prust vnd scheub In Indes mit dem stich stark
zu ruck etc

49r *No text.*

49v *No text.*

50r Item aber ein pruch fertt er mit dem gehultz als vor vnd wil
dich stechen In das gesicht als vor So setz den stich ab mit
lerer handt vnd laß deyn messer valen Indes greuff in seynes
messers klingen nohendt bey seynem gehultz mit deyner rechten
handt vnd mit deyner lincken verkertten handt gleych als du
da mit versetz hast greuff Im In dy swech seynes messers
vnd druck oben von dir vnd Reyß mit deyner rechten handt
vntten zwischen seyn payd arme vnd Reyß starck auff dein
rechte seytten etc

50v Hye sagtt der meyster pruch wider dy andren zwen verß
vnd stuck als will er ringen etc

Item den ersten pruch mach alzo fertt er vber deyn messer
mit dem gehultz vnd kumbt mit gebappender handt an deyn
hals vnd will dich drucken vber deyn rechteß oder linckeß
peyn So prich das alzo Grewff mit deyner rechten handt
hynden an seyn olpogen vnd schewb In woll auff deyn rechte
seytten vnd greuff mit deyner lincken handt zwischen seyn
payd arme woll hyn auff an seyn hals seyner rechten seytten
vnd schreytt mit deynem lincken fuß woll hynter seynen
lincken vnd wirff In vber deyn linke huff etc

51r Item aber ein pruch fertt er vber deyn messer mit dem ge=•
hultz als Ee geschriben stett So grewff mit deyner lincken
hand an seyn lincke vnd mit deyner rechten schewb In pey
dem olpogen von dir vnd spring mit deynem rechten fuß für
seyn lincken vnd nymb Im das gewicht etc

51v Item aber eyn pruch fertt er mit dem gehultz als vor
so grewff mit deyner lincken hand In seyn rechten arme nohend
bey der handt vnd ruck den ~~vber~~ ᵛⁿᵗ~~sich~~ [64] vntersich an dich Indes

---

[64] The scribe started to write the correct word above
vbersich, then crossed that out too.

schlag deynen rechten arme ſtarck vber seynen lincken vnd druck
sy beyde arme woll an deyn leyb vnd tritt deynen rechten
fuß fur seyn peyde vnd swing dich auff deyn lincke seytten etc

Item aber eyn pruch Greuff mit deyner lincken handt vber
seyn messer nohend bey der handthab[65] oder pindt vnd reyß
vntersich vnd far mit deyner rechten handt auff zwischen seyn
payd arme In dy swech seynes messers vnd druck gegen Im
vaſt nyder vnd reyß ſtarck auff deyn lincke seytten so nymbſtu
Im aber seyn messer etc

Item aber eyn pruch Greuff oben vber seyn rechten arme
In seynes messers klingen mit deyner rechten handt nohendt
bey dem gehulţ vnd Reyß gegen deyner rechten seytten Indes
Greuff mit deyner lincken verkertten hand In die swech
seynes messers vnd reyß ſtark auff deyn rechte seytten so
schlechſtu Im seyn messer an seyn kopff etc

Item aber eyn pruch fertt er vber deyn messer als vor vnd
hellt dich bey dem hals als vor So far auff mit deyner
lincken handt hynter seyn rechte zwischen seyn hendt vnd
gehulţ vnd druck da mit ſtarck auff deyn rechte seytten
vnd wendt dich geswindt auff dy Eegemelt seytten etc

Item hatt er dich bey dem hals als vor So grewff mit
deyner lincken handt vorne an seyn lincke vnd ſtoß Im
deyn messer vnter seyn rechten arme auff seyn pruſt vnd
vmbker deyn rechte hand vnd druck In faſt an seyn
lincken arme mit dem messer vnd reyb seyn lincke
handt mit deyner lincken wol vmb vnd schreyt mit deynem
rechten fuß fur seyn payd vnd Swing dich auff deyn
lincke seytten vnd wirff In auff das antlitʒ So iſt er besch=
lossen vnd Tritt mit deynem lincken fuß auf das gehulţ
des messers in mäß vnd geſtalt wie das vnden gemalt
ſtat etc

Item aber eyn pruch wyder den wyderpruch will er den
pruch machen wyder das halten mit dem füß vnd iſt sich
vmb werffen auff seynen ruck So grewff In das messer
vnd schlew In Inn dem vmbwenden seynen rechten arme
dar eyn So kann er nicht woll auff ſtan wiltu ƶ so
tritt mit peyden fussen auff das messer mit yţlichem fuß
auff eyn ortt Das ſtuck das dynet dem kampff So eyn

---

[65] *handt hab*

54r (cont.) geharnischter man geworffen wirtt vnd alzo beschlossen
wirtt mag er nicht mer auff sten Das stuck soltu
haben verporgen vnd nyemant weysen wann dy Rosen
sol man nicht fur dy Sew Streuwen etc

54v Item hatt er dich bey dem hals gefast als vor vnd will
dich vber das peyn drucken als vor So far Im mit dem
kurtzen ortt zwischen seyn ped arme vnd setz Im den ortt
an seyn kelen wye vnden gemalt stet etc

55r Item aber eyn ander pruch hatt er dich bey dem hals
gefast als vor So far mit deynem ortt deynes messers
auff deyn lincke seytten vnd Stoß In mit dem gehultz
zwischen seyn ped arme zu dem angesich ader der kelen
wye vnden gemalt stett etc

55v Item aber eyn pruch fertt er dir vber deyn messer mitt seynem
gehultz vnd will dich bey dem hals nemen wye vor
So weyß Im mit deynem gehultz das seyn hyn weck
vnd Tritt mit deynem rechten fuß woll auff seyn rechte
seytten vnd schlag Im mit dem gehultz auff seyn kopff
wye vnden gemalt stett etc

56r Item aber eyn ander pruch fertt er dir vber deyn messer mit
dem gehultz als vor Indes var Im schnel vber seyn
handt nohendt zu seym gehultz In das ~~gehultz~~ gelenck
seyner handt außbendig vnd var mit der lincken handt
In dy sterck seyneß messers vnd Reyß mit payden
henden starck auff deyn lincke seytten So nymbstu Im das
messer etc

56v Item aber ein pruch hatt er dich bey dem hals gefast wie
vor so laß deyn messer vallen Greuff ym mit deyner
rechten handt hinter seynen lincken olpogen vnd scheub
In da mit von dir auff seyn rechte seytten Indes greuff
In mit deyner lincken[66] ~~rechten~~ handt In seyn lincke knyepug vnd
wirff yn auff seyn rechte seytten etc

57r Item aber ein ander pruch hatt er dich bey dem hals gevast
als vor so loß deyn messer vallen vnd var mit deyner rechten
handt hinter seynen lincken olpogen vnd grewff mit deyner
rechten handt In dy mit seynes leybs auff seyner rechten seytten
vnd mit deyner lincken hand In dy mitt seyner lincken seytten
vnd heb In starck auff vnd dree In auff deyner prust vmb

---

[66] The word "left" is inserted in the margin.

vnd wir[f] In mit dem kopff auff dy erden wie vnden gemalt
ſtett etc

57r (cont.)

Item aber ein pruch will er dir mit dem gehultz vber
deyn meſſer faren wie vor So zeuch deyn meſſer Im vnter
deyner rechten seytten ſtee Indes far mit deym gehultz vnd
mit gewappender handt vber seyn rechte handt außben=
dig vnd reyß ſtarck auff deyn rechte seytten So nymbſtu
Im das meſſer

57v

Item Eyn ander pruch will er mit dem gehultz vber deyn
meſſer faren als vor So far mit der langen ſchneyden an
seyn handt zwischen deß gehultz vnd handt Das deyn
meſſer vber dy zwerch fur dich kum Indes fall zu gewapp=
ender handt vnd ſtich Im Inwendigs durch seynen rechten
arme da mitt vnd kum Im außbendigs In seynen olpogen
vnd druck mit der swech mit deynem meſſer vnd deynem
olpogen ſtarck nyder vnd druck mit deyner rechten handt
vnd meſſer auch seyn rechte handt ſtarck vnd leg dich mit
ſterck auff seynen arme alzo haſtu In beschloſſen kunſtenlich
vnd meyſterlich vnd magſt In halten drucken oder werfen etc

58r

Item aber eyn pruch will er dir mit dem gehultz vber faren
wie (sic) far Indes far mit deyner lincken handt Inbendigs
vber seyn meſſer vnd fall zu gewappender handt in dy
mit deyneß meſſers vnd far Im mit der swech außben=
digs In seyn gelenck der handt vnd druck vaſt In das
gelenck auff deyn rechte seytten so nymbſtu Im das meſſer etc

58v

Vberfaren schnel piß remen
vnd mit ortt dy kelen solt nemen

59r

Hie lertt der meyſter aber eyn gutz ſtuck auß dem vberfaren
In dem meſſer vnd sprich vberfaren etc Das soltü alzo verſtan
gee nohendt an In vnd vberlauff seyn meſſer mit deym
gehultz alzo das seyn meſſer vnter deyn rechte vchsen
oder arme kum Indes vall mit deyner lincken handt
In dy mit deyneß meſſers vnd far vnden auff ausbendigs
seyner rechten handt vnd ſtich Im alzo zu dem gesicht
oder kelen will er zu Rück tretten So tritt Im frōlich
nach vnd druck seynen arme mit ſterck nyder etc

Vberfar ſtarck schneyd das glenck
Seyn messer zu erden swenck

59v

Hie lertt der merer der kunſt Ein anders ſtuck auß dem
vberlauffen mit dem gehultz vnd sprich vberfar ſtarck

59v (cont.) etc das soltu alzo versten ferstu Im vber sein messer mit
deym gehultz So far Im vnden durch das seyn messer
vnder deyn rechte vchsen kum vnd far Im außbendigs
In das gelenck seyner handt vnd druck gegen der erden
geswindt nyder So nymbstu Im das messer etc

60r **Messer nemen wilt begynnen**
Vberfar mit cloß ler reyssen Innen

Hie lertt der Merer der kunst aber Ein stuck auß der Ee
gemelten artt vnd spricht alzo Messer nemen etc Das soltu
alzo verstan far Im mit dem gehultz außbendigs vber
seyn messer vnd laß den dy swech sincken Inwendigs nyder
vnd far Im mit dem gehultz außwendigs In das gelenck
seyner handt vnd Swing dich Indes auff deyn rechte seytten
vnd Reyß starck mit dem gehultz dar auff wie vntten gemalt
stett etc so nymbstu Im das messer;

60v **Messer nemen mit eyner hendt**
Mit lerer handt vberfar piß behendt

Hie lertt der merer der kunst wye man aynen eyn messer
nemen solt mit eyner hendt vnd ob eyner halt nicht
mer hett wenn Eyn handt oder lamb wer In der lincken
handt das Er dennich Eynem seyn messer nemen mocht
schnell mit gewallt Das verstee alzo Gee auff mit
der zwirch mit deynem messer auff seyn lincke seytten
nohendt an seyn gehultz Das dy lang schneyd gegen
dem man stee Indes laß deyn messer fallen vber das
haubt vnd var mit deynen rechten handt Im außbendig
vber seyn handt vberbindt dye Das sy wyder außbendig
kum Indes Reyß da mit starck mit der eyn hendt auff deyn
rechte seytten So nymbstu Im gar lyderlich an alle nott
seyn messer etc

61r **Mit lerer handt ler an dich rucken**
Mit peyden armen starck drucken

Hye lertt der meyster vnd der merer der kunst aber eyn gutz
stuck das gett zu auch mit lerer handt So dw zu Im
kumbst wye vor So laß deyn messer fallen vnd far aber
vber seynen arme wye vor vnd windt den arme vmb das
dir seyn rechte handt kum vnter deyn rechte vchsen vnd
deyn rechte handt auff seynen olpogen vnd der linck arme
wer geschlagen vber genß rechten arme vnd druck seynen
rechten arme starck an deynen leyb oder prust vnd halt In

wye vnden gemalt ſtet wiltu In aber werffen so schreyt
mit deynem lincken peyn fur d̶ seyn pede peyn vnd wirff
In dar vber oder vall auff den arß so fellt er auff seyn
antlitz vnd kan nicht auff ſtan den du laſt In gerne
auff ſten etc

Fertt er auch vberfar
Linck vberschlag recht reyssen mit spar

Hye lertt der meyſter Eyn pruch wyder dy vberfaren mit
dem gehultz vnd spricht fertt er etc Das soltu alzo verstan
fertt Er mit dem gehultz vber deyn messer wye vor So
far auch mit deym gehultz Indes vber seyn rechte handt
außbendigs vnd Reyß ſtarck dye handt auff deyn rechte
seytten vnd wendt dich mit dem leyb an seynen lincken
arme vnd reyß Im seyn handt mit dem messer woll zu
deyner pruſt vnd schlag deynen lincken arme vber seynen
rechten olpogen vnd nymb Im alzo das messer oder schreyt
mit dem lincken fuß fur seyn payde vnd wirff In dar
vber etc

Sy absetzen lere
Hew Stich kunſtenlich were
Von vyer Enden
Hew Stich ler ab wenden

Hie sagtt der meyſter wye man soll Im messer absetzen
Es seyn hew oder ſtich Wenn du mit dem zuuechten
zu Im kumbſt ſtelt Er sich den gegen dir als woll Er
ſtechen vnd setz deynen lincken fuß fur leg dich gegen (sic) Im
In dy hut des Ebers zu deyner rechten seytten vnd gib dich
ploß mit der lincken seytten ſticht Er dir denn zu der ploß
So windt mit deynem messer auff deyn lincke seytten gegen
seynen ſtich das dy kurtz schneyd an seyn messer kumb vnd
setz da mit ab das der ortt gegen seynem gesicht ſtee oder gegen
dem leyb vnd ſtich Im zu dem gesicht oder pruſt etc

Item Ein anders ſtück wenn du ſteſt zu deyner rechten scytten

In dem Eber hawtt er dir denn zu der lincken seytten
oben zu der plöß so far auff mit dem messer vnd windt
da mit auff deyn lincke seytten gegen seynen haw Das daß
gehultz fur das haubt kum In der Stiren vnd schreyt
mit deym füß f̶u̶r̶ zu vnd ſtich Im zu dem gesicht
oder pruſt Dy ſtuck magſtu treyben von payden seytten
auß den vyer legren etc

63r (cont.) Durchwechsel das stuck

Dy meyster treybt zu rück

Merck deß artt vnd lere

Von payden seytten stich mit sere

Hye sagtt der meyster von dem durchwechsel vnd spricht
Durchwechsel das etc Nu ist zu wissen das der durch=
wechsel vill sind vnd mancherlay du magst dy machen
auß allen hewen Gegen den vechteren Dy do hawen zu
den messeren vnd nicht zu den zynnen des mannes Nu
den durchwechsel soltu treyben mit grosser fursichtikeyt das
man dir nicht ansetz Dy weyll dw durchwechselt etc

63v Item den durchwechsel Treyb alzo wenn du mit dem
zuuechten zu dem man kumbst So haw Im von oben
lanck Eyn zu dem kopff hawtt er dann gegen dir zu
dem messer vnd nicht Zu dem leyb So loß den ortt
mit dem haw vntten durchwechselen Ee er dir an das
messer pindet vnd stich Im Zu der andren zynnen wirtt
er denn das stichs gewar vnd fertt mit dem messer
dem stich nach mit der versatzung So wechsel aber
durch Zu der andren seytten das treyb albegen wenn Er dir
nach dem messer vertt mit der versatzung Das soltu
treyben zu payden seytten etc

64r Item aber Ein durchwechselen wenn du mit dem zuuechten
zu Im kumbst So setz deynen lincken fuß fur vnd halt
Im den langen ortt gegen dem gesicht haw[67] er dir denn von
oben oder von vnten zu dem messer vnd will dir das weck
schlagen oder dar an pinden Indes laß den ortt vntersich durch
sincken vnd stich Im zu der andren seytten Eyn das magstu thun
gegen allen hewen etc

Item Ein gutte ler[68]

64v Item Ein gutte ler vnd das dritt stuck merck gar eben
wenn Er dir versetzt hatt oder sunst an deyn messer hatt
gepunden helt er den am messer seyn ortt nicht gegen deynem
gesicht oder plöß deynes leybs vnd lest den ortt neben
dir ⁊ beseytz auß gan auff Eyn seytten So wechsel Im
frolich durch pleybt er aber mit dem ortt gegen dem antlitz
oder gegen den plossen So wechsel Im nicht durch Sunder

---

[67] LH, f 34v: *hawt*

[68] This is clearly a copying error. This line belongs to the play on the other side of the page.

pleyb am messer vnd erbeyt Im da mit zu der negſten ploß
So mag er dir nicht nachraysen noch anſetzten etc

Wiltu linck lanck vnter hawen
Durchwechsel w soltu dich frewen

Hye lertt der meyſter wye man durchwechslen soll Im
messer auß langen geraden vnterhewen von payden seytten
zu vor auß von der lincken seytten das der recht fuß vor gee
vnd der linck dem rechten nachvolg alzo das dy fuß Im
schreytten zusammen kummen vnd haw dy hew außgeſtr=
acktem arme lanck Eyn so dw denn In dem zuuechten
zu Im kumbſt so thu gleych als du Im den ortt wolleſt
Eyn schyssen zu seyner rechten seytten zu dem gesicht Indes
wechsel vntten durch bey dem gehultz vnd ſtich Im In
seyn angesich seyner lincken seytten

Linck lanck laß recht eyn schyessen
Windt ſtich wirtt In verdyessen[69]

Hie lertt der meyſter aber Eyn gutt ſtuck auß den Eegemelten[70]
vnterhewen vnd spricht linck lanck etc Das soltu alzo
verſtan Gee auff den man In vnterhewen von deyner lincken
seytten zu deyner rechten seytten mit geſtracktem arme vnd mit
Eynem zwiuachen tritt als vor das der dawm oben ſtee
auff der flech deß messers vnd dy ſtumpff schneyd vor
gee So du denn đ zu dem man kumpſt So scheuß Im
den langen ortt geradt Eyn zu dem gesicht Das aber der
dawmen oben ſtee haſtu In denn mit dem ortt nicht getroffen
So pleyb alzo ſten mit dem messer vnd windt deyn messer
auff deyn lincke seytten auff seyn messer das dy kurtz schneyd
vntten ſtee vnd dy recht schneyd oben ſtee vnd Tritt woll
auff seyn rechte seytten mit deynem lincken fuß schreytt woll
vmb In vnd ſtich Im zu dem gesicht đ wiltu denn so
mach das dupliren zu seyner rechten seytten hinter seynes
messers klingen auch magſtu den treffer dar Eyn machen;

Wer das radt kann machen
Den wechsel mag Er zwiuachen

Hie sagtt der meyſter von dem ſtuck Das do heysset das radt
vnd spricht wer das Radt etc Das soltu alzo verſtan wiltu
das radt machen So haw zu Im von deyner rechten achsel

---

[69] LH, f 35v: *verdryessen*

[70] Based on LH, it could also be *Ergemelten*, since this word is written *er gemelten* there.

66r (cont.)
Eyn geraden treyb haw mit geſtrackten arme vnd ſtee mit
dem lincken füß fur vnd far mit dem ortt vbersich das dy
ſtumpff schneyd oben ſtee vnd laß oben von deyner rechten achsel
woll hintersich abgen vnd von vntten wider auff alzo das albeg
der ortt gegen dem man gee vnd thu als dw Im den ortt zu dem
gesicht wollteſt schyssen zu seyner lincken seytten Indes far auff
mit dem ortt vnd wechsel durch zu seyner rechten seytten zu
dem gesicht etc

66v
Item ſtee mit deinem lincken fuß fur vnd mach das radt von
deyner lincken seytten vnd schreyt mit (sic) deynenem (sic) rechten fuß
fur vnd mit dem lincken albegen dem rechten nach vntz dw
zu dem kumpſt So du denn In Erraychen magſt mit dem
ortt so thu als dw Im den ortt wolleſt Eynschyssen von
deyner lincken seytten zu seyner rechten Indes laß den ortt
bey seynem gehultz vntten durchwischen vnd ſtich Im zu
dem gesicht seyner rechten seytten alzo magſtu auch mit
den andren hewen durchwechselen als ausß den ſtreychewen[71]
Treybhewen halphewen vnd zwiuachen vnterhewen

67r
Zuck dy treffen
Den meyſtren wiltu sy effen
Will er auff dich pinden
Zuck schnel so wirſtu In finden

Hie sagtt der meyſter wye man Im messer zucken soll vnd
wiß das daß zucken iſt auch der haubtſtuck Eynes vnd
iſt vill vnd manchlay vnd gehoren zu treyben gegen den
meyſtren dy ſtarck an das messer pinden vnd ſtarck In der
versatzung sindt wiltu dy selbigen meyſter effen vnd Teuschen
So mach dy zucken gegen In also pindt mit dem Entruſt=
haw an seyn messer auff seyner linckn seytten Indes haw
mit dem selben haw zu seyner rechten seytten vnd wenn
dy messer zusammen klitzen Indes zuck mit dem arme
Eyn benigs zu ruck vnd mach auß dem schlag Eyn
ſtich vnd ſtich Im zucken auff seyn lincke seytten zu
dem gesicht etc

67v
Item wiltu so laß das messer durchgen vnd schlag Im wider
Ein mit der ſtumpffen oder scharpffen schneyden zu seyner
rechten seytten auff dy obren zynnen vnd waß das durchgen
iſt das wirſtu hin nach vnterricht vnd sich eben das du

---

[71] LH, f 36v: *steych hewen*

dy durchgen machest nach meyner maynung wann Ich dy 67v (cont.)
In dem messer anders bedewtten pin dann Im swertt etc

Item haw Im von der rechten seytten oben starck zu dem kopff 68r
fertt er dar vnd will verstetzen So zuck dein messer an dich
Ee wenn Er dar an pindet vnd stich Im zu der andren zynnen
der andren seytten das thu In allen treffen das messers etc sequir
aliud notabile[72]

Item hatt er an deyn messer gepunden vnd stet er dann gegen dir 68v
vnd wartt ob du dich wollest ab zyhen vnd als du wollest
zucken vnd pleyb am messer vnd zuck deyn messer vntz
an dy mit der klingen an dich vnd stich am messer wider Eyn
pald zu dem gesicht oder prust vnd So du wilt zucken
So mustu deyn messer albeg winden gegen dem seynen das
dy lang schneyd oben stee als du vor gehörtt hast etc

Item Trifstu In nicht recht mit dem stich so erbeyt mit dem
dupliren oder mit andren gutten stucken etc

Item Ein anders zuckn steestu In der pastey vnd hawt 69r
er denn von oben auff dich auff deyn lincke seytten so
schreyt ziuach[73] auff deyn rechte seytten auß dem haw
vnd zuck den leyb auß dem haw vnd haw Im nach
der obren ploß vnd loß dir nicht an das messer pinden;

Item haut Er dir von dem luginslandt zu so du auch dar In 69v
stest so zuck In seynem haw das messer an dich vnd laß In
nicht dar an komen Sunder hawt Er von seyner rechten
seytten zu deyner lincken So zuck deyn messer an dich vnd
schreyt auff seyn rechte seytten Ein zwiuachen Trit vnd
schlag Im zu den öbren zynnen das magstu Treyben
von payden seytten etc

Item hätt er dir an das messer gepunden auff deyner lincken 70r
seytten vnd will Er vmb schlahen zu der andren zynnen
so versetz Im nicht Sunder schlag zu der andren zynnen
vnd Tritt zwiuach auß auff seyn rechte seytten vnd
schlag Im zu dem nacken oder auff das haubtt etc

Item aber gar Ein gutz zucken wiltu das machen so thu 70v
Im also stestu In der hut luginslandt vnd hatt er dir an
deyn messer gepunden auff deyner lincken seytten vnd Er

---

[72] Latin: *aliud notabile* = "Further noteworthy things follow"

[73] LH, f 38r: *zwivach*

70v (cont.) will vmb schlagen zu deyner rechten seytten So zuck
deyn messer an dich vnd Erhöch das gehultz fur das
haubtt das dy kurtz schneyd gegen dir stee vnd laß den
ortt nyder sinken an deyner lincken seytten ab vnd setz
Im den ortt In seyn kelen oder prust vnd scheub In da
mit zu ruck etc.

71r ## Stet offen dy pfortt
Im feler gee dar eyn mit ortt

Hie lertt der meyster wye man mit dem ortt Erbeytten vnd
zucken soll In dy feler vnd spricht stet offen etc Das sol
dw also verstan Gee mit der zwirch nohendt an den man
zu seyner linckn seytten vnd Thu samb[74] dw wolst schlagen
zu seyner rechten seytten vnd far vber deyn haubtt so dw
kumbst vber dy mit deynes haubtz das der ortt hinden hin
auß gett vnd das gehultz gegen dem man stee So schlag
wyder zu seyner lincken seytten So du schir zu seyner
lincken seytten zu dem messer kumbst so haw nicht zu
seynem messer das dw das Treffest Sunder Thu samb
dw es Treffen wollest Indes zuck nohendt bey seynem
kreutz deyn messer an dich vnd stich Im zu seynem
antlitz auff seyner rechten seytten das
dy kurtz schneyd vntten stee vnd
der dawmen auff der flech lig etc

71v ## Haw deyn flech zum luginslandt
Durchzuck den ortt zu handt
hatt er den stich versetzt
Indes fell tzwiuach Er wirtt geletzt

Senesu[75] dy maynung des Textz soltu alzo verstan stett er
In der hutt luginslandt so haw von deyner rechten achsel
mit der flech gegen seynem messer das dy lang schneyd
gegen dir stee vnd Im haw So schreyt mit deynem
rechten fuß woll auff seyn rechte seytten zu seynem rechten
fuß vnd haw gleych zu seynem messer vnd So dy messer
Schir zu sammen wollen klitzen Indes zuck deyn messer
an dich mit dem gehultz gegen deyner lincken seytten vnd stich
Im den ortt starck an alle versatzung grob zu dem gesicht seyner
rechten seytten vnd merck das du Im Stich ym nachschreytest

---

[74] Word could come from *sambalde = alsbald* (HEN-NIG 2001, p 276) or *sam = als* (HENNIG 2001, p 275)

[75] Could come from *senen = ersehnen* (LEXER 1885, p 225)

deyne͞n lincke͞n fuß mit deynem rechte͞n das der schritt
zwiuach wer <u>etc</u>

71v (cont.)

It<u>e</u>m haſtu In mit dem ſtich getroffen oder nicht So windt
nach dem ſtich Indes auff dein lincke seytten das der dawm
an dem messer vnde͞n sey vnd dy lang schneyd gegen dem
man gekertt sey vnd far mit deyne͞m messe͞r vbe͞r dy mitte͞n
deynes ~~mes~~[76] haubtz zu deyne͞r rechten seytten vnd thue
samb dw Im zu seyne͞r lincke͞n seytten schlagen wolſt
Indes far Im wide͞r zu seyne͞r rechte͞n seytte͞n Indes
far wide͞r auff mit dem messe͞r das der dawm abe͞r vnde͞n
sey vnd schreytt abe͞r mit deyne͞m rechte͞n fuß woll auff
seyn rechte͞n vnd schlag Im wide͞r Ein zu der selbige͞n
seytte͞n hinte͞r seynes messers klinge͞n dy kurtz schneyd
Im Inn das angesicht vnd schreyt abe͞r mer mit deyne͞m
lincke͞n fuß woll hinte͞r seyne͞n rechte͞n vnd schlag Im In
den nack Indes wiltu magſtu Im den Treffe͞r mache͞n
wie vnde͞n gemalt ſtet <u>etc</u>

72r

Ob er ſtarck iſt
Durchlauff zu alle͞r friſt
Der handt vnd arm͞e soltu nahen
vnd weyslich wartt des vah͞en
Der glider soltu war neme͞n
Der knypug soltu dich remen

72v

Hie sagtt der mere͞r der kunſt abe͞r vo͞n Eyne͞m haubt=
ſtuck als vo͞n dem durchlauffe͞n vnd wye ma͞n dem
man durchlauffe͞n soll Nu merck dy durchlauffe͞n vnd
dy ringe͞n dy gehore͞n den meyſtere͞n zu treybe͞n dy gern͞e
Ein lauffen vnd hoch auff mit den armen vnd
versatzung lige͞n vnd dy dich mit ſterck wellen vbe͞r
dringe͞n <u>etc</u>

It<u>e</u>m dy durchlauffe͞n mach also hautt er gege͞n dir eyne͞n
obe͞rhaw So haw auch gleych mit Im Eyn den Entruſt=
haw nohendt zu Im das dy hendt schir an Eyn
ander rüre͞n Indes greyff mit deyne͞r lincke͞n handt
hinte͞r seyne͞n olpogen außbendig vnd scheub In vo͞n dir
auff seyn lincke seytten Indes laß deyn messe͞r vallen vnd
greyff mit deyne͞r rechte͞n handt oder arm͞e außbendigs

73r

---

[76] The scribe probably wanted to write *messer*. However, LH, f 39r, clearly only says *Haubtz*.

73r (cont.)

In sein rechte knÿpug vnd heb In auff vnd wirff In
furdich <u>etc</u>

73v

It<u>em</u> habt Ir payde an gepund<u>en</u> So pegreyff seyn rechte
handt mit deyn<u>er</u> linck<u>en</u> vnd heb da mit hoch auff
vnd kum deyner linckn handt zu hilff mit deyner
recht<u>en</u> vnd wendt dich mit deyn<u>em</u> ruck g<u>egen</u> Im
vnd prich Im den arm<u>e</u> vbe<u>r</u> deyn rechte achs<u>eln</u> mit
payd<u>en</u> hend<u>en</u> <u>etc</u>

74r

It<u>em</u> habtt Ir an gepund<u>en</u> wy vor So greyff mit deyn<u>er</u>
recht<u>en</u> handt In seyn rechte vnd kum deyner recht<u>en</u> mit
deyn<u>er</u> linck<u>en</u> zu hilff vorn<u>e</u> bey dem gelenck vnd wendt
deyn lincke achsel g<u>egen</u> seym recht<u>en</u> arm<u>e</u> vnd heb den
hoch da<u>r</u> vbe<u>r</u> vnd prich Im den arm<u>e</u> vbe<u>r</u> deyn lincke achsel
vnd maß dich das de<u>r</u> olpog<u>en</u> albeg auff dy achsel kum
vnd mach das wye vntt<u>en</u> gemalt ſtett <u>etc</u>

74v

It<u>em</u> hatt er dir Eyn gel<u>o</u>ffen vnd dw Im wid<u>er</u> so pegreuff
sein rechte hand mit deyn<u>er</u> linck<u>en</u> Inbendigs vnd reyb dye
vo<u>n</u> dir vnd heb Im den arm<u>e</u> woll auff Indes lauff mit
dem haubt durch sein arm<u>e</u> auff seyn<u>er</u> recht<u>en</u> seytt<u>en</u> vnd
ruck Im seyn recht<u>en</u> arm<u>e</u> mit deyn<u>er</u> linck<u>en</u> handt auff
den nack<u>en</u> vnd vbe<u>r</u> deyn lincke achs<u>eln</u> vnd Erheb
dich woll auff mit dem nack vnd mit den schult<u>ern</u>
vnd prich Im den arm<u>e</u> ſtarck da<u>r</u> vber

75r

It<u>em</u> Ein anders ſtuck hatt er dir Ein gel<u>o</u>ff<u>en</u> vnd du
Im auch So pegreyff seyn rechte handt mit deyn<u>er</u> linck<u>en</u>
vnd heb Im seyn arm<u>e</u> woll vbe<u>r</u>sich vnd gee mit dem
haubt durch den arm<u>e</u> vnd greyff mit deyn<u>er</u> recht<u>en</u>
handt In ~~I~~ seyn rechte knypug vnd heb In mit d<u>en</u>
schult<u>ern</u> auff vnd ~~wir~~ greyff mit deyn<u>er</u> recht<u>en</u> handt
In seyn rechte knypug vnd heb In pey dem pein auff
vnd wirff In vbe<u>r</u> den ruck <u>etc</u>

75v

It<u>em</u> abe<u>r</u> Ein ſtuck Greyff mit der linck<u>en</u> hand In seyn
rechte vnd reyb dy vo<u>n</u> dir In dcyn lincke seytt<u>en</u> vnd
schreytt mit deyn<u>em</u> recht<u>en</u> fuß hint<u>er</u> seyne<u>n</u> recht<u>en</u> vnd
greyff mit deyn<u>er</u> recht<u>en</u> handt vorn<u>e</u> umb seyn leyb
vnd wirff In furdich vbe<u>r</u> dy hüff deynes recht<u>en</u> peyns;

76r

It<u>em</u> Ein anders ſtuck wenn Er dir In laufft mit auff
gerackt<u>em</u> arm<u>e</u> vnd du Im wid<u>er</u> so lauff Im durch mit d<u>em</u>
haubt durch seyn rechte vchs<u>en</u> vnt<u>er</u> seyne<u>n</u> recht<u>en</u> arm<u>e</u>
seyn<u>er</u> recht<u>en</u> seytt<u>en</u> vnd schreytt mit deyn<u>em</u> recht<u>en</u> fuß forn<u>e</u>

fur seynen rechten vnd far Im mit deynem rechten arme vnter
seynen rechten durch vnd hinten umb seynen leyb vnd
senk dich Ein benig nyder vnd faß In auff deyn rechte
hüff vnd wirff In hinter dich etc

Item laufft dir Eyner Ein zu deyner rechten seytten vnd iſt
hoch mit dem arme vnd du auch So haltt deyn messer
zu dem seynen vnd far mit dem gehultz vber seyn
rechte handt vnd wendt dich an In mit deyner lincken
seytten an seyn rechte vnd spring mit deynem lincken fuß
fur seynen rechten vnd far mit dem arme hinten woll vmb
seyn leyb vnd senck dich Ein wenig nyder vnd faß In
auff deyn lincke hüff vnd wirff In furdich auff seyn
antlitz ~~vnd~~ oder laß In Ein[77] deyn messer vallen als dw
woll weyſt etc

Item ob er dir zü ſtarck wollt seyn Das du In nicht
leychtiklich mochſt vber deyn hüff werffen So far Im
mit dem gehultz deynes messers ſtarck In seyn genick
vnd reyß da mit auff deyn rechte seytten so wirffſtu In
geruicklich[78] treten

Item wenn dir Eyner Ein lauffet vnd hellt seyn messer
nyder vnd dy handt So greyff mit deyner lincken verkerten
handt In seyn rechte Inbendig vnd spring mit deynem
rechten füß hinter seynen rechten vnd far mit deynem rechten
arme vorne vmb seyn hals vnd wirff In alzo vber dy
rechten huff;

Item haſtu In außbendig gefaſt bey seynem rechten arme mit
deyner rechten handt so zeuch sere seyn rechte mit deyner
rechten vnd far mit der lincken handt aussen vmb seyn
~~leyb~~ hals vnd wirff In vber deyn linckes peyn oder huff;

Item wenn dir Eyner Ein lauffett Im messer vnd iſt nyder
mit der handt so far mit deynem gehultz aussen Im oben
vber sein arme vber dy rechten handt vnd druck do mit
nyder vnd zeuch dy da mit woll auff dcyn pruſt vnd
faß In mit der lincken handt bey seynem rechten olpogen
vnd spring mit dem lincken fuß fur seyn rechten vnd druck
In also dar vber;

---

[77] Should be *In*.

[78] *ge-* can come before all forms of a verb to complete,
strengthen, or reinforce the action (LEXER 2001, p 60),
rucke, rücke, ze ruck = zurück

78r Item Gee an In wye vor vnd scheub In mit deyner lincken
handt auff seyn lincke seytten auff seynen rechten arme vnd
far mit deyner Rechten handt Indes hinter seynen rechten ōlpo=
gen vnd scheub In woll auff seyn lincke seytten vnd far
mit deyner lincken handt Inbendigs In sein rechte knypug
vnd schreytt mit deynem lincken fuß fur seyn lincken vnd
wirff In für dich etc

78v Item der text spricht der glider soltu etc Wenn dw zu Im
kumbſt Im zuuechten So fall Im mit der lincken hend
Inwendigs oben vber seyn messer vnd greyff Indes mit
deyner lincken hand vntten an deyn messer In dy mit vnd
far ſtarck mit deynem messer an das seyn außbendig vnd
spring mit gantʒem leyb Indes auff seyn lincke seytten
so nymbſtu Im das messer

79r Item Gee an In als vor vnd far mit deynem lincken arme
vber seyn messer Indes far mit deyner lincken handt In dy
mitt deynes messers vnd laß den ort auff deyner lincken
seytten hyn auff gan vnd far Im Inwendigs mit dem gehulʒ
an seyn handt vnd spring auff deyn lincke seytten So ny=
mbſtu Im seyn messer;

79v Item hawtt er auff dich So spring mit dem pogen vnter
sein messer vnd greyff mit deyner lincken handt aber
vber seyn messer vnd faß Indes dein messer mit deyner
lincken handt In der mit vnd heb den ortt auff seyn
rechte seytten vnd druck In Inn das gelenck seyner
rechten handt mit der scharpffen schneyden so muß Er das
messer fallen lassen;

80r Item Thu Im als vor fall Im vber seyn rechten arme vnd
kum der rechten zu hilff In massen als vor vnd far mit
dem fodren teyll deß messers zwischen sein handt vnd
messer Inbendigs vnd druck faſt So nymbſtu Im aber
das messer alzo soltu Erbeytten In dy glider des armes vnd
der handt mit dem ortt deß messers etc

Item Du magſt In auch mit dem ortt deß messers Inbendigs
In seyn dener[79] schneyden;

80v Item aber Eyn messer Nemen Thu Im als vor So du
deynem messer zu hilff piß kummen mit deyner lincken handt

---

[79] *tener* or *tenner* refers to the palm that is not protected by armor. (BAUFELD 1996, p 50)

so far vntersich mit dem gehultz vntten durch auff seyn
rechte seytten außbendigs far oben vber seyn rechte handt
vnd reyß starck auff deyn rechte seytten So nymbstu Im
aber seyn messer etc

80v (cont.)

Item Gee an man wye vor vnd far vber seyn messer
mit deyner lincken handt vnd fall In dy mitten deynes
messers wye vor vnd far mit dem gehultz Inbendigs
vber seyn handt vnd Reyß auff deyn rechte seytten etc

81r

Item haw Im den Entrusthaw zu seyner rechten seytten Indes
fall zu gewappender hant zu der selbigen seytten vnd far
mit deynem gehultz oben vber seyn arme vnd far Im aussen vber
seyn arme vnd windt mit deym gehültz seyn handt starck
auff seyn rechte seytten vbersich so nymbstu Im das messer etc

81v

Item aber Ein stuck fall mit deyner lincken handt vber
seyn rechte Inbendigs vnd ruck dy In deyn lincke seytten
vnd laß dein messer vallen Indes greyff mit der rechten
handt In seyn rechte knypüg vnd wirff In fur dich etc

82r

Item als du dy glenck der handt mit der swech des
messers vnd des ortz ersicht hast alzo magstu dy auch
ersuchen mit dem gehultz des messers außbendigs vnd
Inbendigs vnd das messer Im alzo nemen als du oben
vnterricht pist;

Bewappent durchwind swech zu handt
Messer nymb deyn schneyd sey aussen bebandt

82v

Hie lertt der meyster aber Eyn gutz stuck vnd Eyn messer
nemen vnd spricht Bewappent durchwindt etc Das
soll du alzo verstan haw mit dem Entrusthaw nohent
zu Im Eyn zu seynen lincken seytten vnd schlag vmb zu
seyner rechten seytten Indes fall zu gewappender hant vnd
wind dy swech oben hyn Eyn Inbendigs in das gelenck
seyner handt vnd windt dürch vber seyn handt vnd druck
dy fest auff deyn rechte seytten so nymbstu Im seyn messer etc

Dy swech gewappendt vber handt ist winden
Windt gesenckt mit kurtz linck solt pinden

83r

Das soltu alzo versten fertt Er dir In deyn hant vnd will
windten wye vor mit der swech seynes messers gewappend
Indes senck dy spitzen deynes messers gegen der erden vnd
Gee mit dem messer auff seyn lincke seytten vnd schlag Im
dy kurtz schneyd auff seyn kopff etc

83v   **Kurtz schlecht er vnd gett mit luſt**
                 Piß schnell gewappend ſtößt dy pruſt

                 Saend[80] Das der mezer der kunſt setz Ein wider pruch
                 wider den Eegemelten pruch vnd spricht kurtz ~~schnel~~
                 schlecht er etc das soltu also verſtan Gett er mit
                 hangenttem messer auf deyn lincke seytten vnd will dich
                 mit der kurtzen schneyden auff deyn kopff schlagen So
                 fall Indes schnel In dy gewappendt handt vnd ſtich
                 Im zu dem gesicht oder pruſt vnd Gee vaſt nohend zu
                 Im etc

84r   **Durchlauff entruſt**
                 Den olpogen reyb handt zu pruſt
                 Recht wiltu umb spring
                 Auff recht arme prich gering

                 Hie lertt der meyſter Eyn Eynlauffen mit dem Entruſthaw
                 vnd thu Im also Gee mit dem Entruſthaw an den man
                 ſtarck das der dawm vntten ſtee am messer auf der
                 flech vnd haw Im zu der rechten seytten zu dem kopff mit
                 der ſtumpffen schneyden vnd ruck faſt an In mit dem
                 gehultz das dy hendt an Eyn ander ruren Indes laß deyn
                 messer fallen vnd pegreyff seyn handt bey dem ~~gehultz~~
                 gelenck Dar In Er das messer hatt vnd Reyb dy vmb vnd
                 nymb das gewicht bey dem olpogen vnd spring In dy
                 wag vnd wirff In vber deyn linckes peyn oder huff

84v   Item wiltu das ander ſtuck machen So laß deyn lincken
                 arme hangen So du In gefaſt haſt seynen rechten arme auff
                 deyn pruſt so ſtoß mit deyner lincken achsel vnd druck mit
                 deyner rechten handt seyn rechten arme faſt vber deyn pruſt
                 vnd spring ſtarck auff deyn rechte seytten etc

85r   Item aber Ein ſtuck laufft dir einer Eyn vnd iſt hoch
                 mit dem arme So lauff auch ſtarck auff In das dy hendt
                 zusammen ruren vnd dring hoch vbersich mit dem arme
                 vnd pregreuff do mit seyn peyde peyn In dy glenck der
                 knypug vnd heb In auff dy achselen deyner rechten oder
                 lincken seytten vnd wirff In vber ruck auff den kopff

85v   Item Greyff mit deyner lincken handt dein rechte vnd laß
                 deyn messer vallen vnd windt mitt deyner lincken seyn

---

[80] *Sa* from *sar* = "immediately afterward", "then im-
mediately" (either alone or reinforced by related   expressions) (LEXER 1885, p 205), *end = ehe*, "before"
(LEXER 1885, p 43)

rechte woll vmb In deyn lincke seytt<u>en</u> vnd schreytt
mit deyne<u>m</u> recht<u>en</u> füß fur seyn<u>en</u> recht<u>en</u> vnd schlag deyn<u>en</u>
recht<u>en</u> arm<u>e</u> ſtarck vber seynen recht<u>en</u> vnd druck ſtarck
Indes In dein rechte seytt<u>en</u> vnd swing dich auff dein
lincke seytt<u>en</u> vnd wirfſt yn vb<u>er</u> das recht peyn <u>etc</u>

85v (cont.)

It<u>em</u> aber eyn ſtuck laufft er dir eyn zu deyn<u>er</u> recht<u>en</u> seytt<u>en</u>
so pegreuff seyn rechte handt mit deyn<u>er</u> linck<u>en</u> verkerten
handt Indes schreytt mit deyne<u>m</u> recht<u>en</u> füß woll auf seyn
recht<u>en</u> vnd far mit deyn<u>er</u> recht<u>en</u> handt vnt<u>er</u> seyn recht<u>en</u>
olpogen außbendigs vb<u>er</u> seyn arm<u>e</u> vnd zeuch mit peyd<u>en</u>
hend<u>en</u> seyn recht<u>en</u> arm<u>e</u> auff deyn pruſt vnd swing dich
ſtarck auff deyn lincke seytt<u>en</u> vnd schreytt mit deynem
recht<u>en</u> fuß fur seyn<u>en</u> recht<u>en</u> vnd wurff In vb<u>er</u> deyn rechtes
peyn <u>etc</u>

86r

It<u>em</u> Greyff mit deyn<u>er</u> recht<u>en</u> handt In seyn rechte auß=
bendigs vnd Tritt mit deyn<u>en</u> linck<u>en</u> füß fur seyn linck<u>en</u>
Indes schlag deyn linck<u>en</u> arm<u>e</u> In seyn recht<u>en</u> heb mit
deyn<u>er</u> recht<u>en</u> handt auff seyn rechte vb<u>er</u> deyn linck<u>en</u> arm<u>e</u>
vnd druck deyn linck<u>en</u> arm<u>e</u> faſt zu deyn<u>er</u> pruſt vnd
druck mit der recht<u>en</u> vnt<u>er</u>sich vnd wirff In furdich
~~auff das~~ vb<u>er</u> deyn linckes peyn <u>etc</u>

86v

Nymbt er das gewicht
Wyd<u>er</u> nym macht zu nicht
Indes erbeyt mit schnellikaytt
So wirtt seyn kunſt nyd<u>er</u> geleytt.

87r

Hye sagtt der meyſt<u>er</u> Eyn pruch wyd<u>er</u> dy verß
als gesproch<u>en</u> iſt Druchlauff Entruſt <u>etc</u> Vnd spricht
nymb das gewicht <u>etc</u> Den pruch soltu also mach<u>en</u>
Vellt er dir In deyn rechte handt mit seyn<u>er</u> recht<u>en</u>
verkertt<u>en</u> handt vnd will dir das gewicht mit der
linck<u>en</u> pey dem olpog<u>en</u> neme<u>n</u> Indes weyl du empfindeſt
Das er deyn rechte handt Greyfft mit seyn<u>er</u> recht<u>en</u>
Indes greyff behentlich mit deyn<u>er</u> recht<u>en</u> auch In seyn
rechte vnd Greyff mit deyn<u>er</u> linck<u>en</u> handt auch
an seyn<u>en</u> recht<u>en</u> olpogen vnd spring mit deyne<u>m</u> linck<u>en</u>
fuß fur seyn recht<u>en</u> vnd mach das ſtuck das er dir hatt
woll<u>en</u> mach<u>en</u> vnd piß behendt <u>etc</u>

It<u>em</u> aber eyn pruch will er dich fassen als so far mit deyne<u>m</u>
gehultz vb<u>er</u> seyn handt vnd fall mit der linck<u>en</u> handt
In dy mit deynes messers vnd leg dich mit dem linck<u>en</u> arm<u>e</u>
vnd messer gewappend ſtarck auff seyn arm<u>e</u> vnd druck vnd

87v

87v (cont.) reyß ſtarck auff deyn rechte seytten vnd schreytt mit deynem
lincken fuß fur seyn rechten etc

88r Item hatt er dir deyn rechte handt gefaſt mit seyner
rechten verkertten vnd will dir pey dem olpogen das gewicht
nemen Indes laß deyn gehultz vber seyn handt sincken
gegen der erden vnd Greyff mit deyner lincken handt vnter
seyn handt In deyn gehultz des messers deyner rechten zu
hilff vnd reyß mit payden henden ſtarck vndersich vnd
haw Im durch seyn kopff etc

Item aber Eyn pruch hatt er dich gefatt[81] als vor Indes
will er dir das gebicht nemen bey dem olpogen etc Das ſtück
vindeſt das daß F[82] ſtett

88v **Greuff linck wydersyns nymb gewicht**
Bey olpogen Trifſtu es / arme pricht

Hye lertt der meyſter Eyn hubsch ſtuck Do mit man eynen
den arme prechen mag Wenn Es anders recht gemacht
wirtt vnd thü Im also Greyff mit deyner lincken handt
In seyn rechte vnd ruck Da mit In dy höch auff deyn lincke
seytten Indes Greyff mit deyner rechten handt vntten an seynen
olpogen vnd druck mit deyner lincken nyder seyn handt gegen
deyner lincken seytten vnd ſtreck Im den arme nicht gantz
auch laß Im den arme nicht zu krump vnd heb Indes
mit deyner rechten handt seynen olpogen woll vbersich vnd
schreytt mit deynem rechten fuß hinter seynen rechten etc

89r Item Eyn pruch wyder das gewicht nemen als gesprochen iſt
durchlauff Entruſt etc Hatt er dich gefaſt als vor vnd
will dir Indes das gewicht Nemen bey dem olpogen so
far geschbing[83] nyder mit dem olpogen ſtarck In deyn rechte
seytten woll nyder an deynen leyp vnd schnell Im dy kurtz
schneyd deynes messers auff seyn kopff etc

89v Item hatt er dich gefaſt bey deyner rechten handt mit seyner
rechten vnd will dir mit seyner lincken das gewicht pey
dem olpogen nemen Indes greyff Im wider mit deyner rechten
ſtarck seyn rechte handt vnd ruck dye fur dich vnd
schlag deyn lincken arme vber seyn ped arme vnd schreytt

---

[81] From LH, f 45v: *gefast*
[82] The symbol could stand for a reference to another page on which this F can be seen. The scribe has

already used this method on other pages. The description in LH is significantly longer.
[83] LH: *geswyns*

mit deynem rechten fuß fur seyn rechten vnd swing dich
auff deyn rechte seytten vnd wurff In fur dich etc

Scheuß recht mit rucken
Vber messer soltu arme drucken

Hie lertt der meyster aber Eyn besunder stuck vnd spricht
Scheuß recht etc Das soltu also verstan pegreyff seyn
rechte handt mit deyner lincken verkertten handt vnd ruck
In da mit Eyn wenig gegen dir vnd scheuß deyn messer
Im vntter seynen arme vnd far Im aussen an den olpogen
vnd vber den arme vnd kum Im mit der kurtzen schneyden
Im vntter seyn hals an dy kelen vnd thu deyn gehultz
Indes aussen an deyn nacken vnd loß dy rechten handt von
deynem messer faren vnd kum deyner lincken zu hilff
vorne mit deynem rechten vorne an seyn rechte handt vnd
schreytt mit deynem fuß hinter seynen rechten vnd reyß starck
vntersich seyn arme etc

Ler arme vberschyssen
Ja kunstenlich denn beschlyssen
Da mit In magst furen
Notten zu lauffen oder nicht zu ruren

Hye sagtt der meyster von Eym kunstlichen stuck da mit
man Eynem pyntten[84] beschlyssen vnd halten mag wenne
er will so nott er den man mit dysem stuck das er still
muß sten ser oder ~~lauf~~ melich lauffen[85] müß vnd dw In
notest das er In Eynen sack krichen müß will du es
anders gehabtt haben dar vmb wirtt das stuck von
den kostenlichen[86] meysteren genandt Der vngenandt
auff[87] das daß stuck nicht gemayn soll werden Man
soll auch das nicht gemayn machen vnd nyemant
wissen lassen denn Es werd Im wol pezaltt etc

Item du solt das Ee gemalt stuck nicht offenbaren leychfertigen
lewtten oder meystern dye durch Römß[88] willen das stuck
gemayn mochten machen auff das/das sy von andren gelobt
sollen berden wanne man soll dy edlen margaritten oder
rosen nicht fur dy sweyn strewen Das sy durch dy selbigen
nich vngeertt werden vnd getretten In das kott Nu will

---

[84] *pinden = binden* (BAUFELD 1996, p 34)
[85] LH, f 46r: *gmelich söl lauffen*. In LM, the *g* cannot be seen.
gemelich = gemechlich (HENNIG 2001, p 108)

[86] *kostlich = kostspielig* (HENNIG 2001, p 190)
[87] In LH, f 46r, it's called *der verporgengriff*
[88] LH, f46v: *römeß* comes from *rom = Ruhm* (BAUFELD 1996, p 197)

91r (cont.)   du das stuck machen So thu Im alzo wenn Er dir Eyn
lauffett In dem messer So laß deyn messer vallen oder behaltz
In deyner rechten handt vnd Greiff mit deyner rechten verk
erten handt In seyn rechte außbendig vnd reyb dy handt
vntersich vmb vnd mit deyner lincken hand faß In bey
seynem rechten olpogen vnd spring mit deynem lincken fuß fur
seynen rechten vnd heb seyn rechte handt auff mit deyner rechten
handt hastu aber deyn messer noch In deyner hendt so heb dy
selbig seyn rechte handt auff mit dem kreutz deynes messers
vnd stoß dye vber deyn lincken arme vnd heb In da mit
vbersich auff vnd druck starck nyder mit deyner lincken handt
Ob aber er so starck wer So greyff mit deyner rechten handt
oben auff seyn rechte achsel vnd Gebyn Im dy wag an so
magstu In furen werffen schlagen halten nötten wye du willt
alzo hastu das stuck wye oben vor gemalt stett etc

91v   Item So du zu dem man kumbst so faß In wye Ee
geschriben stett vnd So du In also gefast hast So fall
Indes neben In auff deyn ars vnd Im fall wendt deynen
ruck an seyn rechte seytten vnd hallt In vest auch
magstu Im pretspilen Essen vnd Trincken das er nicht
auff mag stan denne du lost In gerne auff Sunst must
er vntter dir erfaulen etc

92r   Item wiltu In Inn eynen Sack schyben mit dem Eegemelten
stuck So pestell heymlich Etlich[89] dy Eynen sack verporgen
pey In haben auff der schull Dy hintter dem volck stenn vnd
faß In als Ee gemelt ist vnd So du In Inn dy wag gepricht
hast So für In mit gewallt Da dye mit dem sack stenn
vnd hayß In dar Eyn krychen will er das nicht williklich
thun So denn dy den sack auff halten So greyff mit deyner
rechten handt In seyn rechteß peyn nohent bey dem knochen[90]
außbendigs vnd heb In da mit auff auff deyn rechte seytten
vnd schreytt mit deynem lincken fuß fur seyn lincken vnd
wurff In In nammen gottes In den sack vnd Thu Im
darnach wye du willt etc

92v   Item Eyn ander beschlyssen So Ir Im zuuechten zu sammen
kumpt So greyff Im vber seyn rechten arme mit deiner
lincken hantt In dem far mit deynem messer vnter seyn
rechten arme vnd fall mit der lincken handt In dy

---

[89] LH, f 47r: *zwen*

[90] The word actually looks like *knorren*. In LH, f 47r, the opponent is caught by the *knypug*.

mit deynes messers klingen vnd far do mit Im In

seinen rechten olpogen hyntten vnd druck mit der ſterck
nyder vnd Tritt genaw an In vnd heb mit deyner
rechten handt ser vbersich mit dein messer alzo iſt er
aber beschlossen etc

Item Eyn ander beschlyssen fall Im vber seyn rechten arme

mit deynem lincken vnd greyff Indes aber vntten In dy
mitten deynes messers vnd var mit dem gehulß vntten durch
seyn arme In das gelenck des rechten olpogens außbendigs
vnd wendt deyn lincke seytten an seyn rechte vnd druck
mit dem gehulß zu gewappender hantt faſt oben nyder
vnd Tritt nohendt an In das er auß dem pandt nicht
kummen mig[91] alzo iſt er verigelt vnd verschlossen etc

Ler handt zu pruſt reyben

Wiltu den vngenantten den ſtarcken treyben
Spring vnd gach
Den lincken arme vberschlag

Hye lertt der meyſter Eyn koſtenlich halten vnd ſtuck vnd
So du felſt In dem ſtuck auff deyn ars So magſtu den
man halten das er nicht auff mag ſten vnd das ſtuck
heysset der vngenant vnd iſt der peſten ſtuck eyns vnd
gehortt zu Treyben gegen den ſtarcken wye hernach geschriben
ſtett etc

Item wiltu das Eegemelt ſtuck machen So thu Im also

haw zu seynen lincken seytten mit dem Entruſthaw
Indes ruck nohent an In an seyn gehulß mit deynem
messer Indes loß deyn messer fallen vnd Greyff mit
deyner rechten verkertten handt In seyn rechte vnd
wendt dich vmb mit der lincken seytten anseyn
rechte vnd Tritt mit deynem lincken fuß fur seynen
rechten vnd Ruck seyn arme auff deyn pruſt vnd schlag
deynen lincken arme vber seyn rechten arme vnd leg dich
mit gantzer ſterck dar auff vnd spring auff deyn
rechte seytten So prichſtu Im den arme etc wye danne
vor gemalt ſtett;

Item wiltu In den werffen [vnd] So hallten das er nicht auff
kann ſtan Indes so dw dich piſt wenden auff deyn rechte
seytten Indes vall auff deyn ars auff dy erden So müß

94r (cont.) er mit dir fallen vnd fellt auff seyn antlitz nebe̱n dich
also kan er nicht auff ſten den du loſt In gerne̱ auff vnd
Thu Im wye vntte̱n gemalt ſtett <u>etc</u>

94v **Vach linck seyne̱n rechte̱n**
Dy achsele̱n ſtoß hinte̱r spring Im vechte̱n

Ite̱m das ſtuck Treyb also wenn du mit de̱m zuuechte̱n zu dem ma̱n
kumpſt so vach Im seyn rechte handt mit deyne̱r lincke̱n verkertte̱n handt
vnd Reyb dye vmb In deyn lincke seytte̱n vnd zeuch dy woll fur dich hyn
vnd Thu als du In wolleſt mit dem gehultz In das angesicht stossen vnd
schreytt Indes mit deynem rechte̱n fuß hinte̱r seyne̱n rechte̱n vnd ſtoß In
mit dem gehultz ode̱r mit deyne̱m rechte̱n arme̱ ſtarck vorne in seyn achsel vnd reyb dich Eyn benig mit dem leyp auff deyn lincke seytte̱n vnd
wurff In vbe̱r deyn rechtes peyn <u>etc</u>

95r **Dy achsele̱n will er ſtosse̱n**
Den olpoge̱n solt ve̱rdrossen

Hye sagtt der meyſte̱r Eyn pruch wyde̱r das obgemelt ſtuck
vnd spricht also Dye achsele̱n <u>etc</u> Das soltu also verſtan
hatt er deyne̱n rechte̱n arme̱ gefange̱n mit seyne̱m lincke̱n vnd
iſt mit dem rechte̱n fuß geschritte̱n hinte̱r deyne̱n rechte̱n vnd
will deyn achsel ſtosse̱n Indes so er ſtosse̱n will So nymb
deyn lincke handt vnd far Im da mitt hyntte̱n an seyn
olpoge̱n vnd scheub In vo̱n dir auf seyn lincke seytte̱n ode̱r
schlag Im deyn lincke̱n arme̱ ſtarck obe̱n vbe̱r seyne̱n rechte̱n
vnd wendt dich ſtarck auff deyn rechte seytte̱n <u>etc</u>

95v Ite̱m abe̱r Eyn ande̱r pruch hatt er dich gefasſt wye vor
So schlag deyne̱n lincke̱n arme̱ ſtarck vbe̱r seyne̱n rechte̱n
vnd schreytt mit deyne̱m lincke̱n fuß fur seyne̱n rechte̱n
vnd druck obe̱n mit deyne̱m lincke̱n arme̱ faſt nyde̱r seyn
payd arme̱ vnd wendt dich ſtarck auff deyn rechte seytte̱n
wye vnde̱n gemalt ſtett <u>etc</u>

96r **Haut er auff dich rechte̱ns**
Entrußt vnd wartt vechtens
Mitt linck seyne̱n rechte̱n vahe̱n
Recht vbe̱rschlag so mag dir keyn schad nahe̱n

Hye sagtt de̱r meyſte̱r vo̱n Eyne̱m gutte̱n ſtuck vnd spricht
hautt er auff dich <u>etc</u> Das soltu also verſtan Gee
geſtrackt auff In mit Eyne̱n Entrußthaw vnd vach seyne̱n
rechte̱n arme̱ mit deyne̱m lincke̱n arme̱ oder handt vnd
ker dy vmb vnd laß deyn messe̱r valle̱n vnd far Im vntte̱n
durch den arme̱ vnd schlag Im deyne̱n rechte̱n arme̱ außbendig

In das gelenck des olpogens vnd ruck seyn olpogen an deyn
prust vnd leg dich starck mit dem leypp dar auff wiltu
In werffen so schreytt mit deynem rechten füß hinter seynen
rechten etc

Will er ym vechten
Vberschlagen den rechten
Den soltu durchfaren
Er ist gefangen als In eynen garen

Hye sagtt der meyster Eyn pruch wyder das Eegemelt
stuck vnd spricht will er etc Das soltu alzo verstan
Fellt er dir mit seyner lincken handt an deyn rechte
vnd will mit seynem rechten vntten durch faren durch
deyn rechten arme vnd will dir den außbendigs
schlahen In das gelenck deynes olpogens Indes laß
deyn messer fallen vnd reyb deyn rechte hand auß
seyner rechten vntersich vnd far vnter seynen rechten arme
vntten durch vnd far hintten vber seyn rechten arme
vntter deyn rechte vchsen vnd druck seyn handt an
deyn leyb vnd wenck deyn lincke seytten an seyn
rechte vnd fall mit deynem lincken arme oben ✠ oder
handt oben vber seyn lincke achselen vnd swing dich
auff deyn rechte seytten oder fall auff deyn ars vnd halt
In vnd sitz neben Im wye vntten gemalt stett

Item das vntten gemalt stett Ist nichtz wedeutten etc

Hic nihil ad implevit

Item Greyfft er dir mit seyner lincken verkertte handt
In deyn rechte vnd will seynen rechten schlagen vber
deyn rechten In das gelenck des olpogenß als vor Indes
piß behendt vnd loß deyn messer fallen vnd reyß
deyn rechte handt Inwendigs auß seyner lincken vnd
far vntten durch seyn lincken arme mit dem schlag
hintten mit deynem rechten arme vber seynen lincken starck
von oben nyder vnd wendt deyn rechte seytten an
seyn lincke vnd wurff In starck vber deyn rechte
huff etc

Item Thu Im wye Ee gemelt ist So du In auff dy huff
gepracht hast so greyff mit deyner lincken handt
hinter dich In seyn lincke knypug vnd heb In woll
auff deyn schulteren vnd wurff In furdich auff deyn
lincke seytten etc

98v   Item aber Eyn ander pruch Greyfft er dir mit seyner lincken
handt als vnd will seynen rechten schlagen vber deyner
rechten außbendigs Indes wendt deyn messers mit
der kurtzen schneyden gegen Im vnd loß Indes dy Ee
gemelten schneyden mit gesencktem ortt an seynen lincken
olpogen schnell ableytten vnd far mit dem messer zwi
schen Euch payd mit der klingen vnden durch seyn
rechten arme vnd kum mit deyner lincken hand In dy
mitten deynes messers vnd druck do mit außbendigs
seynen rechten arme faſt nyder vnd wendt dich mitt
deyner lincken seytten an seyn rechte vnd wirff In
vber deyn linckes peyn etc

99r   Item hatt er dich gefaſt mit seyner lincken handt als
vor Indes senck deyn gehultz woll vber seyn lincke
handt vnd kum mit deyner lincken handt deyner
rechten zu hilff vntten an dem gehultz vnd reyß
mit deynen peyden henden ſtarck vntersich vnd haw Im
zu dem kopff vnd schreytt Indes mit deynem lincken
fuß zu ruck Du solt auch wissen das dw mit dysen
Eegemelten bruchen schnell seyn sollt das er dich
nit In dy wag pring etc

99v   **Recht mit linck ler arme beschlyssen**
Hallt In vaſt zu verdyssen
Mitt messer ler arme Tauchen
Wiltu der beschlyssen prauchen

       Hye sagtt der meyſter aber von eynem beschlyssen vnd spricht
Recht mitt linck etc Das soltu also verſtan haut er auff
dich Eynen oberhaw So haw gegen Im den Entruſthaw vnd
kum nohent an In das dy hend zusammen ruren Indes
halt den rechten arme ſtarck an seynen rechten das deyn handt
Ewchling an seyner ſtee vnd schlag Indes deyn lincken arme
In das gelenck seynes olpogens seynes rechten armes außbendigs
vnd far mitt der lincken handt Inwendigs auff deynen
rechten arme vnd leg dich mit der pruſt auff den selbigen
arme also das seyn olpogen an deyn pruſt kum vnd druck
vaſt mit deynen peyden henden also iſt er aber
beschlossen etc

100r   Item aber eyn beschlyssen Greyff mit deyner lincken handt
vber seynen rechten arme woll zu dem olpogen vnd mit
der selben lincken handt far hyn ab außbendigs hinter
seyn arme vnd nym mit deyner lincken deyn messer In

der mitten. vnd laß den ortt beseyt neben deyner lincken
seytten hyn auß gan vnd schreytt mit deynem rechten fuß
hynter seynen rechten vnd ſtoß In also mit deynem messer
oder arme auff seyn rechte achselen vnd druck mit deyner
lincken handt seynen rechten arme faſt In deyn lincke seytten
vnd wurff In vber deyn rechtes peyn oder leg Im das
gehultz an den hals In seyn genick auff seyner rechten
seytten etc

100r (cont.)

Item gee Im aber durch mit dem messer als vor das dir seyn
messer kum vnter deyn lincke vchsen vnd far Im mit dem
gehultz vnter seynen hals auff seyner lincken seytten vnd schreyt
mit deynem rechten fuß hinter seynen rechten vnd wendt dich
Eyn wenig gegen deyner lincken seytten vnd druck In vber
deyn rechtes peyn etc

100v

Beschlossen so er dich hatt
Mitt drucken er dir macht natt[92]
Deyn messer solt sencken
Auff linck ſtarck ler swencken

101r

Hye lertt der meyſter Ein pruch wyder dy Eegemelten beschlissen
vnd spricht beschlossen so er dich etc Das soltu alzo verſtan
hatt er dir griffen mit dem lincken arme In das Glenck
deynes rechten olpogen außbendigs vnd mit seyner rechten vnter
deynen rechten vnd hatt deyn arme auff deyn pruſt gefaſt vnd
will dich da mit vber ruck drucken Indes ſtee ſtreyteß
gegen Im vnd loß deyn messer sincken nyder In deyn lincke
vchsen vnd Greyff mit deyner lincken handt In seyn arme
nohendt bey der achsel vnd schreytt mit deynem rechten fuß
hinter seynen rechten vnd reyb dich vmb auff deyn lincke seytten
vnd wirff In vber deyn rechten fuß etc

Item aber eyn pruch hatt er dich gefaſt mit seynem rechten
vntter deynen rechten vnd mit seyner lincken geschlagen
yn deynen rechten olpogen als vor So fall In dy mit deynes
messers mit deyner lincken handt vnd far auff mit der
klingen In dy höch vnd far mit deynem gehultz vber seyn
peyde arme vnd schreyt mit dem lincken fuß fur seynen
lincken vnd druck mit dem gehultz ſtarck von oben
nyder vnd mit deynem lincken arme scheub ſtarck seyn
achselen nyder vnd wurff yn vber deyn linckes peyn

101v

---

[92] *nott*

102r Item hatt er dich gefast als so stoß mit deyner lincken handt
seynen rechten olpogen woll auff seyn rechte seytten vnd schlag
In mit deynem messer auff seyn kopff etc

102v **Nicht vergyß der schnytt**
Zwen vntten zwen oben mitt

Hye sagtt der meyster von den schnytten vnd von Iren
eygenschafften vnd spricht das Ir vyer seyn dy zwen
ersten als dy zwen obren gehoren zu Treyben gegen den
vechternn dy auß dem pandt gen vnd dy mit dem messern
gerne vmb schlagen zu den andren zynnen das prichtt
der schnytt also wenn Er dir mit dem versetzen oder sunst
an deyn messer pyndett zu deyner lincken seytten vnd
schlecht er denne d paldt vmb mitt dem entrusthaw oder
sunst zu deyner rechten seytten so spring auß dem haw
mit dem lincken fuß auff seyn rechte seytten vnd fall
Im mit der scharphen schneyden vber seyn armm vnd
druck In mit dem schnyt von dir das soltu albogen
treyben wenne er auß der versatzung schlecht etc

103r Item ligtt er auff deyner rechten seytten bey dem oren vnd du
ligest an seynen messer vntten ader oben so er denn will vmb
schlagen zu der andren zynnen deynes lincken orens so schneyd
In mit der kurtzen schneyden In seyn rechten arme Inbendigs
das ist der ander schnydt etc

103v Item dy andren zwen schnytt gehoren zu treyben gegen den
vechtren dy do Eyn lauffen mit auff geracktem arme vnd
dy schnytt Treyb also wenn Er dir an das messer pindt
Es sey mitt versatzung oder sunst fertt er denn hoch auff
mit dem arme vnd laufft dir Eyn Zu deyner lincken seytten
so verbent deyn messer das der dawm vntten kum vnd mit
der scharpffen vntter seyn gehultz In seyn arme vnd druck mit
dem schnytt vbersich etc

104r Item laufft er dir Ey auff deyn rechte seytten so verbendt deyn
messer das der dawm vntten kum mit der kurtzen schneyden vnter
seyn gehultz In seyn arme vnd druck mit dem schnytt vbersich
also hastu dy vyer schnytt vnd wenne du dich recht dareyn
schicken kanst so magstu dye treyben In allen an pynden etc

104v **Will er auß dem pandt rucken**
Dy handt soltu Im drucken

Hye sagtt der meyster wye man Eyn schnyt In den andren
wechselen solt vnd spricht will er auß etc Das soltu alzo

verstan wenne dw eynem an gepunden hast auff seyner
rechten seytten vnd Ir ligett ped Im hangenden ortt auff der
selbigen seytten Indes so er vmb schlech schneyd In an seyn
arme vbersich mit der kurtzen schneyden Indes wendt deyn messer
vmb auff seynen arme vnd schneyd Im den obren schnyd vber
den selbigen arme etc

104v (cont.)

Item laufft dir eyner zu deyner rechten seytten Eyn mit auffgeracktem
arme So windt Im deyn messer mit der kurtzen schneyden vnter
seyn gehultz In seyn arme vnd druck vast vbersich vnd schreyt
auff seyn lincke seytten vnd loß das gehultz vntten durch gen
vnd wendt das messer mit der langen schneyden vber seyn arme
In den obren schnytt Druck von dir auch magstu schneyden
den man auff seyn hals an seyn prust vnd dy schnytt treyben
darnach als dir der man stett etc

105r

Dy handt will er schneyden
Messer nemen solt nicht meyden

105v

Hye sagtt der Merer der kunst wye man dy schnytt prechen
soltt vnd spricht dy handt etc Das soltu also verstan
Schneydt er dir In deyn rechten arme außbendigs Indes laß
deyn gehultz vnd handt auff seyn swertt vnd greyff mit
deyner lincken handt oben In seyn gehultz hynter seyner handt
vnd reyß starck auff deyn rechte seytten etc

Schneydt er dir dy handt
Olpogen stoß sey dir bekandt

106r

Hye sagtt der meyster Eyn ander pruch wyder dy schnytt vnd
das soltu alzo versten wiltu vmb schlagen zu seyner rechten
seytten Indes schneytt er dich aussen In deyn arme oder handt
So stoß Indes mit deynem rechten olpogen auff seyn messer auff
der selbigen seytten do er dich schneydt vnd schlag Im Euchling
dy kurtz schneyd auff seyn kopff etc

Schneydt er dy handt ploß
Vberlauff gewappend stoß

106v

Hye lertt der meyster aber Eyn pruch wyder dy schnytt
vnd mach den also schneytt er dich außbendig vber deyn
arme wye vor so far mit dem gehultz auff seyn messer
vnd fall mit deyner lincken handt In dy mitt deynes
messers vnd stich In zu dem gesicht gewappendt etc

Schneydt Inwendig / swech durchge
Vberbindt haw stich Im geschicht we

107r

107r (cont.)

Das soltu also verstan schneydt Er dich Inwendigs
In deynen rechten arme so wind deyn messer gegen dem seynen
das dy lang schneyd oben stee vnd dy kurtz gegen dir vnd
senck den ortt gegen der erden vnd gee mit der swech vntten
durch auff seyn rechte seytten mit der kurtzen schneyden
auff seyn messer vnd haw In Indes Zu seynen kopff der
selbigen seytten mit der langen schneyden etc

107v

No text.

108r

Dy handt ist er schneyden ynnen
Stoß handt haw mag gewynnen

Hye lertt der meyster aber Eyn pruch wyder den Indren schnyt
vnd mach den also Schneydt er dich Inwendigs vber deyn
handt oder arme Indes stoß seynen rechten arme nyder mit deyner
lincken handt vnd haw Im zu dem kopff mit der langen sch
neyden wye vntten gemalt stett etc

108v

Inwendig handt mit schneyden wirtt
Durchge auff recht ortt lanck stoßt hartt

Hye lertt der meyster aber Eyn pruch wyder den Inbendigen
schnytt vnd mach den also So er dich schneyt In deynen
arme Inbendigs Indes laß deyn messer Sincken vnd wechsel
Im den langen ortt durch auff seyn rechte seytten vnd stich
Im zu dem gesicht der selbigen seytten Trifstu In nicht
mit dem langen ortt so tritt nohendt In Inn vnd windt Im
den hangenden ort Zu seynem gesicht der selbigen seytten etc

109r

Deyn schneyd ler wenden
Ortt do mit senden

Hye sagtt der meyster Eyn stuck auß den schnytten vnd spricht
deyn schneyden etc das soltu also verstan So du den schnytt auff
seyner rechten seytten gemacht hast Indes windt deyn messer
gegen dem seynen vnd stich In mit dem hangenden ortt zu dem
gesicht oben zu der selbigen seytten vber seynen arme etc

109v

In dy andren schnytt
Ortt erbeyt albeg mitt

Hye lertt der meyster wye man / ~~dy andrenschnytt prechen~~
~~Soltt vnd spricht~~ / den ortt In dy andren schnyt prauchen solt
vnd spricht In dy andren etc Das soltu also verstan
Schneydestu In In seyn rechten arme Inbendigs Indes
wenndt deyn messer gegen seynen arme vnd stich Im den

hangenden ortt zu dem gesicht wye vntten gemalt ſtett

also haſtu den ſtich auß dem andren schnytt etc

109v (cont.)

Item auß dem dritten schnytt mach den ſtich also Su dw

mit der langen schneyden von vnden auff schneydeſt dy

seytten oder den arme auff seyner rechten seytten Indes

so du auff das hochſt kumeſt mit dem gehultz deyneß

messers Indes senck Im den ortt vntter seyn vchsen oder

pruſt etc

110r

Item auß dem vyerden schnydt mach den ſtich also So dw

In geschnytten haſt auff seyn pruſt oder Inwendigs In

seynen rechten arme Indes far Im mit dem langen geraden

ortt zu seynen gesich seyner lincken seytten etc

110v

Das sunnen zaygen
mit dem messer wiltu naygen
Dy achselen taſt
Gegen nack druck vaſt

111r

Hye sagtt der meyſter von Eynen ſtuck das heyſt das sunnenzaygen

das mach also hawtt er zu dir oder du Zu Im vnd so dy

messer zu sammen klitzen so far mit deym gehultz deynes messers

vber seyn messer auff seyner rechten seytten vnd schreytt mit

deynem lincken füß hinter seynen rechten vnd greyff mit deyner

lincken handt hyntten an seyn achselen seyner lincken seytten

vnd ruck In vaſt zu dir vnd far mit dem gehultz vnd

kreutz deynes messers vntter seynen hals vnd druck Im seyn

kelen ser gegen dem nack das heysset Im messer das

sunnenzaygen etc

Ob er daſt[93]
Vnd gegen nack druckt vaſt
den rechten arme vberschlag
vnd bey dem olpogen nym dy wag

111v

Hye sagtt der meyſter Eyn pruch wyder das sunnezaygen vnd

mach den also iſt er dir mit dem gehultz gefaren vber deyn

messer vntter deynen hals vnd mit der lincken hyntten an

deyn lincke achselen so laß deyn messer fallen vnd far mit

deyner rechten handt durch seyn peyde arme oder der zwischen

seyn payd arme vnd far mit der rechten handt auff seyn rechten

arme hyntten an den olpogen vnd zeuch do mit ſtarck gegen

---

[93] *dast* could come from *dasten = tasten* (HENNIG 2001,
p 51). In LH, f 56r: *tast*

111v (cont.) deyner rechten seytten vnd far mit deyner lincken handt außben
digs vntter seynen rechten olpogen deyner rechten handt zu hilff
vnd wendt deyn lincke seytten an seyn rechte vnd scheub mit
payden henden mit gantzer krafft auff deyn rechte seytten vnd ˢᵒ kumt
dir seyn rechter arme beschlossen vber deynen lincken vnd auch
seyn messer wye es vntten gemalt stett etc

112r Item hat er dich mit dem gehultz gefast als vor vnd will
dir aber das sunnenzaygen machen so far mit deynem messer
vntter seyn payd arme auff deyn lincke seytten das der ortt
hynden auß hyn gee vnd fall zu gewappender handt vnd
far auff mit dem gehultz vnd mit dem rechten arme vnd
wenck mit der prust eyn wenig zu ruck Indeß stoß mit
deynem gehultz starck an seynen rechten olpogen außbendig
vnd schreytt mit dem lincken fuß fur seynen rechten vnd
far mit dem gehultz starck vber seynen rechten olpogen
das der ortt oben stee vnd das gehultz gegen der erden Inbendigs
an seynen arme vnd druck starck von oben nyder mit gewappen
der handt auff deyn rechte seytten etc

112v Item aber eyn ander pruch hellt er dich als vor mit dem
sunnenzaygen so far aber vntten seyn beyd arme mit dem
ortt auff deyn lincke seytten vnd far mit dem gehultz
zwischen seyn payd arme auff oben Inbendigs vber seyn
handt vnd reyß starck mit dem gehultz vntersich vnd
druck oben fast nyder mit der klingen deynes messers
vnd swing dich auff deyn rechte seytten etc

113r Item aber eyn ander pruch hatt er dich mit dem sunnenzaygen gefast
als vor So merck Indes dy weyll er fertt mit dem lincken arme
nach deyner lincken achsel so laß In vmb dich mit dem arme nicht
kummen Sunder weyll er tasten will gegen dir vnd seynen arme streckt
Indes far mit dem ortt deynes messers mit der swech zwischen
seyn payd arme auff In das gelenck des olpogens seynes rechten armes
vnd far mit dem gehultz vnd mit der sterck deynes messers von
vntten auff Inbendigs In gelenck seynes lincken ölpogens
vnd wendt dich mit deyner rechten seytten an seyn lincke vnd schreyt
mit deynem rechten fuß fur seynen rechten vnd druck geswindt auff
deyn lincke seytten also hast In aber gefangen vnd verigelt das du
In furen magst wo dw hyn wildt vber seyn danck vnd magst
In auch werffen das er nicht auff mag stan etc

113v Item helt er dich als vor ym sunnenzaygen so far aber mit dem
messer mit gewappender handt auff seyn lincke seytten als vor
nun Ee er deyn achsel pegryffen hatt Indes weyll er den

arme gegen dir ſtreckt Indes far auff mit dem gehultʒ zwischen

seyn payd arme vnd wendt deyn lincke seytten an seyn rechte

vnd schreytt mit deynem lincken füß für seyn rechten vnd reyb

mit dem gehultʒ seyn rechten arme oder handt ſtarck nyder

vnd druck mit deynes messer klingen vnd mit deynem olpogen

seyn rechte achselen ſtarck von dir vnd leg dich mit deynem

arme vnd pruſt auf dy selben ~~seytten~~ achselen etc

Item hatt er dich geuaſt[94] als vor dy weyll er aber greyffen will

nach Deyner achselen als vor so far mit deynem rechten arme aussen

ſtarck vber seyn lincken außbendigs yn das gelenck vnd olpogen

auff seyn lincke seytten vnd schreytt mit dem lincken fuß hinter

seyn rechten vnd druck In dar vber etc

Mit gehultʒ vberfar
Mit swech der gelenck nym war

Hye lertt der meyſter Eyn gutt ſtuck wye man dem man

das messer nemen soll zu gewappender handt mit der swech

vnd thu Im also Gee an den man von deyner rechten seytten

mit dem Entruſthaw auff seyn lincke seytten Indes far mit

deym gehultʒ vber seyn messer auff seyner rechten seytten das

seyn[95] messer vntter das seyn kum Indes far Im mit der swech

deynes messers Inbendigs zwischen seyn gehultʒ vnd handt

vnd reyß ſtarck auff deyn lincke seytten etc

Item aber Eyn gutt ſtuck Gee zu Im als vor vnd far

aber mit deym gehultʒ vber seyn messer auff seyner rechten

seytten vnd laß vntten durch gan dy swech deynes messers

vnd far Im außbendigs In das gelenck seyner rechten handt

vnd druck ſtarck auff deyn rechte seytten So nymbſtu Im

aber das messer etc

Vberfar gewappendt ſticht
Ortt zu der kelen vicht

Hye lertt der meyſter aber Eyn gutt ſtuck wye man den

man zu gewappender hent an setʒen soll den ortt vnd thu

Im also far vber ~~de~~ seyn messer mit dem gehultʒ auff seyner

rechten seytten vnd laß dy swech deynes messers sincken auff

seyn lincke seytten ~~vntten dürch wyder auff seyn rechte seytten~~

---

[94] After LH: *gefaßt* | [95] Should actually be *deyn.*

115v (cont.)   vnd ſtich Im jnbendigs[96] gewappen[97] zu dem gesicht vnd das seyn

messer kum vntter ~~seyn~~ deyn rechte vchsen etc

116r   ### Für schleg er pultz
Auff linck wer nymb mit gehultz

Das soltu also verſtan far Im mitt dem gehultz vber seyn

messer wye vor vnd so deyn swech vntten durch iſt gegangen

auff deyn lincke seytten vnd seyn messer iſt dir kummen vnter

deyn rechte vchsen so kum Im mit dem gehultz Inbendigs In

das gelenck seyner rechten handt vnd druck da mit geswindt

auff deyn lincke seytten so nymbſtu Im das messer

116v   ### Vberfar mit cloß
Mit gehultz hals: vber peyn ſtoß

Das soltu also verſtan Gee an In mit deynem messer wye vor

vnd far Im mit dem gehultz vber seyn messer auch wye vor

vnd kum vntten durch mit der swech auff deyn lincke seytten

vnd far mit deyner lincken handt In dy mitten deynes messers

vnd far mit dem gehultz auff Im an seynen hals vnd schreyt

mit deynem fuß hinter seynen rechten vnd swing dich Eyn

wenig auff deyn lincke seytten vnd wurff In fur dich vber

deyn rechte peyn etc

117r   ### Zu Im soltu rücken
Vnd dy handt ler drucken

Hye lertt der meyſter aber Eyn ſtuck wye man Eynen vahen[98]

solt bey seyner handt vnd thu Im also So du mit deynem

gehultz vberferſſt seyn messer wye vor So zeuch mit deynem

gehultz auff seynem messer zü ruck das dy swech deynes messers

kum auff seyn rechte handt vnd fall yndes mit deyner lincken

In payde messer vnd auch mit deyner rechten oben In payde messer

vnd druck mit peden henden vaſt zü vnd wendt deyn lincke

seytten an deyn rechte vnd mach was du willt etc

117v   ### Wiltu ab lauffen
Von payden seytten soltu wauffen

Hye lertt der meyſter aber Eyn haubtſtuck vnd sagtt wye du Im

messer ab solt lauffen vnd sprich wiltu ab etc das soltu alzo

verſtan wenne du Im zuuechten zu Im kumbſt vnd haſt

---

[96] Prefix of -bendigs effaced and overwritten with jn- in a different hand.

[97] Written in the margin in a different hand with a small cross signifying the insertion point.

[98] vahen = fassen (LEXER 1885, p 310)

nahendt zu Im mit dem anpind̲e̲n wiltu den̲n ablauffen mit
dem messer so soltu da mit bayffen⁹⁹ als dy fraw̲e̲n wan sy das
garn̲n ab haspel̲e̲n¹⁰⁰ von payd̲e̲n seytt̲e̲n vnd thu geleych als dw
nyd̲e̲r wolleſt setz̲en mit dem messer vnd laß das mess̲e̲r von payd̲e̲n
seytt̲e̲n abgen dach¹⁰¹ das der dawm̲e̲n albeg vntt̲e̲n ſtee auff der
flech des messers vnd wenn̲e̲ dw denn Sichſt das er deyn̲e̲m mess̲e̲r
nach gett mit der ve̲r̲satz̲ung Indes mach Eyn feler auff welch
seytt̲e̲n du den man̲ am gewiſt̲e̲n meynſt zu hab̲e̲n so triffeſtu In ¶

117v (cont.)

¶Item od̲e̲r so du ablauffeſt von̲ payd̲e̲n seytt̲e̲n Indes so du abgeloff̲e̲n
piſt alzo pald schlag zu der untter̲e̲n zynn̲e̲n Indes schlag
pald wyd̲e̲r zu der obr̲e̲n zynn̲e̲n od̲e̲r schlag Im den ortt In seyn
pruſt vnd schlag Indes nohendt zu der forig̲e̲n zynn̲e̲n etc

118r

Ligt er ym hang̲e̲nden orcht
Bnym das messer an vorcht
Mit dem kreutz soltu schüb̲e̲n
Mit payd̲e̲n schneyd̲e̲n dich üb̲e̲n

Hye lertt de̲r̲ meyſt̲e̲r wye man sich leg̲e̲n solt In den hang̲e̲nd̲e̲n
ortt vnd dar auß mach̲e̲n sol auff payd̲e̲n leger̲e̲n vnd schick
dich In den hang̲e̲nd̲e̲n ortt auff payd̲e̲n seytt̲e̲n wye vntt̲e̲n gemalt
ſtet etc

Item hye sagtt der meyſt̲e̲r ab̲e̲r von eyn̲e̲m haubtſtück das iſt
von dem pnem̲e̲n Nu merck das pnem̲e̲n Im mess̲e̲r iſt wenn̲e̲
Ir Im zuuecht̲e̲n zusam̲m̲e̲n kumpt vnd so Ir payde ligtt
Im hang̲e̲nd̲e̲n ortt wye ob̲e̲n gemalt iſt ligtt er geg̲e̲n dir
Im hang̲e̲nd̲e̲n ortt auff deyn̲e̲r̲ recht̲e̲n seytt̲e̲n vnd dw ligeſt
auch auff seyn̲e̲r̲ recht̲e̲n seytt̲e̲n Im hang̲e̲nd̲e̲n ortt vnd so Ir
also ligtt geg̲e̲n Eyn ander vnd also seyt Im pandt so far
mit dem kreutz auff seyn lincke seytt̲e̲n an seyn mess̲e̲r vnd
scheub mit deyn̲e̲m mess̲e̲r seyn messer auff seyn rechte seytt̲e̲n
vnd schreytt Indes woll vmb auff seyn lincke seytt̲e̲n
mit deyn̲e̲m recht̲e̲n füß vnd schlag In mit der ſtumpff̲e̲n
schneyd̲e̲n an seyn kopff etc

118v

Item ligſtu auſſ seyn̲e̲r̲ linck̲e̲n seytt̲e̲n an seyn̲e̲m mess̲e̲r Im
hang̲e̲nd̲e̲n ortt so far mit dem kreutz od̲e̲r gehultz auff seyn
rechte seytt̲e̲n in seyns messers kling̲e̲n vnd scheub Im also
seyn messer auff seyn lincke seytt̲e̲n Indes schreytt mit deyn̲e̲m

119r

---

⁹⁹ *weifen = schwingen, schwencken* (GRIMM AND
GRIMM 2004, vol. 28, p 632, 5)

¹⁰⁰ *Haspel* [*Garnwinde*] (HENNIG 2001, p 147)
¹⁰¹ *doch*

119r (cont.) lincken füß fur auff seyn rechte seytten vnd schlag Im zu
dem kopff mit der langen schneyden etc

119v
Der Im hangenden ortt ligtt
Vnd dy pnemen wigtt
Dem soltu durchgan
Wiltu seyn nicht schaden han

Hye sagtt der meyster eyn pruch wider dy pnemen vnd spricht
der Im etc Das soltu also versten ligtt er Im hangenden
ortt auff deyner rechten seytten vnd du auch seyner rechten
seytten Indes dy weyll er deyn messer scheubt auff deyn rechte
seytten vnd will schlagen zu deyner lincken seytten zu dem
oren Indes zuck deyn messer an dich mit gesencktem ortt von
seynem messer nohendt an deyn leyb vnd kum Ee mit dem
schlag vnd schlag Im zu dem oren seyner rechten seytten

120r Item ligstu auff deyner lincken seytten Im hangnden ortt vnd
Er ligett auch gegen dir auff seyner lincken seytten Im hangen
den ortt vnd will dir deyn messer pnemen mit seynem messer
so merck dy weyl er deyn messer scheubt so zuck deyn messer
gegen dir nohendt zu deynem leyb vnd kum Ee mit dem
schlag denne er vnd schlag Im zu seynen lincken oren Indes
windt gegen Im deyn messer vnd stich Im zu dem gesicht
ob dw willt etc vnd der pruch ist ser letz vnd nicht
am pesten vnd so dw nicht weytter kummen magst so magstu
dich dennich do mit behelffen

120v Item aber Eyn pruch vnd ist der pest will er dich pnemen als
vor so merck dy weyll er dir nach deynen lincken oren schlecht
Indes far auff mit deynem messer als vor vntter seyn messer vnd
setz Im den ortt an seyn hals auff seyner rechten seytten vnd
mach Indes waß du willt etc

121r Item Eyn pruch wyder dy zucken Im hangenden ortt willer
mit seynem messer zucken In dem hangenden ortt Indes so er
auff fertt mit seym gehultz vnd will zucken so far mit
deynem ortt starck nach seym gesicht vnd heb deyn arme hoch
auff vnd laß In nicht von dem messer kummen vnd volg
nach der ploß vnd seynem messer mit dem ortt etc

Item ligtt er dir Im hangenden ortt auff deyner rechten seytten
vnd du ligest Im auch auff seyner rechten seytten So thu
sam[102] du Im dringen wollst zu der ploß auff der selbigen

---

[102] *sambalde = alsbald* or *sam = als* (HENNIG, p 276)

seytten Indes zuck vbersich mit deym gehultz vnd messer
vnd stich Im oben eyn zu der ploß seyner lincken seytten
Der pruch der oben gehortt vber ditz stuck etc

Wiltu In pnemen
Mit lērer handt wer nemen
Zuck wechsel ob du wilt
Dy wach hab hinter dem schilt

Hye sagtt der meyster wye man den mann pnemen solt vnd
wiltu In etc Das soltu also verstan wiltu Im seyn messer
nemen so thu Im also So Ir payd ligt Im hangenden ortt auff
seyner rechten seytten vnd Er auch auff deyner rechten seytten So
pnym Im seyn messer woll auf seyner rechten seytten Indes fall
mit deyner leren handt auff seyn handt an das gehultz des
messers vnd reyß vntersich so nymbstu Im das messer etc

Item ligt Ir payd als vor so pnym Inn aber mit lerer
hand vnd thu Im also So Ir payde ligt Im hangenden
ortt so greyff mit deyner lincken auff seyn lincke seytten
durch seyn messer vnd fall außbendigs oben mit der lincken
hand zwischen seyn messer vnd deyner rechten handt alzo
ist er pnummen mit lerer handt etc

Item liget Ir ped Im hangenden ortt so wendt oben deyn
hand Ewchling gegen seyner ~~Ewchlich~~ ewchlingen vnd greyff
Indes mit deyner lincken handt In peyde messers In dy
mit vnd far mit deynem gehultz vntten durch seyn arme
außbendig In das gelenck seyner rechten handt vnd reyß
Indes starck auff deyn rechte seytten mit dem gehultz etc

Item ligstu Im hangenden ortt als vor so magstu auch
durchzucken vnd auch ~~d~~ durchwechselen schneyden etc
Item thu Im also als du wollest stechen zu den vntren zynnen
Indes stich beseytt auff seyn rechte seytten vnd Triff In
nicht mit dem stich Sunder schlag In mit dem Entrußthaw
zu seyner lincken seytten zu den obren zynnen etc

Item So Ir payde ligt als vor Indes windt mit deynem
messer gegen seynem messer das der dawm oben stee vnd
stich Im behentlich zu dem pauch Indes windt wyder
auff deyn lincke seytten das dy stumpff schneyd vntten stee
vnd stich Im zu der vorigen zynnen etc

Item so du stest als vor so thu geleych als du Im zu dem
pauch wollest winden vnd stich Im den langen ortt Indes
zu dem gesicht etc

124r **Im winden piß bericht**
Deyn ortt trifft vnd seynen pricht
Messer nemen solt er weren
Mit reyssen dich zu Im keren

Hye sagtt der meyster Eyn pruch vnd spricht Im winden
etc Das soltu also verstan Stestu Im hang[e]nden ortt gegen
seyner rechten seytten vnd er windt (sic) seym messer gegen deynem
vnd will Indes durchgen mit dem ortt zu deynem gesicht
Indes windt auch gegen seynem messer vnd far da mit
hoch auf zu seynem gesicht so trifft deyn ortt vnd der seyn
wirtt ab gesetzt etc

125r Item hye sagtt der meyster Eyn pruch wyder das stuck das
er dich Im winden an den pauch sticht vnd will du das
prechen so thu Im also So er dir nach dem leyb will faren
vnd winden will gegen deynem messer vnd dich stechen In den
pauch Indes so er winden ist gegen deynem messer so far
auff vnd stich Im den langen ortt zu seym gesicht wye
vnden gemalt stett etc

125v Item Eyn ander pruch wyder das zucken In dem hangenden ortt
so du mercken pist das er seyn messer an sich zuckt vnd
will dir durch zucken Indes so er auff fertt mit dem gehultz
vnd auch arme Indes far Im mit hangendem ortt zu seynem
gesicht vnd loß In nicht ab kummen von deynem messer noch
durchzucken etc

126r Item Eyn pruch wyder das messer nemen Im hangenden ortt
pnymt er dich auff deyner rechten seytten vnd greyfft mit seyner
lincken handt außbendigs auff deyn rechte handt In das gehultz
vnd will dir das messer nemen Indes so ~~far vntersi~~ er vntersich
druckt so fall mit deyner lincken handt vntten an das gehultz
vnd mit deyner rechten far wyder oben an das gehultz vnd reyß
mit payden henden starck vntersich so nymbstu Im das messer
wyder vnd hawest In durch den kopff etc

Item das ander messer nemen prich auch mit disem Eegemelten
pruch der do heysset frey außgezawmbtt[103] etc

126v **Durchge dy zynnen**
Hew stich schnytt lere vinden

---

[103] *zäumen = gefangen nehmen* (BAUFELD 1996, p 254)

Dy ſtuck ſoltu woll bedencken
Vnd do mit dy meyſter krencken

126v (cont.)

Hye ſagtt der meyſter wye man mit dem meſſer durchgen ſollt
vnd das durchgen iſt ſere gutt Im meſſer vnd Im ſvertt
Du ſteſt nohendt oder fern ſo magſtu der prauchen mit ortt
hewen oder ſchnytten Item wenn Ir Im zuuechten zuſammen
kumptt wiltu denn dy durchgen machen So thu Im alſo
haw von deyner rechten achſel gerad Eyn oberhaw zu Im Eyn
auff ſeyn lincke ſeytten Indes Ee dw Triffeſt mit dem haw
So windt dy ſtumpff ſchneyd gegen ſeynem meſſer vnd ſenck den
ortt nyder vnd gee zwiſchen Im vnd dir auff ſeyn rechte ſeytten
vnd ſchlag Im zu dem kopff etc

Item kumbt er zu dir Im zuuechten So haw aber eyn oberhaw ſtarck
auff ſeyn lincke ſeytten zu dem kopff Indes windt gegen ſeynem
meſſer dy ſtumpff ſchneyden vnd loß den ortt nyder ſincken vnd
far zwiſchen dir vnd Im auff ſeyn rechte ſeytten vnd ſtich Im
oben Eyn zu dem geſicht vnd wendt deyn meſſer das dy ſtu[m]pff
ſchneyd vntten ſtee etc

127r

Item So du alſo auff deyn rechten ſeytten geſtochen haſt In dem
durchgen alſo magſtu auch Indes durchzucken vnd wydervmb
ſtechen auff ſeyn lincke ſeytten etc

Item du magſtu dich auch verſuchen ob du Im mugſt durch
payde wang ſtechen vnd thu dem alſo So du zu dem
man kumbſt In dem zuuechten als vor Indes haw
Im ſtarck eyn Eynen oberhaw auff In als vor Indes gee
durch mit dem meſſer auff ſeyn rechte ſeytten vnd ſtich
Im zu dem geſicht Indes zuck das meſſer wider an
dich vnd ſtich Im zu ſeyner lincken ſeytten zu dem geſicht:

127v

Item haw Eyn oberhaw auff ſeyn lincke ſeytten als vor vnd
gee aber durch vnd ſtich Im auff ſeyn rechte ſeytte zu dem
geſicht Indes far auff mit dem meſſer vnd far vber deyn
haubt das der dawmen vntten auff der flech ſey vnd wenn
dw vber dy mitt deynes haubtz kumſt das dy ſpitz hintten
vber den kopff hyn auß ſtee vnd das gehultz gegen dem
man Indes haw oben wyder eyn auff ſeyn rechte ſeytten zu
dem kopff oder ſchneyd Im ſtarck vber ſeyn arme oder
erbeytt mit dem dupliren oder Indes haw zu den vntren
zynnen etc

128r

Pogen zwiuach
Deyn erbeyt do mit mach

128v

128v (cont.) **Von payden seytten**
ziuach ler schreytten

Hye sagtt der meyster aber als von eynem haubtstuck als von dem
pogen vnd von seyner eygenschafft das ist nun zu mercken
vnd zu wissen das auß dem pogen vill gutter stuck vnd
messer nemen kummen vnd vill stuck haben Ir dar pringen auß
dem pogen

Item den pogen mach also hallt deyn messer auf deyner lincken
seytten oberthalb des knys das der ortt gegen dem man stee
So er denne auff dich pindt von seyner rechten seytten auff deyn
lincke seytten so peug mit deynem messer fur deynem leyb auff
deyn lincke seytten das der ortt eyn wenig gesenckt sey vnd
der dawmen vntten stee vnd dy stumpff schneyd gegen dir stee ¶

129r ¶Indes so seyn messer auff das seyn klitzt so schreytt auff seyn
lincke seytten wol auß dem haw mit eynem zwiuachen
tritt vnd haw Im durch seyn hawptt seyner lincken seytten

Item den andren pogen treyb also leg dich mit dem messer auff
deyn rechte seytten als vor vnd wenne er auff dich hawett
auff deyn lincke seytten nach der ploß so peug mit dem messer
auff vntter seyn messer auff deyn rechte seytten das der ortt
gee auff deyn rechte seytten vnd das gehultz stee gegen deyner
lincken vnd der dawm vntten vnd das dy lang schneyd gegen dir
stee schreytt auß dem haw auff deyn rechte seytten eyn zwiuachen
tritt woll vmb In vnd haw Im zu dem kopff seyner rechten seytten etc

129v Item[104] Der dritt pogen der ist also zu machen leg dich mit
deynem messer auff deyn lincke seytten das der dawm oben
lig auff dem messer vnd der ortt des messers auff der erden
vnd dy stumpff schneyd sey gekertt gegen dem man vnd
so er denne auff dich hawtt so peug deyn messer auff
deyn rechte seytten vnter seyn messer vnd hab deyn schillt
das ist dy versatzung woll fur deyn haubtt vnd schreytt
mit deynem lincken fuß woll auß dem haw fur seynen rechten
vnd auff seyn rechte seytten vnd haw Im Indes durch seyn
haubt seyner rechten seytten etc

130r Item der vyerd pogen ist zu machen auf der lincken seytten
vnd thu dem also leg dich mit dem messer auf deyn lincke
seytten das der ortt auff der erden sey vnd der dawm oben stee
auff der flech vnd gib dich ploß mit deyner rechten seytten

---

[104] *Item* is written in the margin.

schlecht er dir denne zu der ploß so peug deyn messer woll
auff deyn rechte seytten Das das gehultz stee auff deyner
rechten seytten vnd der ortt des messers auff deyner lincken vnd
schreytt zwiuach mit deynen payden fussen woll vmb In
auff seyn lincke seytten woll auß dem haw vnd haw Im
durch seyn hawbtt seyner lincken seytten also hastu den pogen
zwiuach von payden seytten etc

Waß vom pogen lanck kumbt
Ortt schnell das abnymbt
Dy kurtz schneyd ler wencken
lanck zu haupp laß sencken

Hye sagtt der meyster Eyn pruch wyder dy geraden hew dy
von dem pogen gehawtt werden vnd den pruch soltu also
machen So du auff In hawest vnd er pewgtt seyn messer
dar vntter vnd versetzt vnd Indes schlecht er oben nyder
mit eynem geraden oberhaw Indes tritt albegen[105] beseytt auß dem
oberhaw vnd haw Im nach dem kopff oder windt Indes deyn
messer gegen seynem vnd stich Im zu dem gesicht Oder so
er auff dich schlecht so wendt deyn messer mit der kutzen
schneyden auff seyn messer Indes wendt deyn handt vnd
messer gegen Im mit der langen schneyden vnd haw Im
durch seyn haubt etc

Mit dem messer nemen
Magstu In beschemen
Mit lerer handt wer nym
Vberwindt auff linck druck geswindt

Hye sagtt der meyster aber eyn haubtstuck als von dem messer
nemen vnd thu Im also Gee starck an In mit dem Entrußthaw
zu seyner lincken seytten vnd Im anpinden so dy messer zusammen
klitzen so far mit deym gehultz auff vnd vber seyn messer
auff seyner rechten seytten vnd far mit den kreutz Inbendigs
In seynen rechten olpogen vnd stoß mit dem kreutz vnd messer
starck von dir vnd mit der lincken handt fall Im an seyn
pindt hinter seyner rechten handt vnd reyß mit deyner lincken
handt starck an dich so nymbstu Im das messer etc

Item hautt er gegen dir eynen oberhaw so haw auch gleych mit
Im Eyn Indes windt deyn messer mit der zwirch Eyn
wenig auff das seyn vnd greyff Indes mit deynem lincken

---

[105] *albeg[en]* = *alleweg* (BAUFELD 1996, p 6)

131v (cont.)   arme vber peyde messer das dy peyd spitzen hynterdich vnd
vntter deyn lincke vchsen hin auß ſten Indes far mit
deynem gehultz vntten durch seyn rechten arme vnd windt
außbendigs deyn gehultz vber seyn rechte handt vnd
reyß ſtarck vbersich auff deyn rechte seytten etc

132r   Item pegreyff seyn messer als vor vnd far mit dem gehultz
Inbendigs oben vber seyn arme vnd reyß ſtarck vntersich auff
deyn rechte seytten etc

132v   Item wiltu Im das messer nemen mit lerer handt als mit deyner
lincken handt so merck hawtt er auff dich gegen deyner lincken
seytten so peug deyn messer vntter seyn messer woll auff deyn
lincke seytten Indes spring woll auff seyn rechte seytten vnter
seyn messer vnd pegreyff seyn arme Inbendigs bey der hand
vnd druck mit dem lincken arme Inbendigs zwischen
seyn handt vnd gehultz ſtarck auff deyn lincke seytten so
nymbſtu Im aber seyn messer etc

133r   Item Greyff aber mit deyner lincken handt vber peyde messer
Indes far mit deynem ~~he~~ gehultz an seyn gehultz das deyn
kreutz kum hintten ader vorne an das seyn vnd windt
do mit gegen seyner rechten seytten vnd scheub mit dem vodren
tayll deynes armes ſtarck auch auff seyn rechte seytten mit
gepogem arme das der olpogen vntten ſtee vnd scheub da mit
faſt ~~nyder~~ vnd reyß Indes oben ſtarck nyder auff deyn rechte
seytten so nymbſtu Im aber das messer etc

133v   **Iſt er das gelenck prellen**
Gleych soltu wyder schnellen
Nymb gehultz ring reyß
Das man dir sag preyß

Hye sagtt der meyſter Eyn pruch wyder das erſt messer nemen
vnd spricht will er das gelenck etc Das soltu also verſtan
fellt er mit dem gehultz vber deyn messer In das gelenck
deynes olpogens vnd will dir mit seyner lincken hand deyn
messer pey dem pindt nemen So merck gleych Indes so
er vberfertt mit dem gehultz vber deyn rechte handt Indes
far Im auch also schnell vber seyn rechte handt vnd far
mit dem kreutz In das gelenck seynes rechten olpogens
vnd scheub den von dir vnd grewff mit deyner lincken
handt an seyn gehultz vnter seyner rechten handt vnd reyß
ſtarck an dich auff deyn lincke seytten also prichſtu Im seyn
ſtuck vnd macheſt selber das er willen hett zu machen

Item aber eyn pruch wyder das messer nemen mit lerer handt
vnd er deynen rechten arme ~~vnd~~ mit seyner lincken handt
gefaſt hatt vnd will dich drucken In seyn lincke seytten
Indes kum deyner rechten handt zu hilff mit deyner lincken
vnd pegreyff da mit deyn gehultz vntten vnd reyß ſtarck von
oben nyder als du vor gehortt haſt etc

134r

Item hatt er pegriffen peyde messer mit seynem lincken arme
vnd fertt vntten mit dem gehultz durch außbendigs In
deyn gelenck der handt vnd will reyssen auff seyn
rechte seytten Indes greyff hinter seyn lincke handt vorne
bey der swech des messers vnd reyß ſtarck auff deyn lincke
seytten etc

134v

Item wyder das vyerd messer nemen mach auch dysen
Eegemelten pruch etc

Linck vnterfar den rechten
Mit gehultz vber reyß Im vechten

135r

Hye lertt der meyſter Eyn messer nemen das iſt sere gutt
vnd abentewrisch[106] wiltu das machen So thu Im also Stee
mit deynem lincken fuß für vnd hallt deyn messer auff
deyner rechten seytten pey dem peyn hawtt er dir denn zu
deyner lincken seytten nach der ploß Indes pewg von deyner
rechten seytten auff deyn lincke deyn messer vnd spring woll
auff In vnd leg deyn lincke handt veſt vntten an seyn
rechte Indes far mit deynem gehultz auch oben vber seyn
rechte handt Inbendigs vber seyn gelenck der handt
vnd reyß ſtarck auff deyne rechte seytten So nymbſtu Im
aber seyn messer etc

Wyder nemen solt nicht sawmen
Vach swech reyß messer müß rawmen

135v

Hye lertt der meyſter Eyn pruch wider das Eegemelt ſtuck
vnd iſt also zu machen So er deyn handt also beschlossen
hatt als oben geschriben ſtett So senck deyn messer
gegen deyner lincken seytten auff das seyn vnd far mit
deynem lincken arme vber peyde messer vnd reyß Indes auff
deyn lincken seytten so nymbſtu Im aber das messer oder reyß
vntersich mit deynem messer als du vor gehortt haſt etc

---

[106] *abenteuerlich = ausgefallen, nicht alltäglich* (BAU-
FELD 1996, p 1)

136r
Geſtu an were
Wirſtu vberlauffen gefere
Ruck an schaden nicht kanſt keren
Standt: dich frolich magſt weren

Hye lertt der meyſter Eyn ſtuck das do heysset Eyn notſtuck
vnd gett zu so du keyn were haſt vnd wirſt vberloffen
vnd kanſt an schaden nicht entflyhen so ſtandt frolich so du
dich weren müſt vnd nicht anders geseyn mag so wendt
dich gegen Im vnd ſtee mit deynem lincken fuß fur vnd so
er danne auff dich hawett von oben nyder so tritt mit deynem
rechten fuß gegen seynen rechten vnd woll beseytt auß vnd
far mit geſtackem woll aussen an seyn messer an dy flech
außbendig vnd far Inbendigs mit deynem rechten arme vber seyn
messer oder arme vnd druck den arme woll zu deym leyb vnd
wendt deyn handt an seyn messer vnd fall mit deyner lincken
hand an seyn gehultz oder hant vnd reyß mit deynem leyb ſtarck
auff deyn rechte seytten so nymbſtu Im das messer etc

136v
Item hautt er auff dich als vor So ſtee mit deynem rechten
fuß für vnd schreytt mit deynem lincken auff seyn lincken
woll auß dem haw vnd far mit deynem lincken vber
seyn messer vnd schreytt mit deynem rechten füß gegen
seynen rechten vnd far mit deyner rechten handt auff seyn
gehultz bey der handt vnter seyn messer ~~auff deyn rechte~~
vnd kum mit deyner lincken handt vnter seyn messer
auff deyner rechten seytten vnd swing dich ſtarck auff
deyn lincke seytten etc

137r
Item haütt er auff dich als vor so schreytt aber auff seyn rechte
seytten ~~so si~~ mit deynem lincken fuß vnd fall aber vber seyn messer
mit dem lincken arme vnd fall mit deyner rechten handt auff
seyn gehultz vnd reyb ſtarck auff deyn rechte seytten etc

137v
Item hautt er aber auff dich als vor so ſtee mit deynem lincken fuß
fur vnd schreytt mit deynem rechten fuß auf seyn rechten wol
ausß dem haw vnd far vber seyn messer mit deynem rechten
arme vnd mit deyner lincken handt auff seyn gehultz vnd
reyß ſtarck auff deyne lincke seytten

138r
Ob er frisch iſt
Vnd ſtett In kunſtenlicher liſt
Piſt gefangen messer solt wenden
Ruck: ortt thut In enden

Hye sagtt der meyster Eyn pruch wyder dy obgemelten messer   **138r (cont.)**
nemen so du auff eynen gehawen hast der keyn messer hatt vnd
er versetz dir mit dem arme als oben geschriben stet vnd er
fertt dir vber deyn messer Indes windt deyn messer vmb das
dy kurtz schneyd vntten stee vnd far auff mit deynem gehultz
vnd setz Im den ortt auff seyn prust

Item fertt er aber vber deyn arme den pruch vindestu da
der text spricht will er praugen da wirtt der pruch gelertt etc

Bewappend wiltu wer strauffen[107]   **138v**
Mitt gehultz soltu vberlauffen
Reyß starck auff deyn rechte seytten
Mit dem fuß zu ruck ler schreytten

Hye lertt der meyster wye man das messer mit gewappender
handt eynem soll nemen vnd spricht gewappend wiltu etc Das
soltu also verstan hawtt er auff dich von seyner rechten seytten
auff deyn lincke So peug auff mit deynem messer vnd mit
deyner lincken handt far Indes In dy mitt deynes messers
vnd vach den schlag zwischen deyn peyd hend auff das
messer Indes far auff mit deynem gehultz Inbendigs vnd
von oben nyder vber seyn handt In das gelenck außbendig
vnd reyß starck auff deyn rechte seytten So nymbstu Im
das messer etc

Item du magst gewappend felen mit dem ortt vnd thu Im also   **139r**
so du Im gewappend an gepunden hast vnd ligest Im auff seyner
rechten seytten mit dem ortt vor dem gesicht Indes thu sam
du mit deynem gehultz wollest schlagen zu deyner lincken
seytten vnd So du kumpst mit dem gehultz Eyner span weytt
gegen seyner lincken seytten so far Im Indes mit dem ortt wyder
zu seyner rechten seytten zu dem gesicht etc

Item ligestu auff seyner lincken seytten gewappend also das der   **139v**
knopff dem man gegen dem gesicht stett Indes far aber
auff mit dem knopff oder gehultz vntz vber dy mitt deyns
haubtz Indes mit dem gehultz schnell wyder vmb zu der
forigen lincken seytten schlag In mit dem gehultz zu dem
kopff oder stoß Im zu dem gesicht etc

Item du magst auch durchwesclen[108] mit dem kurtzen ortt vnd thu   **140r**
dem also so du ligest wye Er mit gewappender handt auff seyner

---

[107] *straufen = 1. die Haut abziehen, abstreifen; 2. streuen* | [108] *durchwechseln*
(BAUFELD 1996, p 228)

140r (cont.) lincken seytten So thu gleych samb du mit dem ortt wollest gan
auff seyn rechte seytten vnd stechen auff dy selbigen seytten Indes
far vntersich mit der spitzen vnd wechsel durch auff seyn lincke
seytten vnd stich Im zu dem gesicht seyner lincken seytten etc

Item so du kumpst mit gewappender handt an den man so mach dy ab=
laüffen gewappend zu payden seytten Indes fall In dy stich schnit
oder hew vberlauff seyn messer vach In pey dem hals vnd
mach also mit gewappnder handt was du willt etc

140v

Will er strauffen
Mit dem gehultz vberlauffen
Seyn swech zeuch zu dir
Schreytt recht stoß schir

Hye sagtt der meyster eyn pruch wyder das erst stuck als wyder
das messer nemen zu gewappender handt vnd spricht wil er
strauffen etc Das solt also verstan will er dir deyn messer nemen
zu gewappender handt So merck eben dy weyll er mit seynem
gehultz vber deyn handt fertt vnd will reyssen auff seyn
rechte seytten Indes Greyff mit deyner lincken handt In
dy swech seynes messers vnd wendt mit deyner rechten
handt deyn gehultz an seyn rechte zwischen seyn payd
hend vnd reyß Indes starck auff deyn lincke seytten etc

141r Item aber eyn pruch fertt er mit dem gehultz vber deyn arme
als vor so merck dy weyll er vber deyn messer fertt so fall
mit deynem lincken arme vber payde messer vnd druck dy In deyn
lincke seytten vnd swing dich starck auff dy selbigen seytten etc

141v Item Eyn ander pruch hatt er deyn messer gefast wye vor Indes
greyff mit deyner lincken hand In seyn lincke vnd reyß dy
auff deyn lincke seytten vnd schreytt mit deynem rechten fuß
fur seyn rechten vnd schlag geswindt mit deynem rechten
arme hyntten starck an seyn lincken olpogen etc

142r Item aber eyn pruch fellt er dir mit dem gehultz vber deyn arme
als vor mit gewappender handt so merck Indes dy weyll er vber
deyn handt mit dem gehultz will faren so wendt deyn messer
auff dem seynen vmb das dy kurtz schneyd auff seynem messer stee
vnd schlag In zu dem kopff mit der kurtzen schneyden etc

142v Item so du Im also auff den kopff felst mit der kurtzen schneyden
wye vor Indes far von seynem messer mit deynem ortt gegen
seyner lincken seytten zwischen dich vnd In mit gesencktem ortt
auff deyn lincke seytten vnd senck Im den ortt auff seyn prust

vnd schreytt mit deynem lincken fuß fuß zu ruck vnd haw
Im zu dem kopff etc

Iſt dir kunſt zu runnen

Das messer iſt genummen
Nicht solt dich sawmen
Vnd den man auß zawmen

Hye sagtt der meyſter Eyn ᵛⁱᵈᵉʳ pruch wyder den erſten pruch des
messers vnd spricht iſt dir etc Das soltu also verſtan hatt
er dir deyn messer pegriffen bey der swech vnd reyſt auff
seyn lincke seytten vnd du Empfindeſt das du deyn messer nicht
gehalten magſt so fall mit deyner lincken handt hynter seyn
lincke an das gehultz mit deyner rechten vorne zu seyner rechten
handt vnd wendt das gehultz gegen Im vnd reyß ſtarck
vntersich So nymbſtu Im wyder seyn messer vnd haweſt
Im durch seyn kopff etc

Du magſt dich deß remen

Gewappent recht mit clotz hals nemen

Hye lertt der meyſter Eyn ſtuck wye man den man mit dem
gehultz auff seyner rechten seytten bey dem hals nemen soll vnd
spricht magſtu etc Das soltu also verſtan haſtu Im an gepunden
auff seyn lincke seytten mitt der zwirch so haw Im vmb zu
seyner rechten vnd Ee der haw verpracht wirtt so vberdring
In mit deym gehultz auff seyner lincken seytten vnd far Im an
seyn hals seyner rechten seytten vnd schreytt mit deynem rechten
füß hynter seynen lincken vnd druck In dar vber etc

Iſt es gelungen

klotz hals hatt gebünnen
Mit klotz recht vberfar
Vach arme handt gewynnſtu In gar

Hye sagtt der meyſter Eyn pruch wyder das ſtuck So man eynen
bey dem hals hellt mit dem clotz vnd iſt das der g̶ erſt pruch
wann der pruch seyn vill vnd manigerlay Item Der erſt pruch
so dich Eyner gewappendt mit dem klotz bey dem hals ſtarck
hellt so nym deyn messer auch zu gewappender hantt vnd far
mit dem klotz vber seyn lincke handt Inbendig auff aussen
vber das gelenck vnd druck vaſt vntersich mit dem gehultz
vnd heb mit der lincken handt vnd mit der klingen des messers
vaſt vbersich auff vnd schreytt mit dem lincken fuß woll
fur seynen rechten vnd druck mit dem pindt seyn lincke handt
vaſt an dich auff deyn rechte seytten vnd scheub In oben mit

144r (cont.)   deynem messer vast von dir vnd wurff In furdich nyder auff
das antlitz vber deynen rechten fuß etc

144v   Item aber eyn pruch helltt er dich pey dem hals mit dem pindt
als vor so loß deyn messer fallen vnd schlag deyn lincken
arme vber seyn rechten oben starck nyder vnd greyff mit deyner
rechten handt In seyn glenck der handt vnd spring mit deynem
lincken fuß fur deyn[109] rechten vnd druck seyn lincke handt
woll In deyn rechte seytten vnd swing dich mit dem leyb
auff deyn rechte seytten vnd wirff In vber deyn linckes peyn
fur dich etc

145-146r   Item aber Eyn pruch So er dich bey dem hals hatt als vor
so nymb deyn rechte handt vnd far da mit auff seynes
messers klingen auff zwischen seyn payd hend nahent zu
seyner lincken vnd reyß starck an dich auff seyn rechte seytten
Indes wirtt der kloß ledig von dem hals Indes begreyff
mit deyner lincken handt vber seyn rechte In das gehultz vnd
reyß hintersich auff deyn rechte seytten so nymbstu Im das messer
Merck wenne du das messer nemen machen willt so soltu
albeg In dy swech des messers fallen mit deyner rechten handt;

145-146v   Item aber eyn pruch scheub In von dir aber pey seynem ~~lincken~~ rechten olpogen
vnd greyff mit deyner rechten hand an seyn rechtes peyn nohent
bey dem fuß vnd schreyt mit deynem lincken fuß fur seynen
lincken Inbendigs seyner peyn vnd wurff In auff seyn antlitz
wye vnden gemalt stett etc

147r   Item eyn ander pruch scheub In von dir mit deyner lincken handt
vnd nymb das gewicht bey dem olpogen vnd greyff Im forne
mit der rechten handt seyn rechte vnd schreytt mit dem lincken
fuß fur seynen rechten etc

Item Eyn ander pruch scheub In starck von dir mit seynem
olpogen mit deyner lincken handt scheub In also auff seyn
rechte seytten Indes laß deyn messer vallen vnd greyff Im
außbendig In seyn gelenck seynes rechten peyns heb In
auff vnd wurff In auff seyn lincke seytten etc

Item Eyn ander pruch hatt er dich gefast als vor so
scheub In von dir auff seyn lincke seytten mit deyner
lincken handt vnd far mit der rechten handt oder arme
vmb seyn leyb vorne vnd schreytt mit deynem

---

[109] Following LH, *seyn*

rechten fuß hinter seyn rechten vnd wurff In vber deyn
rechte huff auff seyn ruck oder kopff etc

147r (cont.)

Item scheub In aber mit seynen rechten olpogen auff seyn lincke seytten
vnd pegreyff Indes seyn rechten füß mit deyner rechten handt
wye vor vnd ruck den selbigen fuß woll vnter deyn rechte vchsen
vnd halt In also da pey das er nicht auff kan stan wo er auff
eyner seytten sich auff wollt richten So wurff In auff dy andren
seytten vnd stee mit deynem lincken peyn vnd leyb woll zwischen
seyn peyde auff das das er dich mit dem lincken peyn nicht
stossen müg vnd haltt In also vnd wurff In von eyner seytten
zu der andren wye du willt etc

147v

Item aber Eyn pruch hellt er dich pey dem hals gewappent
als vor so druck vast zu ruck mit dem nack Indes greyff
auch zu gewappender hantt In deynes messers klingen vnter
seyn messer vnd loß plupffling¹¹⁰ dy recht hand von dem pindt vnd
schreytt hinter In auff seyn rechte seytten vnd schlag In mit dem gehultz
In seyn pauch vnd schreytt woll hinter In Indes schlag In Inn seyn nacken
vnd schreytt aber paß hyn vmb als du denne woll weyst etc¹¹¹

148r

Item hatt er dir gelegtt den klotz an deyn hals als vor So far
auff mit deynem messer vnd fall zu gewapp[e]nder hantt vnd far
hoch auff mit dem gehultz vnd stich Im zu dem gesich oben
wye vntten gemalt stett etc

148v

Item Eyn ander prüch hatt er dich bey deym hals gefast
aber als vor mit dem gehultz Indes laß deyn messer fallen vnd
far mit deyner rechten handt hynter seyn lincke zwischen seyn
handt vnd dy swech seynes messers vnd wendt dich Indes
starck auff deyn lincke seytten vnd reyß starck mit deyner
rechten handt auff dy selbigen seytten etc

149r

Item hatt er dich bey deynem hals mit dem gehultz aber
als vor Indes greyff mit deyner lincken handt In dy mitt seynes
messers vnd far auff mit deyner rechten handt vnd thu
sam du Im das gehultz In das angesicht wollest stossen Indes
stoß In dy mitt seynes messers starck mit deyner handt oder
arme vnd swing dich auff deyn lincke seytten etc

149v

*wen man fechten wil*¹¹²

---

¹¹⁰ *plüpflich = plötzlich* (SCHMELLER 1837, vol I, p 460)
¹¹¹ There is a short note in a different hand at the bottom, but it is too smudged to read.

¹¹² Written in the margin in a different hand. First word may be *zwen*.

☰ 89 ☰

150r Item aber eyn pruch hatt er dich gefaſt mit deyne_m_[113] gehultʒ an
deym hals wye vor So greyff mit deyne_r_ lincke_n_ handt In
seyn lincke nohendt zu der handt vnd reyß dy auff deyn
lincke seytte_n_ vnd schlag Indes deyn rechte_n_ arm_e_ vber seyn
ped arm_e_ fluchs ⱴ nyder vnd schreyt mit deyne_m_ rechte_n_
uß hinte_r_ seyne_n_ rechte_n_ vnd wurff In dar vbe_r_ ete vnd
swing dich Eyn wenig auff deyn lincke seytte_n_ etc

150v Item hatt er dich bey dem hals als vo_r_ vnd will dich vber
das peyn drucke_n_ wye vor Indes fall zu gewappe_n_de_r_ handt
vnd ſtich Im vntte_n_ zwische_n_ seyne_n_ payde_n_ hende_n_ mitt
gewappe_n_de_r_ handt auff In seyn kele_n_ vnd wenne_e_ du zu
gewappe_n_de_r_ hant felſt so far mit deyne_m_ lincke_n_ arm_e_ obe_n_
vbe_r_ seyne_n_ rechte_n_ etc

151r Item hatt er dich gefaſt pey dem hals als vor Indes far
mit deyne_m_ messer zwische_n_ Euch payd mit der swech
auff deyn lincke seytte_n_ vnd[114] fall zu gewappe_n_de_r_ handt vnd
ſtoß In mit dem gehultʒ an seyn kele_n_ als vntte_n_ gemalt
ſtett etc

151v **Auß vber durch ſteche_n_**
Bewappend ler wer auß preche_n_
Auch magſtu arm_e_ beschlyssen
Will gluck des kanſt genyssen

Hye sagtt der meyſte_r_ von dem vbe_r_durch ſteche_n_ vnd dy
selbe_n_ gehore_n_ zu treybe_n_ mit grosser fursichtikaytt vnd
nohent bey dem man_n_ Das du dar ob nit nyde_r_ligeſt nun
thu Im also Gee zu dem man_n_ mit deyne_m_ Entruſthaw
zu seyne_r_ lincke_n_ seytte_n_ vnd schlag da_r_nach vmb zu der
rechte_n_ seytte_n_ vnd windt dy swech seynes messers vbe_r_ seyn
messer gegen seyne_r_ lincke_n_ seytte_n_ vnd far vntte_n_durch seyn
arm_e_ wyde_r_ auff seyn rechte seytte_n_ vnd fall In dy mitte_n_
mitt deyne_r_ lincke_n_ handt In dy klinge_n_ deynes messers
vnd druck da mit außbendigs In das gelenck seyns arm_e_s
vnd olpogens vnd wendt deyn lincke seytte_n_ an seyn rechte
vnd druck mit der lincke_n_ handt obe_n_ faſt nyde_r_ vnd mit de_r_
rechte_n_ pey dem pindt vnd heb seyn handt vbe_r_sich also iſt
er beschlossen etc

152r Item Stich Im abe_r_ vber seyn ~~linck~~ rechte handt vnd fall abe_r_
zu gewappe_n_de_r_ hantt vnd kum Im nohent das seyn rechte

---

[113] Should be *seynem*. [114] Probably should be *zu*.

handt kum vntter deyn rechte vchsen vnd fall zu gewappender   **152r (cont.)**
handt mit deynem messer ✠ hintten an seynen rechten olpogen vnd
layn dich mit der prust vast dar an vnd ker deyn lincke
seytten an seyn rechte vnd spring mit deynem lincken fuß fur seyn
rechten vnd swing dich starck auff deyn rechte seytten

Item aber eyn stuck Im durchstechen Thu Im als vor vnd   **152v**
Indes mit dem gehultz auff seyn rechte achselen an seyn hals
woll hyn vmb In seyn nacken vnd heb In mit der lincken
handt woll vbersich vnd verbendt deyn handt auff deyn
messer das du dester stercker hebst vnd schreytt mit deynem[115]
rechten fuß hynter seynen rechten vnd druck In dar vber etc

Item far aber Im dürch mit der swech vber seynen rechten arme wye   **153r**
vor vnd far vntten auff deyn lincke seytten mit der swech
seynes messers alzo das dir seyn messer vntter deyn rechte vchsen
kum vnd stoß In mit dem gehultz zu seyner kelen oder
angesicht etc

Item aber eyn stuck ist er dir zu starck das dw außbendigs   **153v**
den arme nicht beschlyssen magst als das erst stuck lautt
So thu Im also du außbendig In wolst beschlissen Indes far
Inbendigs oben vber seyn arme an dy achsel vnd schreytt mit
deynem rechten fuß hinter seynen rechten vnd druck In dar vber

Item nota wenne du Im also Inbendigs In seyn arme ferst mit
dem messer so tritt albegen mit deynem rechten fuß woll hinter
seynen rechten das seyn arme zwischen dich vnd In kum so gett
eß recht zu etc

Item oder leg Im das pindt an seyn hals seyner lincken seytten
vnd schreytt mit deynem rechten füß hinter seynen rechten vnd
druck dy klingen deynes messers an seyn arme fast an deyn
leyb das der ortt auff deyner lincken seytten peseytt hyn auß
stee etc

Wer vber durchist stechen   **154r**
Kunstenlich ler auß rechen
Windt recht piß behendt
Ortt zu gesicht sendt

Hye sagt der meyster Eyn pruch wyder dy ~~du~~ vberdurchstechen
vnd spricht wer vber etc Nu wiltu dy vberdurchstechen Im messer
prechen so thu Im also wenne er dir vber deyn messer windt

---

[115] Corrected from *deynem*.

154r (cont.) vnd will dir mit dem ortt vntter deym arme durchgen Indes windt
gegen Im auff seyn rechte seytten deyn messer an das seyn das dy
lang schneyd oben sey vnd ſtich Im zu dem gesicht also kann
er zu den obgemelten ſtucken nicht kummen etc[116]

154v Item aber Eyn pruch auff das vberdurchſtechen vnd merck
dy weyll er vntter deyn arme durchſtechen will vnd dy
weyll er nyder nach der mit seynes messers will greuffen
Indes far auff mit dem gehultz vnd fall zu gewappender
hant vnd setz Im den ortt an dy kelen etc

155r Item leg Im deyn messer an den hals auff seyner lincken seytten
zu gewappender hantt dy swech des messers vnd druck Im
den olpogen woll vnter seynen hals vnd thu als du vor vnter=
richt pist also hastu aber eyn pruch wyder das durchſtechen;

155v Item aber Ey pruch so er durch ſtechen will In dem dy weyll
er deyn messer nyder druckt Indes far auff mit dem gehultz
vnd haw oder schneyd Im durch seyn hals auff seyner lincken seytten
vnd das gett als leychtiklich zu der es recht treyben kan;

156r **Will er dir nahen**
Recht mit linck ler vahen
Setz an dy kelen vnd üchsen
Bewappend thut das püssen

Hye sagtt der meyſter Eyn ſtuck wye man das zu gewapp
ender handt treyben soltt vnd spricht will er etc Das soltu
also verſtan So er Im zuuechten zu Ir dir kumbt so ſtee
vor mit deynem lincken füß vnd halt deyn messer auff deyner
rechten seytten vnd ob deynem rechten kny vnd das der dawmen
oben ſtee vnd der ortt ſtee gegen dem mann vnd gib dich mit
der lincken seytten ploß schlecht er dir denne der ploß nach so
pewg auch vntter seyn messer deyn lincke seytten vnd spring
nohent auff In mit deynem gantzen lincken arme Greyff
ſtarck Indes vber seyn rechten nohent bey dem olpogen vnd
druck seyn arme feſt In deyn lincke seytten Indes greyff vntter ¶

156v ¶ seyn arme In dy mitt deynes messers vnd setz Im den ortt
an seyn kelen oder vntter seyn rechte vchsen oder pehallt
deyn messer In der lincken hand In der mitt vnd laß deyn
rechte handt von dem pindt des messers vnd greyff

---

[116] Between the two figures there is an illegible note
that was written by another hand. "Man..."

Im mit der rechten handt In seyn rechte knypüg vnd
wirff In furdich etc

156v (cont.)

Item oder wendt deyn lincke seytten an seyn rechte vnd
schreytt mit deynem lincken fuß fur seyn rechten vnd far[117]
mit deynem gehultz deynes messers außbendigs an seyn
hals vnd schreytt mit deynem lincken fuß fur seyn rechten
vnd wurff In fur dich etc

Item Eyn ander pruch so dw In pey dem arme gefast hast so
schreytt mit deynem rechten fuß hinter seynen rechten vnd far mit
deyner rechten handt vmb seyn hals Inbendigs auff seyner
lincken seytten vnd wurff In furdich vber deyn rechte huff etc

157r

B¶[118] Item Eyn pessers stuck so dw Empfindest das er will vber
deyn arme fallen gleych Indes so er fellt mit seynem arme vber
deynen Indes heb Im seynen lincken arme außbendigs starck
auff vbersich vnd schreytt mit deynem rechten fuß fur seynen
lincken vnd druck mit deynem rechten arme fast nyder seynen
lincken das dy gelenck der olpogen zu samen ruren oder
kummen ist er starck so kum deym rechten arme oder handt
zu hilff mit deyner lincken vnd wurff In auff das antlitz
vber deyn rechtes peyn etc

157v

Item oder greyff Inn dy mitt deynes messers mit deyner lincken
handt der rechten zu hilff vnd reyß starck auff deyn
lincke seytten vnd druck oben nyder mit der lincken handt
vnd spring mit deynem rechten fuß für seynen lincken vnd wurff
In dar vber also hastu den pruch wyder das erst stuck

A¶ **Will er prangen**[119]
So der arme ist gefangen
Arme vberfar dy kelen
Mit drucken soltu nicht velen

158r

Hye sagtt der meyster Eyn pruch wyder das stuck so eyner
deyn rechten arme gefangen hatt mit seynen lincken vnd
spricht will er ~~fa~~ prangen etc Das soltu also verstan ist deyn
arme gefangen mit deynem messer so far mit deynem messer
aussen vber seyne arme vnd leg Im das messer mit der
schneyden forne an den hals In dy kelen vnd druck fast
von dir In mit dem messer mit gewappender hantt etc

---

[117] Corrected from *gar*.
[118] The letters B and A indicate that ff 157v and 158r
are in reverse order. LH confirms this.

[119] *prangen = prangen, glänzen* (HENNIG 2001, p 256)

158v Mit messer will[120] er kelen drucken
Dy were soltu rucken
Nit lang thu peytten[121]
Anseƶ oder recht piß schreytten

Hye sagtt der meyſter Eyn widerpruch wyder den gemelten
pruch vnd spricht mit messer etc Das soltu also verſtan
wenn du Eynem vber seyn messer piſt gevallen vnd Er will
dir deyn kelen drucken als oben geschriben ſtett Indes nymb
seyn messer bey der klingen mit deyner rechten handt oder
gehulƶ vnd druck dy klingen seynes messers vaſt an dich
In deyn rechte seytten vnd far mit deyner lincken handt
oben auff seyn rechte Inbendigs an seyn pindt seynes messers
vnd reyb ſtarck auff deyn rechte seytten so nymbſtu Im das
messer etc

159r Item aber eyn ander pruch will er In dy kelen drucken mit
seynem messer als vor Indes dy weyll seyn messer In dy kelen
will fallen so fall zwischen dich vnd In mit deynem messer
an das seyn vnd loß deyn lincke handt vnd fall In dy mitt
deynes messers vnd seƶ Im den ortt an dy kelen etc

159v Item Eyn anders ſtuck so dw seyn arme haſt gefangen als vor
vnd er will dir nach der kelen faren Indes swing dich auff
deyn rechte seytten woll vmb so kan er aber nichƶ schaffen;

160r Item aber Eyn ander pruch will er zu dem hals mit
seynem messer als vor Indes senck deyn lincken arme mit
deynem leyb woll nyder vnd druck den faſt an deyn leyb vnd
peug dich mit dem ruck eyn so iſt seyn will aber vmb Sunſt

160v Item hatt er dich mitt dem messer gefaſt hinter dy kelen wye vor
so greyff mit deyner rechten handt In dy swech des messers vnd
druck vaſt auff deyn lincke seytten Indes greyff mit deyner
lincken verkertten handt oben auff das gehulƶ hinter seyn rechte
handt vnd reyß ſtarck vntersich auff deyn lincke seytten etc

161r Item aber eyn pruch greyff aber als vor mit deyner rechten handt
seyn rechte reyß dye ſtarck auff deyn lincke achselen seynen
rechten arme Indes far mit deynem lincken arme ſtarck vmb seyn
hals außbendigs vnd schreytt mit deynem lincken fuß fur
seynen rechten vnd swing dich auff deyn rechte seytten etc

---

[120] From LH, f 80r: *wil* | [121] *beiten = warten, zögern* (HENNIG 2001, p 24)

Item greyff (sic) seynnen rechten arme als vor vnd reyß ſtarck vber   **161v**
deyn lincke achselen vnd druck vaſt nyder Indes druck auch
seynen lincken arme dar Innen du seynen rechten arme haſt vnd
druck mit deynem lincken arme vaſt In deyn lincke seytten vnd
kum mit deyner lincken vntten In dy mitt deynes messers
vnd schreytt mit deynem fuß fur seyn lincken vnd swing
dich auff deyn rechte seytten etc

Will er dich beschemen   **162r**
Das messer pey gehulʒ nemen
Seyn handt ler rucken
Den ölpogen soltu drucken

Hye sagtt der meyſter Eyn pruch wider den wyderpruch
das soltu also verſtan will er dir das messer nemen vnd
greyffen mit seyner lincken handt oben an deyn pindt vnd
scheubt dir deyn messer nyder auff deyn lincke seytten vnd
will dir das messer nemen Indes greyff mit deyner lincken
handt In seyn lincke vnd reyß ſtarck vntersich auff deyn
lincke seytten vnd schlag deyn rechten arme oben vber
seyn lincken vnd far vntten durch den selben arme mit deyner
rechten handt vnd swing dich woll auff deyn lincke seytten
vnd wurff In vber deyn rechten fuß wye vntten gemalt
ſtett etc

Item Eyn ander pruch wyder den wider pruch thu Im also   **162v**
hatt er deyn messer gefaſt als vor so fall mit deyner
lincken handt hyntten In seynen lincken olpogen vnd reyß
mit deynem gehulʒ vnd rechten handt ſtarck nyder vber
seyn lincke achsel vnd schreyt mit deynem lincken fuß hinter
seynen lincken vnd swing dich auff deyn rechte seytten etc

Item aber eyn pruch wyder den wyderpruch hatt er dir   **163r**
gefaſt deyn messer als vor Indes so er will schyben deyn
swech auff seyn lincke seytten so fall Indes mit deyner lincken
handt In dy swech deynes messers für seyn handt In
dy mitten vnd scheub Im deyn messer ſtarck In seyn hals
vnd druck mit deynem gehulʒ wyder auff seyner lincken achselen
vnd schreytt mit deynem lincken füß hinter seynen lincken
vnd swing dich auff deyn rechte seytten etc

Will du dich rechen   **163v**
Recht mit linck prechen
Recht auff linck wer gesprungen
Linck zu hillff recht magſt kummen

163v (cont.) Hye sagtt der meyster von eynem arme pruch als er denn oben
spricht wiltu dich etc Das soltu also verstan vnd wye
du den arme prechen solt ste mit deynem lincken fuß für vnd
halt deyn messer auff deyner rechten seytten als auff dem lincken
peyn Das der ortt auff ste gegen dem man schlecht er dir denn
nach der ploß auff deyn lincke seytten so spring mit dey
nem rechten fuß fur auch mit dem lincken woll zu Im
mit auff gepognen messer Indes fall mit deyner lincken handt
vber seynen rechten olpogen hintten an den vnd druck mit
gantzer sterck In deyn lincke seytten Indes wendt dich mit
deynem leyb von Im vnd spring mit deynem lincken fuß starck
auff deyn rechte seytten mit gantzem leyb vnd krafft so prichstu
Im den arme etc

164r Item ste als oben geschriben stett vnd greyff seyn rechte handt mit
deyner lincken Indes schlag deynen rechten arme starck vber seynen
rechten vnd heb mit deyner lincken handt auff seyn rechten arme
vber deyn rechten arme vnd druck deyn peyd arme fast zu sammen
vnd schreytt mit deynem lincken füß hintter seyn rechten vnd
wirff In dar vber etc[122]

164v Item Eyn ander armepruch wenne du Im zuuechten zu Im
kumst so ste als oben geschriben stett hawtt er dir denne
zu deyner lincken seytten so far gepogen mit dem messer vnter
seyn messer vnd ~~mit~~ spring zwiuach zu Im als vor
Indes pegreyff seyn rechte handt mit deyner lincken In
bendigs vnd windt Im dye vmb auff deyn lincke seytten
an deyn prust vnd far mit deynem rechten arme vntten an
seynen rechten olpogen vnd prich starck vbersich alzo prich
estu Im seyn rechten vber deyn ~~lincken~~ rechten etc

165r **Auff recht will er wencken**
Vnd arme rencken
Linck hinter linck ler springen
Dy achsel recht solt dringen

Hye sagtt der merer der kunst Eyn pruch wyder dem armepruch
vnd spricht also auff recht will etc Das soltu also verstan
hatt er dir mit seynem lincken arme vber deyn rechten gefaren
vnd will dir den prechen vber seyn rechte seytten Indes so
er sich vmb wenden will von dir auff seyn rechte seytten
so spring mit deynem lincken fuß hinter seyn lincken vnd

---

[122] The letters B and A indicate that ff 164r and 164v
are in reverse order. LH confirms this.

druck mit deynem lincken arme vnd olpogen seyn lincke
achselen vnd wenne du seyn achselen stossen willt so greyff
mit deyner lincken handt vber seyn arme In dy mit deynes
messers vnd druck oben vnd vnden mit gantzer macht auff
deyn rechte seytten vnd wurff In auff seyn ruck etc

165r (cont.)

Item aber eyn pruch wyder daß ander stuck vnd armepruch
hatt er dir deyn rechten arme aber pegriffen mit seyner
lincken vnd hatt seyn messer lassen fallen als vor vnd
schlecht seyn rechten arme vber deyn rechten Inbendigs
In deyn olpogen ~~vber~~ vnd will dich werffen vber deyn
rechtes peyn Indes schreytt mit deynem lincken fuß fur
seynen rechten vnd far Im mit dem gehultz oben vber
seyn rechte handt vnd reyß starck vntersich vnd schlag
deynen lincken arme oben vber seyn arme vnd vber
seyn lincke handt vnd swing dich auff deyn rechte
seytten vnd wurff In aber auff seyn antlitz etc

165v

Item aber eyn pruch hatt er dir mit seyner lincken handt griffen
In deyn rechte vnd hatt das messer lassen fallen vnd fertt
mit seynem rechten arme vnter deyn rechten olpogen vnd will
dir deyn rechten arme also darvber prechen Indes laß deyn
messer auch fallen vnd pricht deyn rechte handt auß
seyner lincken vntersich vnd far mit deyner rechten handt
vntten durch seyn lincken arme vnd far Im Indes vntter
seynen rechten olpogen vnd windt deyn handt vnd arme
starck vber seyn rechten arme vnd druck seyn rechte handt
an deyn prust oder vntter deyn rechte vchsen vnd spring
mit deynem lincken fuß fur seynen rechten vnd schlag
deyn lincken arme auch starck vber seyn rechte achselen
vnd druck Im starck auff deyn rechte seytten vnd wirff
In fur dich auff seyn angesicht etc

166r

Hutt dich vor den trewffen
Indes hant begreyffen
Ruck: druck: ablaytt
Schlach stich schnell piß peraytt

166v

Hye lertt der merer Der kunst wye man soll ledig
werden so man eynem seyn handt gefangen hatt als mit
der lincken handt seyn rechte etc vnd Thu Im also
hatt er deyn handt gefangen vnd er will machen
armepruch oder waß eß sey so loß In dar zu nicht kummen
prich Im deyn handt auß seyner zum Ersten also
wendt dy kurtz schneydt außbendigs an seynen arme

166v (cont.) das dy flech dar an layn Indes reyß ſtarck vntersich mit
gesencktem ortt auff deyn lincke seytten so wirsſtu ledig
Indes schlag In durch den kopff oder arme etc

167r Item Eyn ander ledigung far mit deynem gehultz ſtarck vber
seyn lincken arme oder handt vnd reyß ſtarck vntersich auff
deyn rechte seytten vnd schlag In Indes auff seyn kopff etc

167v Item iſt er aber ſtarck das dw In nicht nyder reyssen kanſt
mit dem gehultz oder das er mit seyner rechten handt
seyner lincken zu hilff kumptt Indes greyff auch
mit deyner lincken handt In das pindt deyner rechten
zu hilff vnd reyß ſtarck vntersich auff deyn lincke
seytten vnd schlag In durch den kopff etc

168r Item hatt er dir deyn rechte handt gefaßt mit seyner lincken
wye vor Indes windt aber aussen an seyn arme mit gesencktem
ortt Gee Im vntten durch auff den selbigen arme vnd faß
In gewappendt pey dem hals schreytt mit deynem lincken fuß
hinter seynen rechten vnd wurff In dar vber etc

168v Item hatt er dir begriffen deyn rechte handt mit seyner
rechten so reyß aber mit deynem gehultz ſtarck vntersich
vnd thu Im als vor etc

169r Item hatt er deyn hant gefaſt als vor vnd du willt vntersich
reyssen mit deynem gehultz vnd er iſt dir zu ſtarck so kum
deyner rechten aber zu hilff mit deyner lincken an das gehultz
vntten vnd reyß vntersich als vor etc

169v Item hatt er dir deyn rechte handt gefaſt als vor so
windt deyn messer Inbendigs an seynen arme vnd gee
vntten durch außbendigs auff seynen arme etc

170r Item hatt er dir deyn rechte handt gefaſt aber als vor so windt
Im abcr Inbendigs durch auff seyn rechte seytten außbendigs an
seyn arme vnd tritt Indes nohendt In Inn vnd nymb In ge=
wappendt pey dem hals auff seyner lincken seytten wye vor etc

170v **Dy regl merck**
Begreyfft er handt arme ſterck
Erbeyt schnell piß besunnen
Zu ſtucken mag er hartt kummen

Hye sagtt der meyster eyn gutte ler vnd spricht dy regel
etc Das soltu also verſtan das du Im messer solt wyssen
dy rechten zeytt zu eynem ydlichen ſtuck vnd bruch
du solt auch wyssen wenn wy vnd wo vnd gegen welchen

vnd haben vermist vnd dich nicht lassen erzornen das du

170v (cont.)

der kunst nicht vergest Du solt auch nich allayn auff
dy pruch gedencken wye du dye recht treyben mugst
sunder auch auff dy stuck vnd das du wist wye du eynen
ydlichen pegegnen solt vnd er mach waß er woll das
du eyn stuck oder eyn pruch vindest der do zu füglich
sey kanstu des nicht so laß dich eß vntterrichten eynen
der es kann vnd kunstu aller meyster kunst vnd west
nich wenne wye oder wo vnd gegen welchen versetzen
hewen stucken pruchen schnytten dytz oder geneß du das
treyben solt so ist dir das keyn nütz Du müst auch In
eynen ydlichen pruch grosse schnellikayt haben vnd
rechte anlegung swech vnd sterck das ist wenn dir
eyner deyn messer hat vntterloffen vnd hatt dir deyn handt
oder arme pegryffen peyn oder fuß hastu denne dy kunst
gemerckt so helt er dich nymer mer hastu dy kunst recht
verstanden so peleybest woll lang meyster
Nota Etliche stuck vnd etlich pruch dy genen[123] zu Er[124] der
man In dy handt greyfft vnd dy selbigen soll man
machen gleych Indes so er will In dy handt greyffen
Indes so er greyffen will In den arme oder farem Indes
so er will schlagen Indes so er will peschlyssen Indes
stechen hawen schneyden etc

Item wann Etliche stuck synd als pald er dich In dy wag

171r

pringtt so kanstü hartt ledig werden eß ist poß[125] teding[126]
wenne eyner In dem stock ligtt vnd nyndertt[127] mag kummen
darnach wiß dich zu richten etc

Item etliche stuck gen zu so der man gantz gefast hatt
vnd wye dy pruch felen wenne du zu lang harest
oder paytest also felen etlich pruch so du sy zu schnell
machest also mustu wissen dy recht artt dyser ding etc

Item du hast manigerlay stuck vnd pruch Schimpfflich
vnd ernnstlich gefellt dir eynes nicht so nym eyn
anders wer dy ding recht verstett vnd ytlichs zu
seyner zeytt treyben kan dem gefallen dyse ding etc

171v

[123] *gen = geben* (GRIMM AND GRIMM 2004, vol 5, p 3342, 21)

[124] *gereicht zur Ehre*

[125] *poß = böse* (BAUFELD 1996, p 39)

[126] *teding* comes from *tedinc = (Zwei-)kampf* (HENNING 2001, p 236), *dinc = ding* (BAUFELD 1996, p 52), *tedingen = strafen, büßen* (BAUFELD 1996, p 49)

[127] *nindert = nirgendwo* (HENNIG 2001, p 243)

171v (cont.)

Wiltu dich mössen

Mit gehultz handt nyder ſtossen

Lerer handt vergyß nicht

Aussen vnd Innen ſtoß mitt

Hye sagtt der meyſter wye man dy ſtuck machen soll
wye man den arme mit dem gehultz nyder ſtoß vnd
wye man auch des geleychen thun mag mit lerer
hand Nu zum erſten iſt zu wissen wen er Im
zuuechten zu dir kumptt so far gegen Im mit dem
Entruſthaw ſtark auff seyn lincke seytten Indes far mit
dem gehultz vber seyn rechte handt außbendigs vnd
ſtoß dy sere von oben nyder gegen der erden Indes far
auff mit deynem messer vnd haw In durch das angesicht

172r

oder schneyd In dar durch etc

Item haueſtu Im auf seyn rechte seytten mit eynem entruſt=
haw oder sunſt Indes so dy messer zu sammen klytzen so ſtoß
Inbendigs mit dem gehultz vber seyn rechte handt oder

172v

arme nyder gegen der erden Indes schlag In durch den kopff;

Item iſt er dir mit dem gehultz vber deyn rechte handt gefaren
vnd will dir dy nyder drucken Indes fall mit deynem
gehultz vber seyn rechte handt außbendigs Indes fall
mit deyner lincken handt In seyn gehultz vntter seyner
handt vnd reyß ſtarck mit peyden henden vnttersich

173r

auff deyn rechte seytten so nymbſtu Im aber das messer;

Item Eyn ander pruch ſtoß Im seyn handt mit dem gehultz
Inbendigs nyder vnd greyff mit deyner lincken handt vntten

173v

an seyn gehultz vnd reyß wye vor etc

Item Eyn ander pruch fellt er dir vber deyn rechte handt
als vor vmit (sic) dem gehultz so fall auch mit deynem
gehultz vber seyn arme vnd reyß faſt da mit nyder Indes
greyff mit deyner lincken handt vnter seyn rechte In seyn
messer vorne pey den[128] handt In dy klingen vnd reyß
ſtarck mit payden henden auff deyn lincke seytten so nymb=

174r

ſtu Im aber seyn messer etc

|Item will er dir deyn messer nemen wye |Ee gemelt iſt
vnd du Empfindeſt das du deyn messer nymmer haben
magſt Indes greyff mit deyner rechten hendt oben an
seyn messer oder an seyn rechte handt vnd far mit deyner

---

[128] Following LH: *der*

lincken handt vntten an seyn pindt vnd windt das vber

174r (cont.)

sich vber seyne rechte handt außbendig vnd reyß starck

vntersich auff deyn lincke seytten so nymbstu Im aber

das messer vnd der pruch heyst außgezawmt[129] etc

174v

|Item fertt er dich mit dem gehultz Inbendigs vber deyn

handt Indes far mit deym gehultz auch Inbendigs

vber seyn rechte handt vnd far mit der lincken handt

Inbendigs vber seyn messer In dy mitt deynes messers

vnd reyß starck auff deyn rechte seytten so nymbstu

175r

Im aber seyn messer etc

|Item fellt er dir Inbendigs vber deyn handt mit dem gehultz

vnd will er dir deyn rechte handt nyder stossen Indes

fall Im auch vber seyn handt Inbendigs mit dem gehultz

vnd greyff[130] mit deyner lincken handt In seyn klingen vnd

reyß starck auff deyn lincke seytten so nymbstu Im aber

175v

das messer

|Item Eyn pruch wyder den wyder pruch felstu Im mit dem

gehultz vber seyn rechte handt außbendigs vnd er felt

dir wyder vber deyn handt auch also vnd er greyfft

mit seyner lincken handt an deyn messer hinden an

das gehultz vnd reyßt mit seyner rechten starck auff

seyn lincke seytten vnd nymbt dir also deyn messer

vnd ligtt deyn rechte handt vor vntter seyner rechten

handt so du denne Empfindest das du deyn messer nicht

mer magst halten mit deyner rechten handt so fall mit

deyner lincken handt In seynes messers klingen nahent

pey der hant hab vnd druck mit deyner rechten handt

seyn gehultz gegen Im vnd reyß mit deyner lincken

seynes messers klingen starck ~~sey~~ auff deyn lincke

176r

seytten so nymbstu Im das messer etc

|Item ferstu Im vber seyn arme mit deynem gehultz vnd er

dir wyder vnd er reyst nyder deyn handt vnd greyfft mit

seyner lincken vntter Ewrer payd hendt In dy klingen deß

messers vnd reyß starck auff seyn lincke seytten Indes so

du Emfindest das dw deyn messer nicht mer halten magst

so fall mit deyner lincken handt In seyns messers klingen

vnd so ist deyn rechte handt vor an deyner rechten |Indes

druck mit deyner rechten handt auff seyn rechte oder gegen

seyner rechten achsel vnttersich vnd reyß starck seyn

---

[129] *zäumen = gefangen nehmen* (BAUFELD 1996, p 254)   [130] At the end of the word, a letter was blacked out.

176r (cont.) gehultz auff deyn rechte seytten so nymbstu Im auch seyn
messer etc

176v |Item fertt dir eyner nach dem arme Inbendigs mit dem
gehultz vnd will dir deyn handt nyder stossen Indes
weyll er fertt mitt dem gehultz nach der handt so
far mit deyner lincken handt vnter seyne rechte handt vnd
mit deym gehultz auch vber seyn rechte auch Inbendigs
vnd druck deyn peyd hendt vnd gehultz fast zu sammen
vnd reyß starck Indes auff deyn rechte seytten so nymbstu
Im das messer etc

177r |Item will er dir mit dem gehultz Inbendigs deyn handt
nyder stossen Indes wendt deyn schneyd vbersich vnd sch=
neyd In dar eyn vnd far mit dem schnytt vbersich auff
deyn lincke seytten alzo schneyd In auch außbendigs In seyn
handt etc

177v |Item habent Ir payde an gepunden als vor vnd er will
dir außbendigs vber deyn handt mit dem gehultz fallen
Indes windt deyn messer mit dem gehultz vntter deyn
arme vnd senck den ortt Im In seyn vchsen oder seytten
oder windt Im Indes dy kurtz schneyd auff seyn kopff
seyner rechten seytten

178r |Item will er dir deyn handt aber nyder stossen Inbendigs mit dem
gehultz So wendt gegen Im auff mit deynem messer vnd
erhoch deyn gehultz vnd fall zu gewappender hant vnd
setz Im den ortt an seyn hals seyner rechten seytten will er
dir das weren so erbeytt mit dem gehultz auff seyn lincke
seytten Inbendigs vber seyn haubt vnd nym Im das messer
als du vor gelertt pist etc

178v |Item fellt dir eyner vber deyn gehultz außbendigs vber deyn
handt vnd du Im wyder vnd druckt er nyder vnd greyfft
mit seyner lincken handt vntter Ewrer peyd hendt In dy
klingen deynes messers vnd will dir es also nemen Indes
fall mit deyner lincken verkertten handt auch In deynes
messers klingen ob seyner lincken handt vnd schreytt mit
deynem lincken fuß hintter seynen lincken oder fur seynen lincken
vnd druckt mit deyner lincken handt vnd mit der klingen
seynes messers starck auff seyn lincke seytten etc

179r |Item hawestu Im oder er dir von oben zu deyner lincken seytten
So haw auch geleych mit ym eyn zu seyner lincken seytten
Indes stoß seyn rechten arme mit deyner lincken handt

außbendigs nyder gegen der erden vnd haw Im mit deyner
rechten durch seyn kopff auff seyner rechten seytten;

|Item hautt er dir von oben eyn eyn oberhaw auff deyn lincke seytten
Indes pewg deyn messer vntter das seyn Indes ſtoß mit
deyner lincken handt auff seyn rechte Inbendigs oben nyder
vnd haw In durch den kopff etc

No text.

No text.

Wen man fechten wil [131]

No text.

|Item fertt er dir vber deyn rechte handt außbendigs so fall Im
auch vber seyn rechte handt außbendigs vnd mach Indes
was du willt etc

|Item fertt er dir mit seyn gehultz vber deyn rechte handt
außbendigs wye vor Indes far Im auch wyder dar vber
außbendigs vnd reyß woll nyder vnd ſtoß Indes das
gehultz seynes messers mitt deyner lincken handt ſtarck
auff deyn rechte seytten so nymbſtu Im das messer etc

|**Den ölpogen ler fassen**
|**Scheub linck haw recht zu der taschen**

|Hye sagtt der meyſter von eynem ſtuck das iſt schimpflich zu
treyben wiltu das machen so thu Im also so du auff In
pindeſt auff seyn lincke seytten Indes schlag vmb zu
seyner rechten seytten vnd Im schlag far hintten an seyn
olpogen vnd ſtoß In vaſt von dir auff seyn rechte seytten
vnd Indes gibtt er dir den ruck so schlag In auff seyn
huff hintten oder hatt er eyn taschen auff seyn ars geschoben
so schlag In dar auff etc

|**Linck vberfar**
|**Indes nymb war**
|**Ruck gegen pauch wendt**
|**Durch peyde peyn ſtich behendt**

|Das soltu also verſtan So Ir Im zuuechten zu sammen
kumptt auff seyn lincke seytten so schlag Im zu der
rechten seytten Indes wendt deyn lincke seytten an seyn

---

[131] Written at the top in a different hand.

183v (cont.) rechte seytten vnd far mit deynem lincken arme aussen
vber seyn messer vnd wend deyn ruck gegen Im Indes
fall mit deyner lincken In deyn messer In dy mitt vnd
stich In durch deyn peyde peyn vnd hab dich woll
mit dem ruck an In das er seynes messers nicht mach
auß dem arme rucken vnd senck dich vorne nyder vnd
stich Im Inn den pauch vnd zeuch den ortt offt wyder
eyn In deyn lincke handt das er dich hintten mitt dem
ortt deynes messers nicht pegreyffen mag vnd stupff
In also eynes oder vyer mall behentlich In den pauch
oder auff das gemecht vnd das stuck ist lecherlich zu
treyben auff der schull etc

184r |Das messer zu rechtem peyn hallt
|Dy wer prich mit gewalt
|Den lincken fuß fur setz
|Mit schritten hew stich letz
|Von payden seytten
|Triff dy lem[132] wiltu schreytten

|Hye sagtt der meyster von etlichen stucken dey dy lem stuck
heyssen da mit man eynem lam hawen mag wanne dy stuck
soll man machen zu den glidern des arme vnd der hant etc

184v |Item wiltu dyse stuck machen so schick dich dar eyn also
Setz deynen lincken fuß fur vnd hallt deyn messer auff deyner
rechten seytten vnd auff deynem rechten peyn das der/ortt gegen
dem man stee vnd der dawm oben sey auff dem messer vnd stee
also als In eyner hutt hawtt er dir danne zu der ploß
auff deyner lincken seytten Indes schreytt mit deynem rechten
fuß woll auff seyn lincken seytten Eyn zwiuachen tritt
woll auß dem haw Im zu der lem seyner rechten handt
oder arme Inbendigs oder wo du zu der lem am negsten
hast als oben gemalt stett etc

|Item leg dich auff deyn lincke seytten mit deynem messer das der
ortt auff der erden sey vnd der dawm vntten auf dem messer
vnd dy recht schneyd stee gegen den man vnd stee mit
deynem rechten füs fur hawtt er denne auff dich oben zu
dem kopff so versetz Im nicht Sunder Tritt Im mit deynem
lincken fuß woll auff seyn rechte seytten woll auß dem
haw vnd schreytt mit deynem lincken fuß dem rechten

---

[132] *leme = Lähmung* (HENNIG 2001, p 204)

nach Indes haw Im nach der andren Ewsren lem seynes
rechten armeß als vntten gemalt ſtett etc

184v (cont.)

|Item merck gleych als du mit dem messer auff der erden
ligſt also far gerichtes gleych auff mit der schneyden
In seyn arme etc

185r

|Item ſte mit deynem lincken fuß fur vnd hallt deyn messer
auff deyner rechten seytten das der ortt gegen dem man ſtee
vnd der dawm oben lig auff dem messer vnd gib dich
ploß mit deyner lincken seytten hawtt er dir dann nach
der ploß so zuck deyn lincken fuß zu ruck auff das
weyteſt vnd haw Im nach der handt In das glid etc

185v

|Item ſtee mit deynem rechten fuß fur vnd leg deyn messer
auff deyn lincke seytten mit dem ortt auff der erden das der
dawmen vntten sey oder auff dem ruck des messers vnd dy (sic)
scarpff[133] schneyd sey gegen den man hawtt er dir auff deyn
rechte seytten nach der plöß so zuck deynen rechten fuß
an dich woll zu ruck beseyt auß dem schlag Im gerichtz
gleych als du gelegen piſt mit dem messer mit der scharpffen
schneyden außbendigs nach der handt Indas glyd vnn
leg dich wyder In das leger auff deyner rechten seytten das
denne dy linck seytt vor ſtee hawtt er dir denne aber auff deyn
lincke seytten so thu als vor geschriben iſt vnd haw Im
wyder In das gelenck Inbendigs vnn leg dich wyder
In das leger auff deyner lincken seytten alzo magſtu dich
eynes ydlichen weren In schimpff oder ernſte etc

186r

|Item lig aber auff deyner rechten seytten als vor hawtt er dir
nach deyner lincken seytten so schreytt auß dem haw mit deynem
lincken fuß woll auf seyn rechte vnd In seynen haw so
haw Im nach seynem rechten arme außbendigs nach der
lemm etc

186v

|Item |Sticht er dir zu dem gesicht mit hangendem ortt auff
welcher seytten du denne ligeſt mit dem messer so zeuch den
lincken fuß schnell zu ruck oder den rechten welchen den
vor ſtett der selbig wer zu ruck gezogen Indes haw Im
außbendigs oder Inbendigs nach der handt vnd fall Im
schlag Eyn wenig mit dem leyp beseytt auß etc

187r

|**Kurtz dich mell**
|**Im pandt piß schnell**

187v

---

[133] *scharpff*

187v (cont.)

|Haw ſtarck zu seyner lincken
|Der rechten handt soltu wincken

|Hye sagtt der meyſter von eynem ſtuck wye man eynem auff
seyn handt oder arme hawen solt das soltu also verſtan so du
das ſtuck machen willt so müſtu gar schnell seyn Im an
pinden vnd thu Im also haw Im zu seyner lincken seytten
ſtarck von oben eyn eynen ōberhaw Indes dy weyll dy messer
zu sammen klitzen oben so haw Im schnell auff seyn rechte
hantt oder arme außbendigs seyner rechten seytten etc

188r

Nach der handt will er hawen
Des orttes soltu dich frawen
Windt ortt zu gesicht vnd messer
Indes erbeytt iſt das pesser

Hye lertt der meyſter eyn pruch wyder das Eegemellt ſtuck
so man eynem zu seyner rechten handt hawett als oben
ſtett vnd spricht also nach der handt etc Das soltu also
verſtan will er dir nach deyner rechten handt hawen
das soltu also prechen wenne er dir hatt an gepunden
auff deyner lincken seytten vnd will dir Indes schnell
nach der handt oder arme hawen vmb zu deyner rechten
seytten So wendt Indes deyn messer gegen dem seynen auff
deyn lincke seytten das dy kurtz schneyd vntten ſtee Indes
ſtich Im zu dem gesicht vnd erbeyt furpaß[134] zu der negſten
ploß etc

188v

Lanck scheuß. von handt
Ortt schlecht er von luginslandt

Das soltu also verſtan wenne dü ſteſt mit deynem lincken
fuß fur vnd haſt deyn messer auff deyner rechten seytten
an dem peyn also das der ortt gegen dem man ſtee vnd der
dawm lig oben auff der flech deß messers hawtt er
denne zu auff deyn lincke seytten nach der ploß So
versetz Im nicht sunder scheub Im mit geſtracktem
arme dy weill er Im schlag iſt den ortt geradt mit
geſtracktem arme ſtarck In das gesicht vnd Im schuß
schreytt Im auff seyn lincke seytten mit deynem rechten
fuß etc

---

[134] *vürbaz = darüber hinaus* (HENNIG 2001, p 448)

Will er auß arme schyssen linck
Scheuß geradt ortt wyndt vnd winck

Item wyder das eynschyssen lertt der meyster Eyn pruch
dar wyder vnd spricht will er etc das soltu also verstan
stett er In dem leger das er seyn messer hatt auff seyner
rechten seytten auff seynem rechten peyn das der ortt gegen
dem man stee vnd so er dir denne den ortt Eyn scheußt Indes
scheuß geradt mit Im Eyn geradt zu seynem angesicht
auch mit gestracktem arme Indes windt deyn messer an
das seyn den ortt zu dem gesicht also das dy recht
schneyd oben stee vnd dy linck stumpff vntten vnd
hallt mit deynem gehultz woll hyn dan auff deyn
lincke seytten vnd stich Im zu dem gesicht seyner lincken
seytten vnd erbeytt Indes etc

Den Storchschnabel soltu erlengen
Das furpayn zu ruck ler prengen

Hye sagtt der meyster wye man das stuck das do heysset
der Storchschnabel machen soll vnd thu Im also stee
mit deynem lincken fuß fur vnd halt deyn messer auff
deyner rechten seytten pey deynem rechten fuß oder ob deynem
rechten kny das der ortt gegen dem man stee vnd gib dich
ploß mit deyner lincken seytten hawtt er denne also nach
der ploß so ruck deynen lincken fuß woll zu ruck
auff das weyttest vnd heb deyn messer mit gestracktem
arme gegen seyner prust vnd loß In an den ortt lauffen;

Item stee als vor so er auff dich hawett aber eyn oberhaw vnd
zeuch deynen lincken fuß zu ruck vnd far auff mit dem
messer geradt vnd mit dem gestrackten arme vnd far Im
mit dem ortt nach der handt oder arme vnd loß In dar eyn
schlagen

Item dyse stuck gehören zu treyben wyder dye dy do Ein lauffen
vnd dy gerne hoch fechten vnd kurtz oder dy verborffen[135] hew
gegen den man machen vnd auch springen mit verborffen hewen
zu dem man also hastu den storch schnabel vnd gegen welchem
stuck oder eygenschafft du das machen solt pist du yetznid vntter
richt etc

Wer dir will eyn lauffen
Mitt dem ortt soltu In gauffen

[135] *verworfen*

190r (cont.) **Zuespringent will er schlagen**
Ortt zu ruck lertt gagen[136]

Hye lertt der meyster Eyn stuck wyder das eynlauffen vnd
wyder dy vechtter dy gerne Ey lauffen schick dich Gegen In
also leg dich In dy hutt der pasteyen vnd wenne er dir denne
eyn lauffen will so hallt deyn messer mit dem ortt furdich
vnd loß In dar an lauffen will er dir denne versetzen den
ortt vnd fertt nach deynem messer so wechsel Indes
durch zu der negsten ploß etc

190v Item leg dich In dy hutt luginslandt wenne dir denne
eyn lauffen will so far hoch auff mit dem arme vnd
stuck Im den ortt In das gesicht oder auff dy prust vnd
scheub In also von dir etc

191r Item nota In dysem stuck so du mit dem messer auff ferst
so soltu winden das gehultz vntter deyn arme woll vbersich
fur das hauptt Merck fellt er dir dann nach dem ortt
so schlag In zu dem haubtt etc

191v **Will er auß dem pandt schlagen**
Ortt macht In verzagen
Sucht er dy plöß
Mitt ortt In verdröß

Den textt soltu also verstan wenne er dir an gepunden
hatt an deyn messer auff deyner lincken seytten vnd er will
schlagen zu deyner rechten seytten Indes far auff mit dem
gehultz vntter deyn arme vnd senck den ortt nyder vnd stich
Im zu dem hals oder zu der prust etc

192r Item hatt er dir an gepunden auff deyner rechten seytten vnd
er will schlagen zu deyner rechten Indes thu als vor vnd
setz Im den ortt an etc

192v Item nota was auff dich wirtt gepunden das du albeg
Indes den ortt zu dem gesicht solt winden er versetz frey
oder krump So magstu albeg den ortt prauchen vntten
oder oben schleg oder stich mitt dem ortt werenn

193r |Scorpian mit seynem kar
|Dem antlitz ist gevar

---

[136] *gagen* = *gegen* (LEXER 1885, p 59/p 63)

|Dy kurtz schneyd gegen haup laß fallen

|Schlag recht zwiuach las prallen

Hye sagtt der merer der kunst von eynem stuck das heysset der
scorpian vnd das stuck treyb also haw Im von deyner
rechten achsel zu seyner lincken starck eyn zu dem haubtt
mit gestracktem arme eyn eynen oberhaw Indes so du gehawen
hast so laß dy kurtz schneyd deynes messers sincken gleych
gegen deynen haubtt dy kurtz schneyd vntten auff deyn lincke
achselen Indes stich Im wider zu dem gesicht seyner lincken seytten
hastu In danne nicht troffen mit dem stich so ruck deyn messer
mitt dem ortt nyder gegen dir doch daß der ortt gesenckt sey
auff deyn lincke seytten vnd schlag In mit der kurtzen schneyd
auff seyn kopff seyner rechten seytten Indes wendt deyn gehultz
vntter deyn rechten arme vnd schreytt mit deynem lincken fuß
hintter seynen rechten vnd schlag Im dy langen schneyd vber
seyn haubtt vnn daß stuck heyst der scorpian etc

Wasilistus mit seynem gesicht

Dy meyster macht vergifft
Vergist dy prüst an vorcht
Vnd versertt das haubt mit schlagendem orcht

Nu den wasilistun mach also so Ir payde an gepunden habt an dy
messer vnn er ist In der krummen versatzung oder pist so leg dich
auff seyn messer mit der langen schneyden vnd wartt wenne er dir
zu deyner rechten seytten schlagen will so er denne nohent bey dir
ist Im pandt Merck dy weyl er Im schlag ist so fall zu ge=
wappender handt vnd stich starck auff seyn lincken seytten zu den oberen
zynnen zu den oren so er denn versetzen will vnd fertt auff mit dem
messer so far hoch auff mit deynem arme vnd stich In mit gewapp
ender handt starck auff seyn prust will er denne auff deyn messer fallen
mit der versatzung so schlag Im das gehultz des messers auff
seyn kopff Indes loß deyn rechte handt wyder von der mitt des messers
vnd greyff deyn gehultz wyder In dy rechten handt vnd setz ~~Im~~
Im den ortt wyder In dy prust also hastu den wasilistun gemacht
also hastu den wasilistun gemacht etc

Klotz mit seyner wartt

Zu den oren stost er hartt
Indes piß gehendt
Bewappendt ortt zu prust wendt

Hye sagtt der meyster von eynem stuck vnd spricht der klotz
mit etc Das soltu also verstan wind Im gerad auff seyn
messer vnd wartt wenne er mit der krum will hawen

194r (cont.) zu deyner lincken seytten wanne mit dem lincken fuß soltu vor
Sten weyll er dir denne hawett zu deyner lincken seytten Indes
fall zu gewappender handt vnd schreytt mit deynem rechten
fuß woll zu seynem rechten fuß vnd Stoß In mit dem gehultz
des messers zu seynen rechten oren Indes zuck mit gewappender
handt den ortt gegen deyner lincken seytten peseyt ab vnd setz
Im den ortt In dy pruSt versetz er denne So erbeyt Indes
mit dem schlagnden ortt zu dem kopff aussen oder far mitt
dem gehultz oben vber seyn handt vnd mach das messer
nemen etc

195r **Gewappend mit dem kloß**
Zu der prüSt Stoß
Gewappend oder sünSt
Versetz recht eß gibt dir luSt

Hye sagtt der meySter wye du mit dem cloß daß iSt mit
dem gehultz soltu den man In dy pruSt Stossen vnd thu
Im also Schlecht er dir zu den oren oder zu dem kopff
deyner rechten oder lincken seytten so fall zu gewappender
handt vnd senck Im das gehultz auf seyn pruSt Indes
erbeytt mit der klingen deynes messers oben Starck zu
dem haubtt oder far mit gewappender handt beseytt ab
vnd setz Im den In das gesicht oder prüSt etc

195v Item schlett er dir zu deyner rechten seytten so ir payd seyt nahent
pey eyn ander In dem so er schlecht zu deyner rechten seytten so
far auff mit dem arme vnd setz Im den ortt In seyn pruSt mitt
gewappender handt wiltu denne so magStu mitt dem ortt
forne vntter seyn gehultz vnd magSt Im den ortt setzen In das
gesicht Gewappendt etc

196r Item ligtt er Im pandt mit dir vnd will dir aber
schlagen zu deyner rechten seytten zu den obren zynnen
Indes far auff ~~mit deynen rechten~~ seyn pruSt mit gewappendem
ortt das iSt mit gewapp[e]nder handt Indes far mit deynem
rechten arme vber seyn messer vnd mitt dem fodren teyll
auch gewappend far Im zwischen das gehultz oder
handt vnd Stoß seyn messer mit deynem lincken olpogen
auff deyn rechte seytten So nymbStu Im das messer etc

196v **Will er mit klossen**
Zu der pruSt Stossen
Versetzen soltu dringen
Vnd dy wer angebynnen

Hye sagtt der meyster ey prüch So man eynen mit dem kloß
oder mit dem gehultz In dy prust will stossen vnd spricht
also will er mit dem klossen etc

196v (cont.)

Item den pruch mach also wiltu schlagen zu seynen rechten
oren Indes fertt er dir zu gewappender hantt mit dem gehultz
auff deyn prust Indes greyff mit deyner lincken handt forne an
seyn gehultz hintter seyn rechte handt vnd stoß mit deynem
rechten arme starck an seynes messers klingen vnd schreytt
mit deynem rechten fuß zu seynem rechten vnd stoß starck auff
deyn lincke seytten etc

Item schlechstu Im zu deyner lincken seytten vnd er versetz
das mit hangendem ortt Indes windt Im mit deynem
messer mit dem mutiren an seynem messer zu seyner prust
oder an gesicht etc

197r

Item schlegstu eynem zu seynen lincken oren mit dem entrust=
haw vnd er fertt dir mit dem gehultz gewappendt auff
deyn prust Indes far mit deym gehultz gegen seyner rechten
seytten vber seyns messers klingen vnd greyff mit deyner
lincken handt In seyn gehultz starck In deyn rechte seytten
nyder vnd mitt deyner lincken handt scheub starck von
dir auff deyn rechte seytten nyder vnd mit deyner lincken
handt scheub starck von dir auff deyn rechte seytten so nymbstu
Im aber das messer etc

197v

Vyer seyn der hengen
Dar auß ler: haw: stich: schnyt: erkennen
In allem gefertt
Leger weych oder hertt

198r

Hye sagtt der meyster wye du dy vyer hengen machen solt
Im messer vnd wye dy genant seyn also nendt er eyn ydlichs
mit namen eyns nach dem anderen;

Item das erst heyst der eber von peyden seytten Das ander der
stir auch von payden seytten vnd dar Innen soltu gar woll
ge vbett seyn vnd also das du erbeyt dar auß gar wol treyben
mügst auch soltu wissen das vyer hengen seyn zway
vntten vnd zway oben vnd wye sy genandt seyn hastu
oben gehortt Nun auß den vyer hengen soltu ach winden
pringen vnd dy selbigen acht winden soltu also wegen
Das du auß ydlichem winden solt machen vnd treyben eyn
stich vnd eyn haw vnd eyn schnytt das seyn dy vyer ob gesch
riben winden etc

198v

198v (cont.) Item Eyn ander oberhengen mit zwayen winden das treyb
also wenne du mit dem zuuechten zu Im kumbst so stee von
deyner lincken seytten In dem stiren hawtt er dir denne oben
zu deyner rechten seytten so windt gegen seynen haw dy lang
schneyden an seyn messer vnd stich Im oben eyn zu dem gesicht
das ist aber eyn wynden setzt er denne den stich ab gegen
deyner rechten seytten so pleyb Im am messer vnd windt
wyder auff deyn lincke seytten wyder In den stiren dy flech
an seyn messer vnd stich Im oben zu dem gesicht seyner
lincken seytten das In dem stich dy stumpff schneyd vntten
stee also hastu zway winden an seynem messer wye
oben gemalt stett etc

199r Item Nu merck wye du auß den vyer hengen solt treyben
acht winden das erst vberhengen hatt zway winden vnd
treyb das also wenne du mit dem zuuechten zu Im kumst
so stee von deyner rechten seytten In dem Stiren schlecht er denn
oben eyn zu deyner lincken seytten so windt gegen seynen haw
dy kurtz schneyd an seyn messer wyder In den Stiren daß
ist eyn winden Setz er denne den stich ab von seyner lincken
seytten so pleyb am messer vnd windt wyder auf deyn rechte
seytten wyder In den stiren dy flech an seyn messer vnd stich
Im oben zu dem gesicht das ist eyn hengen von der rechten seytten
mit zwayen winden an seynem messer etc

199v Item du solt auch wissen das auß den untren zweyen hengen
das ist der Eber von payden seytten treyben solt auch vyer hengen
auch vyer winden mit allen Iren gefertten als auß den obren
das seyn dy ach winden vnd als offt du windest so gedenck
In Eynen ydlichen winden an den haw an den stich vnd an den
schnydt also kummen auß den acht wynden vyerundzwaynzig
vnd auß welchen vnd gegen welchen stucken vnd gegen welchen
hewen stichen oder schnytten du den haw vnd den schnytt
treyben solt das pistu zu gutter maß vor vntterrich werden etc

200r Item will du dich auß der pastey weren so thu Im also
wenne du mit dem zuuechten schir zu Im kumbst
so setz deynen lincken fuß fur vnd hallt deyn messer
mit dem ort gegen der erden als du woll vnterricht
pist kumptt er dir denne Entgegen als oben geschriben ist
so far mit deynem gestrackten arme auff mit dem ortt
Im zu seynem gesicht oder In dy prust hastu In getroffen
mit dem ortt vnd er erschrickt vnd zuckt den leyb
oder kopff zu ruck Indes fall wyder her ab mit deynem
messer In dy hutt der pasteyn vnd zeuch deynen

rechten fuß wyder zu ruck vnd loß dich an schaden   **200r (cont.)**
nit auß der hutt paſtey pringen vnd also thu Im
auch auff deyner lincken seytten etc

Item ligſtu In der hütt paſtey Es sey auff der lincken seytten   **200v**
oder auff der rechten seytten Sichſtu das er dar auff will krummen
mit dem messer oder sunſt dar auff will fallen so wechsel
vntten durch als du woll weyſt etc

Item iſt er hoch Im pandt so far mit dem ortt zu dem
gesicht oder pruſt Iſt er aber nyder mit dem arme so wartt
der ringen etc

Dy zynnen will er ſteygen   **201r**
Paſtey ortt thutt In ab treyben
Waß auff lugisland wirtt geschlagen
Paſteyn ortt iſt das ab tragen

Nota dy maynung des text iſt so du ligeſt In der hutt
der paſtey will er dir Im zuuechten zu den vyer zynnen
schlagen das soltu mit dem ortt als oben schriben ſtet
weren etc

Item dy ander meynung des text iſt waß den obren zynnen
zu geschlagen wirtt das soltu mit dem ortt versetzen
auß der hutt der paſtey doch zu merer Erklerung soltu
wissen das man sich mit dem langen ortt gegen allen anpinden
weren solt vnd den ortt da wyder prauchen Er haw oder
ſtech so pricht das der lang ortt gar wanne der lang ortt iſt
dy peſt werr Im messe[r] vnd Im swert vnd wer dar auß
vechten kan der zwinget den man das er sich vber seyn
danck schlagen müß lassen vnd ob er das vor wollt gewynnen
so magſtu Im das nemen mit dem langen ortt Indes erbeytt
mit deynem messer nach dem nach etc

Item setz deynen lincken fuß für vnd leg dich In dy hutt   **201v**
paſtey schlech er dir den zu von lugislandt zu deyner
lincken seytten zu dem kopff so far mit deynem messer auff
vnd wechsel auff seyner rechten seytten durch vnd ſtoß
seyn messer mit deynem messer oder ruck des messers Eyn
benig beseytt vnd haw Im mit der langen schneyden
durch seyn angesicht oder schneyd In durch seyn rechten
arme vnd leg dich wyder In dy obgemelten hutt etc

Item lig als vor hautt er aber auff von der hutt lugis   **202r**
landt so gee gerad auff mit geſtracktem arme vntter
seyn messer auff seyner rechten seytten vnd Im durchwechslen

202r (cont.) Schreytt mit deynem rechten fuß für gegen seynen rechten vnd
stich Im mit gestracktem messer zu dem gesicht seyner rechten
seytten wiltu denne so magstu machen dy dupliren oder
durchgen wiltu aber nicht so leg dich wyder abher In dy
pastey etc

202v Item gleych als du dy ding machest so du ligst in der hutt
pastey auff deyner rechten seytten so der linck fuß vor stett
also magstu durchwechselen so du stest in der hutt der
pastey auff deyner lincken seytten so der recht fuß vor stett
vnd dar auß hawen stechen vnd schneyden denne das eß
von der lincken seytten etzwaß letzer zu gett denne auff der
rech vnd eynem vngeubtten In dem messer gar vnpequem
dunckt etc

203r **Wer der pastey zu setzt**
Vom luginslandt wirtt er geletzt
Dringtt er Im anpinden
Ortt zu gesicht solt winden

Hye ist zu wissen das daß dy maynung des textz ist also
was den vntren zynnen wirtt zu geschlagen (sic) ader gestochen
das soltu weren auß der hutt luginslandt auch mit dem
langen ortt vnd ob du mit dem ortt gefelt hast so windt
albegen deyn messer mit dem seynen In den stiren von peyden
seytten gett das zu etc

203v Item ligstu In der hutt lugislandt so stee mit deynem peyn
woll In dy wag vnd hallt deyn messer gestrackt vnd
laß In machen was er will kanstu Im recht thun
so kann Er nichtz schaffen Er wechsel durch oder mach
waß er woll du magst dich auch In dyse hutt legen
auff seyn messer also gestrackt vnd dar auß machen waß
dw wild vnd alle ploß da mit süchen vnd wo du daß
fenster sichst offen stan so sich[137] fröhlich dar eyn mit deynem
ortt oder mit deynem haw vnd mach Indes waß du vor
untterricht pist etc

204r Item du magst dich Er weren aller vnterhew auß der hutt lugis
landt mit dem ortt vnd gestrackt In von oben eyn schyssen
den ortt zu dem gesicht vnd Indes ob du willt durchwechslen;

204v **Krump wer versetzt**
Windt stich er wirtt geletzt

---

[137] *stich*

Der winden soltu dich remen
An der krumb beyslich ler abnemen

204v (cont.)

Hye sagtt der meyster wenn dir ymant mit der krum
versetz wye du dich gegen Im solt halten so ir payde
angepunden habtt vnd du ligest auch In seynem messer
vnd er ligtt vntter deynem Thu Im also ligtt er auff
deyner lincken seytten gegen dir oder an dir gegen deyner rechten
seytten vnd hast auff seyn messer auch gepunden das dy stumpff
schneyd gegen den man stett Indes windt deyn messer mit
der langen schneyden Im auff seyn haubtt etc

Item fertt er auff vnd ist hoch mit dem arme das du zu
dem haubt mit dem messer nicht magst kummen vnd hast
vernummen oder pyst das er hoch auff fertt so windt hoch
auff vber seyn messer gegen deyner lincken seytten auff seyn
rechte vnd fall Im mit dem ortt von seynem messer Im
In dy kelen oder auff dy prust etc

205r

Item ligtt aber in der krummen versatzung als vor auff deyner
rechten seytten so windt dy lang schneyd gegen seynem
haubtt fertt er denne aber hoch auff mit dem arme vnd
will versetzen als vor so merck Im winden weyll du
gegen dem haubtt pist winden auff seyn messer auff seyner
lincken seytten vnd schlag Im dy kurtz schneyd auff seyn
haubtt das der dawmen vntten sey vnd lig auff der flech dey
nes messers versetz er denn den schlag. fellt dar auff mit
seynem messer so reyß wyder behendtlich an seyn messer
zu dem haubtt aber mit der kurtzen schneyden hatt er
das versetzt so windt wyder auff dy lang schneyd gegen
seynem haubt vnd loß den ortt aber fallen auff deyn
lincke seytten ~~Ind~~ vnd zu seyner rechten seytten In dy kelen
oder prust etc

205v

In windinden wilt verfuren
Zwiuach soltu dupliren
Wiltu dich paß rechen
Durch dy wang ler stechen

206r

Hye sagtt der meyster wye man auß der krüm soll dupliren
wiltu Im das dupliren machen so thu Im also So Ir payde
habt an gepunden vnd er ligtt dir auff deyner rechten seytten
vnd du Im auff seyner lincken seytten vnd du ligest mit deynem
messer auff dem seynen vnd das dir der dawm vntten an deynem
messer stee Indes windt dy lang schneyd auff seyn haubtt
vnd Indes schlag Im zu seynen lincken oren mit dem Entrust=

206r (cont.) haw auff seynes messers klingen mit der ſtumpffen schney=
den woll vmb zu dem haubtt will du denne abnemem
am messer das magſtu auch thun etc

206v Item ligtt er auff deyner rechten seytten gegen dir als oben ge=
schriben ſtett Thu samb du Im zu dem gesicht wolleſt
winden auff seyn lincke seytten den ortt nach seynem
lincken wang vnd ſtrack den arme woll von dir auff
deyn seytten vnd windt Im den ortt zu dem gesicht etc

207r Item als du dy ſtuck gemacht haſt auff der rechten seytten
also magſtu dy machen auff der lincken seytten vnd waß
dy winden eygenschafft gehabt haben von der rechten seytten
dy selbigen haben dy auch von der lincken seytten dy ab nemen
magſtu auch machen vnd zu dem kopff schlagen denne
wenne du auff der lincken seytten gegen deynem haupt win=
deſt so magſtu mit der ſtumpffen schneyden schlagen auch
magſtu zu seynen rechten wang ſtechen etc

207v Item wiltu Im das messe[r] nemen so er ligtt Im hangenden
ortt auff seyner lincken seytten vnd du ligeſt auch auff
seynem so windt Indes vber seyn messer dy lang schneyd
den ortt In seyn seytten vnd. mach das mutiren Indes
Dritt behentlich auff In mit deynem lincken fuß das
seyn messer neben deyner rechten seytten beseytt mit dem
ortt hyn auß gett Indes fall mit deyner lincken handt
hintter seyn rechte vnd reyß ſtarck damit pey dem pindt
auff deyn rechte seytten vnd ſtoß In mitt dem ortt deyneß
messers ſtarck In seyn leyb etc

208r Item ligtt er Im hangenden ortt auff seyner rechten seytten vnd
du ligeſt vntter dem seynen Indes fall mit deyner lincken handt
vber seyn messer Indes fall In dy mitt deynes messers vnd far
Im also mit dem ortt an seyn kelen oder leg Im deynes messers
klingen an seyn hals auff seyner lincken seytten vnd schreytt
mitt deynem lincken fuß hintter seyn rechten vnd druck In
dar vber etc

208v Item ligeſtu auff seynem messer vnd er ligtt als vor Indes far
mit deynem lincken arme vber seyn messer vnd ſtoß mit deynem
messer Im an seyn messer mit dem kreutz ſtarck oben an
seyn gehultz oder an seyns messers klingen nohent bey der
handt ſtarck auff deyn lincke seytten so nymbſtu Im aber das
messer etc

Item ligestu aber auff seyner rechten seytten vnd auff seynem messer 209r
vnd er ligtt Im hangenden ortt als vor Indes windt
gegen seynen hals vnd druck Im das messer an seyn nack
vnd ob er dir zu starck wollt seyn das du Im seyn messer
nicht nyder drucken kunst so kum mit deyner lincken handt
hinden an ~~deym~~ seyn gehultz der rechten handt zu hilff
Indes fall mit deyner lincken handt Im oben In seyn messer
In des gehultz ob seyner handt vnd druck Im das vberruck
vnd vber daß deyn so nymbstu Im aber das messer etc

Item ligtt er auff seyner rechten seytten In dem hangenden ortt 209v
als vor Indes fall mit deynem lincken arme vber seyn messer
vnd thu als du Im mit dem gehultz deynes messers
wollest stechen In seyn angesicht so er denne auff fertt
mit seynem gehultz vnd will dir den stoß versetzen
Indes stoß mit deynem gehultz starck auff deyn lincke
seytten so nymbstu Im aber das messer vnd hawest In
da mit durch das gesicht etc

Vberfar arme will er possen 210r
Messer nym mit gehultz solt stossen

Item das soltu also verstan kumstu zu dem mann mit deynem
Entrusthaw auff seyn lincke seytten so schlag Im zu
seyner rechten seytten Indes far mit deynem lincken
arme vber seyn messer vber dy klingen vnd thu als du
Im mit dem gehultz In das angesicht wollest stossen
Indes stoß mit dem gehultz In das gelenck seyner
rechten handt außbendigs starck auff deyn rechte seytten;

Item fall Im vber seyn messer als vor Grafft[138] Im denne das 210v
gehultz Inbendigs In der handt vbersich so stoß aber mit
deynem gehultz zwischen seyn handt vnd gehultz starck
auff deyn lincke seytten so nymbstu Im ab das messer wye
vntten gemalt stett etc

Entrust swech vberwindt 211r
Lanck das haubt gewindt

Hye lertt der meyster wye man den man mit der swech deß
messers vber wintten soll vnd thu Im also Schlag Im mit
eynem Entrusthaw zu seyner rechten seytten Indes wendt
deyn handt vmb vnd windt In dy swech deynes messers

---

[138] *graffeln = greifen* (HENNIG 2001, p 138)

211r (cont.)   auff sey̲n messer vnd schlag Im Indes mit der lange̲n
schneyd̲en auff sey̲n kopff. e̲t̲c

211v   **Wiltu In betryge̲n**

So du pi∫t vnt̲erlige̲n

Linck windt durchgee In winde̲n

Hew ∫tich schnytt ler vinde̲n

Zu kopff zu leyb

Was du peger∫t das teyb

In allem gefertt

Prüff weych od̲er hertt

Hye endt de̲r mey∫t̲er vnd de̲r mere̲r de̲r kun∫t das let̲zt ∫tuck
mit seyne̲r Eyge̲nschafft vnd spricht wiltu In betrig̲en
das i∫t wenne̲ Ir payde an gepunde̲n habtt ligtt er dir
de̲nn auff deyne̲r rechte̲n seytte̲n mit der krumme̲n versat̲zung
vnd du lige∫t vntt̲er seyne̲m mess̲er auch In de̲r selbe̲n ver∫at̲zung
Indes windt dey̲n mess̲er auff sey̲n lincke seytte̲n gege̲n seyne̲r
plöß vnd winck woll gege̲n de̲r selbe̲n ¶

212r   ¶Das i∫t als vill gesproche̲n das du nach der ploß
sollt greyffe̲n ey̲n benig vnd dich vo̲n de̲r selbige̲n lincke̲n
seytte̲n schnell durchge̲n mit de̲m messer fur dey̲n leyb mit
gesenckte̲m ortt auf dey̲n lincke seytte̲n vnd schlag
In zu de̲m haubtt seyne̲r rechte̲n seytte̲n od̲er set̲z Im de̲n
ortt zu de̲m gesicht od̲er zu der pru∫t e̲t̲c

212v   Ite̲m lig∫tu also auff seyne̲r rechte̲n seytte̲n vnd er auch auff d̲er
selbe̲n als vo̲r Indes windt vntt̲er seyne̲m messer dey̲n messer gar
woll vmb ~~vntt̲er~~ also das du In auff seyne̲r rechte̲n seytte̲n mit
de̲r kurt̲ze̲n schncyde̲n od̲er flech sey̲n rechte achsel rure∫t Ein
wenig Indes gee durch mit de̲m messer mit gesenckte̲m ortt
zwische̲n Ewre̲r payd seytte̲n od̲er leyb das dy lang schneyd
vor gee Indes set̲z Im de̲n ortt In sey̲n pru∫t vnd od̲er schlag
Im mit de̲r kurt̲ze̲n schneyde̲n / zu seyne̲n lincke̲n ore̲n vnd
halt dey̲n messer albeg fur das haupt mit de̲m gehult̲z woll
vbe̲rsich das de̲r dawm vntt̲en ∫tee auf de̲r flech deß messers e̲t̲c

213r   Ite̲m lige∫tu auff seyne̲m mess̲er auff seyne̲r rechte̲n seytte̲n so er
abe̲r ligtt Im hangnden ortt mit der zwirch Indes windt
dy kurt̲ze̲n schneyde̲n ∫ta̲rck auff sey̲n kopff vnd far hoch
auff mit de̲m gehult̲z will er mit de̲r versat̲zung nach fare̲n
so schlag Indes mit de̲r zwirch zu seyne̲n ore̲n seyne̲r rechte̲n
seytte̲n mit de̲r kurt̲ze̲n schneyde̲n e̲t̲c

213v   Ite̲m lig∫tu auff seyne̲m mess̲er seyne̲r rechte̲n seytte̲n wye
vo̲r Indes greyff mit deyne̲m lincke̲n arme̲ vber seyns messers

klingen vnd greyff vntten zu gewappender hantt das der
ortt deynes messers stee auff deyner lincken seytten vnd far
auff mit dem gehultz vnd stoß starck da mit an seyns
messers klingen nohendt pey der handt auff deyn lincke
seytten so nymbstu Im aber das messer etc

213v (cont.)

Item haw von oben gerad eyn Eynen langen oberhaw mit gestracktem
arme gerad vnd frey zu seynem kopff seyner lincken seytten vnd
so der haw gerad verpracht wirtt ist So wind gegen seynen
messer dy kürtz schneyden dar an das der dawmen vntten stee vnd
schnel Im dy kurtzen schneyden an seyn linckes ören etc

214r

Item haw Eyn oberhaw Im zu seyner rechten seytten lanck vnd
gerad eyn mit gestracktem arme Indes so der haw verpracht
ist so windt deyn messer gegen deyner lincken seytten an seyn
messer vnd schnell In zu seynem kopff oder arme seyner
rechten seytten etc

214v

Item nota dy stuck seyn gutt das du eynem dester leychter magst
verfuren vnd Indes mach dy dupliren dy feler vnd dy treffer

Item du solt auch wyssen das auß dysen winden sullen
gefunden werden hew stich schnytt vnd thu Im also
so du pist durchgangen mit dem messer von seyner lincken
seytten auff deyn rechte so schneyd vbersich an seyner seytten
Indes so du geschnytten hast mit dem messer stich Im auff
seyn prust Indes far auff mit dem messer vnd mach eyn
feler auch auff seyner rechten seytten vnd haw Im zu der
negsten ploß etc

215r

Item liegstu Im vntter seynem messer auff seyner rechten seytten
wye vor Indes windt gegen Im dy kurtz schneytt
das der dawmen auff der flech lig vnd stich Im zu dem
wang der selbigen seytten etc

215v

Item ligestu Im vntter seynem messer auff seyner lincken seytten
wye vor windt Indes gegen Im dy lang schneyden das
dy kurtz oben stee vnd stich Im zu dem wang der selbigen seytten etc[139]

216r

416 par fechter[140]

Item du solt auch gar eben mercken ob er weych oder hertt In
der versatzung Ist vnd merck dy swech vnd dy sterck
vnd zu vor an soltu des wortz Indes nicht vergessen

216v

---

[139] The folio number is underlined, and "HC" written below it.

[140] Added in a later hand.

216v (cont.)

vnd was das vor vnd das nach ist des pistu vor
vnterricht worden Nu soltu wyssen was das wortt
Indes ist ob du Enpfindest ob ir weych oder hertt ist In
der versatzung so merck das du Indes erbeyttest In alle
stuck wye du merckest dy eygenschafft der ding dy auff
dich gepracht vnd gemach werden oder dy du auff eynen
machen pist Indes duplire: Indes mutir: Indes durchlauf:
Indes peschleuß: Indes nymm den schnytt: Indes ring:
Indes messer nym: Indes pnym: Indes durchge: Indes
durchlauff: Indes thu was deyn hertz begertt: Indes ist
Eyn scharpffes wortt da mit dy meyster hartt verschnytten
werden zu vor auß dy meyster dye deß wortz Indes nicht
wyssen noch vernemen also hastu gar genaw In disen
puch den meysten tayl der kunst des messers vnd den
grosen Grundt etc

Alzo hatt Herr Hanns Lecküchner von Nurenberg das püch
gemacht vnd geticht geendt gott Im den hayligen
segen sendt vnd vergeb Im vbel myssetat vnd schuld
vnd pebeyß Im seyn parmmhertzige gottliche gnad
vnd hulf amen

Composita Est materia illa per dominum Johannem Leckuchner
tunc tempore plebanus Jn hertzogaurach Anno domini M° cccc°
septuagesimo octauo sed iste liber scriptum est et completus
Anno 8° secundo Jn vigilia sancti Sebastiani etc[141]

---

[141] Latin: "This material has been composed by Reverend Hans Lecküchner, at the time parish priest in Herzogenaurach, in the year of our Lord 1478, but this book has been written and completed in the 82nd year on the eve of St. Sebastian."

DANIEL BURGER

Hans Lecküchner of Nuremberg and His Fencing Treatise in the Long Knife

This is the reverend Hans Lecküchner of Nuremberg's art and epitome of Messer fencing, of which he himself made and put together both the text and the explanation, for the highborn prince and lord, Duke Philip, Count Palatine of the Rhine, Arch-Steward and Elector, and Duke in Bavaria.

~ Hans Lecküchner, 1482

|Das ist herr hannsen Lecküchner von Nurenberg künst vnd zedel ym messer dy er selbs gemacht vnd getícht hatt Den Text vnd dy auslegung dar über Dem hochgeporen fursten vnd herren hertzogen philippen phaltzgraffen Bey reyn Ertzdruckseß vnd kurfürst vnd hertzog yn Bayern etc

ompared to most other authors of late medieval fencing manuscripts, a great deal is known about Hans Lecküchner († 1482), but still very little is known about his life.[142] In his own words, he came "from Nurenberg", the most powerful imperial city in the Franconian Imperial Circle.[143] Through his mention in the print of Egenolff's Fechtlehre in 1531, his name was handed down as "Hanns Lebkommer of Nuremberg" and at the end of the Holy Roman Empire was even included in the *Nürnbergische Gelehrten-Lexicon*.[144] But this was all known and so one asked oneself: "Wer war dieser... vnd wann lebte er?" ("Who was this... and when did he live?")[145]

The name "Lecküchner" and Nuremberg

The family name refers to a craft once practiced in the family of gingerbread bakers or Lebküchner (Middle High German lebekuoche, leckuche) for whose products the city is still famous today.[146] The name "Her(mann) Leckucher" first appears in Nuremberg sources in 1392,[147] and in 1414 there is evidence

[142] Cf. BARTSCH 1883: p 108 (outdated). — KIST 1965: p 254. — HILS 1985B: pp 641–644. — MÜLLER 1994: pp 355–384 (with a strong second focus on presentation of the Liechtenauer tradition).

For the biography, the sources can be found quoted and partly even as an illustration on the website of the HEMA group *Ochs*: "Johannes Lecküchner – der fechtende Pfarrer" siehe: http://www.schwertkampf-ochs.de/leckuechner.html (accessed 20.09.2015).

The early 19th century assumption that the name of Johannes Liechtenauer was later corrupted into Lecküchner is of only more interest in terms of research history; cf. MASSMANN 1844: p 52.

[143] A highly readable short introduction to the history and culture of Nuremberg in Lecküchner's time is provided in FLEISCHMANN 2012.

[144] WILL 2805: p 280.

[145] KIEFHABER 1793: 103–104.

[146] Cf. EBERT-WOLF 1963/64: pp 491–531.

[147] Cf. SCHEFFLER-ERHARD 1959: p 207. In the original source, no doubt due to a scribal error, the abbreviation symbol for "-er" is additionally appended after the

of an unlawful act by a Stepfan Lekuchner at the Nuremberg market.[148] However, a genealogical connection of the fencer Hans Lecküchner to these persons is not possible, just as nothing at all can be proven with certainty about his family. This should be emphasized at the outset, as the few mentions of the name Lecküchner in the sources too easily lead to connecting them to one and the same person. Due to the derivation from a profession, there were several families of this name in southern Germany in his time.[149]

In the 1430s, the Nuremberg city accounts note that a certain Hans Lebküchler at Heilbronn twice paid 10 and 12 groschen, respectively, under the expenses of recognition duties (i.e., duties for the recognition of duty-free status).[150] Due to the date, it cannot have been the fencing master,[151] but perhaps a relative (such as a father of the same name), who obviously had a relationship with the administration of the imperial city.

In his direction, i.e. a relative of Hans Lecküchner (the Elder), there is also a mention about 10 years later: in 1450, the owner of a benefice at the Nuremberg Church of Our Lady, Johannes Knopfer, gave his consent to the sale of a perpetual sum of 4 gulden from the house of Hans Lebkucher at the fleiſchbänke ("slaughterhouse").[152] In the summer of 1450, a Mich(el) Leckküchner died, followed by his wife a little later (only as "Mich. Lebkucherin" in the death records of the parish church of St. Sebald),[153] but the connection with the family of Hans Lecküchner is unknown. In 1451, Friedrich Ottenfels and Hans Lebküchner donated a benefice at the altar of St. Anthony in St. Mary's Church, which was endowed with three farms and five estates in five villages.[154] In 1463, the "taſchner (purse maker) Hans Lebküchner" received the hereditary right to the bath on the Zotenberg (today the fruit market east of the Frauenkirche, about 200 meters from the former slaughterhouse).[155]

written out "Leckucher", so that "Leckucherer" is written, but certainly not intended.

[148] Staatsarchiv Nürnberg, Reichsstadt Nürnberg, Ratskanzlei, Ratsbücher (1a) 1/a Seite 122 Abschnitt 4. It is probably "Stefan Lecküchner", without whose will Hans von Lochheim "went to his daughter's house" in 1409, for which von Lochheim was punished with imprisonment and banishment from the town; cf. MÜLLNER 1972: p 205.

[149] Whether, for example, there was a relationship to the Bayreuth Lecküchner family (of which Fritz and his wife Margaretha née Riedner as well as their son Hans are documented in the late 15th and early 16th centuries) is not documented. Fritz Lecküchner, who was first mentioned in 1472, owned a house in Bayreuth's Breite Gasse, which was sold after his death in 1514, whereby his son Hans Lecküchner did not accept the inheritance due to high debts. (Stadtarchiv Bayreuth, Stadt- und

Gerichtsbücher B 4/185, B 11/360). That Hans Lecküchner is in Bayreuth still in 1521, 1522, 1529 and 1534 (op. cit. B 11/594, B 11/699, B 11/742, B 16/256, B 16/277, B 16/565).

[150] SANDER 1902: pp 515–516.

[151] Lecküchner is not mentioned anywhere contemporarily as a certified fencing master, as was done later by the Marx brothers, for example. However, in the glosses he refers to himself (albeit anonymously) as a "master", so that this term will nevertheless be used here.

[152] Staatsarchiv Nürnberg, Reichsstadt Nürnberg, Landalmosenamt Urkunden 100 (1450 November 17, Nürnberg).

[153] BURGER 1961: p 29.

[154] WINKLER 1956: pp 186–187.

[155] Stadtarchiv Nürnberg, E 49/III Nr. 8: Sal- Zins- und Gültbuch des Carl Holzschuher, f 14v. I would like to

Fig. 1: Nürnberg in 1598:
1) Fleischhaus and Fleischbrücke
2) Zotenberg
3) Hallerweise
4) Heilsbronner Hof
5) Insel Schütt

Whether the house by the slaughterhouse was the family home muſt remain unproven, but it has a certain probability insofar as that Knopfer had been 𝔚𝔞𝔫𝔨𝔢𝔩𝔥𝔢𝔯𝔯 (a kind of official) at the New Hospital until then, and the deed was witnessed by a prieſt at the New Hospital—Hans (the Younger) Lecküchner's life path would ultimately lead to a parish that was connected to the New Hospital. Nothing of the late medieval situation of this house in Nuremberg has been preserved after the extensive reconſtructions of the Renaissance (Fleischhaus, Fleischbrücke, 1596–1598) and the deſtruction during the Second World War, even though the charming location on the Pegniʒ ſtill invites visitors.[156]

Fencing in Nuremberg

Although Lecküchner's fencing treatises reveal a broad knowledge of this subjeᴄt, there are no further sources on him as a fencer—if we did not have his treatise, the Herzogenaurach prieſt would remain unknown as a fencer. Thus, one muſt rely on references within his texts themselves as well as on the observation of his environment.

thank Dr. Walter Bauernfeind, Nuremberg City Archives, for his kind advice.

[156] The situation before the construction of the new bridge at the slaughterhouse is shown in a 1594 plan by Paul Pfinzing (Staatsarchiv Nürnberg, Reichsstadt Nürnberg, Karten und Pläne 230).

Fig. 2: The imperial town *Nuremberga* in the late 15th century, view from the south (*Schedelsche Weltchronik*, printed in Nürnberg 1493, ff 99v-100r)

Where Hans Lecküchner learned the art of fencing is, like everything else on the subject, unknown. His older fencing manuscript, dated 1478, suggests that he encountered fencing in Nuremberg not as a student, but very probably at a young age.[157]

The imperial city was a major arms producer and itself a considerable military player in southern Germany. In Lecküchner's youth, the First Margrave's War (1449–1450) devastated the Nuremberg countryside, about ten years later another Margrave's War (the so-called 'War of the Princes', 1458–1463) moved into Nuremberg's immediate neighborhood, and in 1474–1475 the imperial city sent troops to the Rhine to support the emperor in his fight against Charles the Bold of Burgundy. On the one hand, Nuremberg was able to rely on the military duty of its citizens and subjects in the event of war, and on the other hand, it kept a number of nobles and mercenaries in permanent service in times of peace.

But even in peaceful times, it was a danger for a Nuremberg citizen to unexpectedly become the victim of one of the numerous noble feuds in Franconia.[158] Therefore, there must have been a fundamentally broad interest in at least elementary knowledge of martial arts.

[157] My sincere thanks go to my HEMA fencing coach Mr. Werner UEBERSCHÄR, Nuremberg, for his numerous references and explanations. Mr. UEBERSCHÄR has been researching the Nuremberg fencing schools and fencing masters for a long time and has made many findings available on the website of the HEMA group "Schwertbund Nurmberg": http://www.schwertbund-nurmberg.de/ (accessed 11.04.2021).

[158] Cf. VOGEL 1998 and ZMORA 1997.

The 'culture of arms' observed in German cities in the early modern period was certainly founded in the late Middle Ages.[159] An interesting light is thrown on the circumstances of the imparting of knowledge in fencing by a council decree of 1479, when the Swiss (𝕾𝔴𝖊𝖞𝖍𝖊𝖗) in Nuremberg service, who fought with spears, were ordered to teach fencing only to citizens(!); one was also allowed to hold a fencing event (𝕾𝖈𝖍𝖎𝖗𝖒𝖘𝖈𝖍𝖚𝖑𝖊), i.e. a public show.[160] There was therefore no lack of 'martial know-how' in Nuremberg. In one documented case, we also know that the Nuremberg merchant Jakob Auer hired such a fighter for an assassination attempt—it was the not-yet-famous Hans Talhoffer, who was involved in some way in the murder of Wilhelm von Villenbach in 1434.[161]

Since the 15th century, public fencing demonstrations are well documented in the imperial city. In 1446, a Nuremberg chronicler reported on an Italian artist (𝖊𝖎𝖓 𝖜𝖆𝖑𝖈𝖍), who was 𝖊𝖎𝖓 𝖌𝖆𝖗 𝖊𝖎𝖓 𝖌𝖚𝖙𝖊𝖗 𝖘𝖈𝖍𝖎𝖗𝖒𝖊𝖗 𝖚𝖓𝖉 𝖊𝖎𝖓 𝖌𝖚𝖙𝖊𝖗 𝖗𝖎𝖓𝖌𝖊𝖗 [𝖜𝖆𝖗] 𝖚𝖓𝖉 𝖜𝖊𝖓𝖓 𝖊𝖗 𝖊𝖎𝖓 𝖕𝖆𝖗𝖆𝖙 𝖒𝖆𝖈𝖍𝖊𝖙 𝖒𝖎𝖙 𝖘𝖊𝖎𝖓𝖊𝖒 𝖕𝖆𝖞𝖗𝖎𝖘𝖈𝖍𝖊𝖓 𝖉𝖚𝖈𝖍𝖊𝖈𝖐𝖓𝖊𝖓 𝖘𝖈𝖍𝖜𝖊𝖗𝖙, 𝖘𝖔 𝖐𝖚𝖓𝖙 𝖎𝖓 𝖓𝖎𝖊𝖒𝖆𝖓𝖙 𝖚𝖓𝖙𝖊𝖗 𝖆𝖚𝖌𝖊𝖓 𝖆𝖓 𝖘𝖊𝖍𝖊𝖓 𝖓𝖔𝖈𝖍 𝖊𝖗𝖐𝖊𝖓𝖓𝖊𝖓...; his feats also consisted of jumps with 𝖟𝖜𝖊𝖓 𝖉𝖊𝖌𝖊𝖓 𝖕𝖑𝖔ß ("two bare daggers") and rope artistry.[162] In the 15th century, these public performances were called 𝕾𝖈𝖍𝖎𝖗𝖒𝖘𝖈𝖍𝖚𝖑𝖊𝖓 (e.g. 1477 as the first mention of this kind)[163] and later 𝕱𝖊𝖈𝖍𝖙𝖘𝖈𝖍𝖚𝖑𝖊𝖓 (for the first time in 1494); such schools are documented several times in Nuremberg.[164] Lecküchner even presents pronounced 'school plays' in his fencing treatise, i.e. these were to be used to amuse an audience, to which he specifically refers: 𝖉𝖆𝖘 𝖘𝖙𝖚𝖈𝖐 𝖎𝖘𝖙 𝖑𝖊𝖈𝖍𝖊𝖗𝖑𝖎𝖈𝖍 𝖟𝖚 𝖙𝖗𝖊𝖞𝖇𝖊𝖓 𝖆𝖚𝖋𝖋 𝖉𝖊𝖗 𝖘𝖈𝖍𝖚𝖑𝖑 ("this play is laughable to make in the school").[165]

The fact that the teachers and participants in the fencing schools were not (only) artists and mercenaries passing through Nuremberg, but fencers who had grown up in Nuremberg—especially from the middle class—is clear from the Nuremberg fencing school permits for craftsmen. A mastery even developed, that is, a certain quality assurance obviously existed in succession from Johannes Liechtenauer, who probably lived in the 14th century.[166] In the fencing manuscript written around 1470 by the fencing master Paul

---

[159] TLUSTY 2011.

[160] Staatsarchiv Nürnberg, Reichsstadt Nürnberg, Ratskanzlei, Ratsbücher (2) 2/ p. 311r.

[161] Cf. BECKER 2020 and the work of Jens P. KLEINAU, which has been published on his blog to date: http://talhoffer.wordpress.com/2011/04/22/1434-the-case-of-wilhelm-of-villach/ (accessed 17.05.2016) and http://talhoffer.wordpress.com/2016/05/18/1434-march-20th-talhoffers-confession-in-salzburg/ (accessed 17.05.2016).

[162] HEGEL 1872: p 166 (2nd edition): 166.

[163] Staatsarchiv Nürnberg, Reichsstadt Nürnberg, Ratskanzlei, Ratsbücher 2, f 115 r; see also Ratsverlass 72, Bl. 11.

[164] Cf. LOCHNER 1860: pp 407–408. This corrects the older opinion that fencing schools in Nuremberg were only documented from the 16th century onwards. For the fencing schools 1553–1698, cf. SIEBENKEES 1794: pp 65–75.

[165] LM, f 183v.

[166] Cf. HILS 1985A.

It is tempting to speculate that Johannes Lichtenauer is related to the town of Lichtenau (Ansbach district), which became a possession of Nuremberg in the early 15th century. Unfortunately, there is no source evidence at all.

Kal, who was in the service of Bavaria-Landshut, several fencing masters of Liechtenauer's apprenticeship and society are named in the introduction, among them a master "Hartman von Nurenberg" (perhaps the same as the Nuremberg pewterer Hans Hartmann, who probably died in 1429)[167] and a master "Hanns Pägniczer" (perhaps the same as the Nuremberg gunsmith Hans Pegnitzer, who was active from 1466 to around 1500).[168] A contemporary of Lecküchner in Nuremberg was Hans Folz (born around 1430/40, died 1513) from Worms, who received Nuremberg citizenship in 1459. The barber and wound surgeon (in 1486 with the rank of master) excelled as a master singer and author of carnival plays, among other things, who also wrote a fencing treatise around 1479.[169]

Fencing with the long sword and other weapons must be excluded at this point. It should only be mentioned that there was probably a separate "Nuremberg fencing tradition"[170] in the 15th century, which included the so-called *Baumann fight book* (B) (c. 1420/c. 1465–1470)[171] and the teachings of the Nuremberg sword polisher Anthoni Rast/Rasch († 1549).[172] Lecküchner restricted himself to the long knife, but it is obvious that he had experience with the long sword. Once he even mistakenly writes 'sword' instead of 'knife'.[173] Lecküchner was aware of the importance of his knife fencing teaching and distinguished it from that of sword fencing:

> …wenn es seyn vill meyster des swertz, dy nicht wissen von der art des messers noch solche lehren recht außsynnen mügen.[174]

> "Namely, there are many masters of the sword who do not know the art of the Messer and cannot comprehend it properly."

He emphasized fundamental differences to the long sword, for example, when it came to Durchwechseln:

---

[167] *Paul Kal-München* (PKM), f 2r. Cf. GRIEB 2007: p 580.

Is he identical with Hanns Hartmann, the captain of the *Marxbrüder*, who is documented in 1500 and 1508? If so, he would have lived to a very old age. Cf. with this noteworthy reference Werner UEBERSCHÄR: Handschriften, Homepage Schwertbund Nurmberg (after Augsburg 1625) (accessed 18.09.2015).

[168] PKM, f 2r. Cf. GRIEB 2007: p 1931.

[169] *Hans Folz* (HF). An annotated transcription was prepared by Andreas MEIER in 2009: http://www.pragmatische-schriftlichkeit.de/transkription/trans_q566.pdf (accessed 22.03.2021).

[170] Cf. "Nuremberg Group". Wiktenauer. Ed. Michael CHIDESTER. http://www.wiktenauer.com/wiki/Nuremberg_Group (accessed 10.04.2021).

[171] *Baumann's fight book* (B). Rainer WELLE presented recently an excellent edition with commentary: WELLE 2014.

[172] Anton Rast (A); cf. http://www.schwertbund-nurmberg.de/Transkription%20Anthony%20Rast.pdf (accessed 10.04.2021).

[173] *Hans Lecküchner-München* (LM), f 105v.

[174] LM, f 1r.

und sich eben, das du dy durchgen machest nach meyner maynung, wann ich dy in dem messer anders bedewtten pin, dann im swertt.[175]

"And make sure that you do the going through according to my meaning, as I have stated it to be different in the Messer than in the sword."

In his texts, Lecküchner anonymously calls himself meister ("master") in the introduction to the glosses of the notes, which are repeated again at the beginning, for example:

Hie lertt der meister aber ein stuck auß dem wecker und spricht: 'recht lere, etc.'

"Here the master teaches another piece of the Waker, saying 'Learn on the right', etc." [176]

Alternatively, he calls himself the merer der kunst ("increaser of the arts") ten times, referring to his inventions of fencing pieces.[177]

The self-confident self-designation as master as well as the dedication to the Palatine Elector make it clear that Lecküchner here wants to join the ranks of the respected fencing masters. This position, partly in princely service, no longer has anything to do with the wage and court fencers of the High Middle Ages, as Martin WIERSCHIN emphasized in his fundamental work on Liechtenauer.[178] The self-designation as master, the recommendation of some pieces for (fencing) schools, the discussion of differences to the sword, all this proves that Hans Lecküchner had contact with fencing communities, even if he never explicitly described himself as a member of one. It is striking in this context that Lecküchner positions himself against the freifechter ("free fencers"; a surprisingly early reference), whom he regards as opponents and recommends his own pieces against their fencing style:

...der wincker... ist der hauptstuck eyns ym messer und geret woll auff dy freyfechter, dy frey versetzen und ist eyn neuer haw und stuck ym messer und ist seltzam und gutt.[179]

"...the Winker... is one of the main pieces of the Messer and strikes well the 'free fighters' who parry freely, and is a new stroke and piece of the Messer, and is peculiar and good."

---

[175] LM, f 67v.

[176] LM, f 15v.

[177] LM, ff 60r, 60v, 61r, 72v, 83v, 105v, 165r, 166v, 193v, 211v.

[178] WIERSCHIN 1965: pp 44–66.

[179] LM, f 30r.

We do not know the location of fencing practices in Nuremberg in the 15th century. The Nuremberg public regulations from the late 15th century forbid **fechten ober schirmen** with sharp swords, daggers (**begen**) or poles on the Hallerwiese, a meadow in city property on the right bank of the Pegnitz directly in front of the western city wall.[180] One may deduce from this that there had been corresponding activities there before which were stopped. It seems as if an 'extraterritorial' site was then sought, but it is not until the 16th century that the Heilsbronner Hof, a property belonging to the Cistercian monastery of Heilsbronn (situated between Nuremberg and Ansbach) opposite St. Lawrence's Church, which lay as a foreign dominion within the Nuremberg city walls, is documented for fencing schools. Still later, in 1627–1628, a municipal fencing house was built on the island Schütt, between the two arms of the Pegnitz within the city.[181]

Studies and clerical ordination

The Lecküchner family must have achieved a certain level of prosperity in Nuremberg—the benefice endowment of 1451 (see above) indicates this—because it was made possible for Hans to pursue a clerical career and even to complete a course of study (which was not necessary for this), which opened the way for him to clearly lucrative positions.

His year of birth can only be approximated and depends on how old he was when he entered his studies in 1455 (see below); if he was a young man, he may have been born in the 1430s. Hans Lecküchner studied in the Saxon university town of Leipzig, where he was enrolled in the summer semester of 1455 as "Iohannes Lechkochner de Nurenberga".[182] Immediately together with him, three other Nurembergers and a student from Schwabach were enrolled, and somewhat later in the same semester another nine Nurembergers, among them sons from the important Stromer and Groland families. In the following year, the Nuremberg humanist Hartmann Schedel also began his studies in Leipzig. In the summer semester of 1457, on September 17th, "Iohannes Lekucheler de Nurembergha" obtained his baccalaureate there[183].

On 22 September 1459, he was ordained an acolyte by the Bishop of Bamberg, who was responsible for the Nuremberg region (i.e., he received the highest of the four lower ordinations and could now assist deacons and priests).[184] As yet, we know little about the following

---

[180] Cf. BAADER 1851: p 52
[181] Cf. BERM 1997: pp 37–62.
[182] ERLER 1895: p 194.
[183] ERLER 1897: p 169.

[184] Cf. KIST 1936: pp 101–111, 136–142, 177–178, 208–209, 243–249, 277–280, 313–317, 341–345, 368–370. 1936. For Lecküchner, p 244, Nr. 840.

period. Lecküchner continued his ecclesiastical career, which apparently allowed him to attend university again due to the financial support he received from the benefice. The time frame of his studies (see below), which covers a total of about 25 years, gives rise to the question of whether we are dealing with two different persons of the same name,[185] but such long periods of study are certainly documented in the 15th century.[186] However this may be decided in future source discoveries, our fencing master Hans Lecküchner was a priest in the end (see below).

The fact that a clergyman was involved in fencing was not all unprecedented in the late Middle Ages. Even the oldest fencing manuscript on fighting with sword and buckler[187] is designed as an instructional work by a priest or monk, and a cleric is generally regarded as its author.[188] Another cleric to be mentioned here is the 𝔓𝔣𝔞𝔣𝔣𝔢 Hanko (Hans) Döbringer at the end of the 14th century. From this point of view, it is less surprising to read in the estate of the chaplain Master Bernhard zu Tramin in 1491, among other things:

Fig. 3: Fencing priests in F1 (f 20r, ca. 1325)

> 𝔄𝔦𝔫 𝔭𝔞𝔯 𝔰𝔭𝔬𝔯𝔢, 𝔞𝔦𝔫 𝔰𝔠𝔥𝔴𝔢𝔯𝔱, 𝔞𝔦𝔫 𝔪𝔢𝔰𝔰𝔢𝔯, 𝔞𝔦𝔫 𝔱𝔦𝔩𝔦𝔠𝔷, 𝔞𝔦𝔫 𝔰𝔱𝔯𝔢𝔶𝔱𝔨𝔬𝔩𝔟𝔢𝔫
>
> "A pair of spurs, a sword, a knife, a 𝔱𝔦𝔩𝔦𝔱𝔷, a mace"[189]

Lecküchner's interests as a clergyman were thus part of a tradition that was not surprising for his contemporaries.

However, Lecküchner makes no reference to his clerical training within his manuscripts. Only twice does a passage in the 1482 manuscript end with *Amen*;[190] this and the request for intercession at the end of the text correspond entirely to contemporary usage and do not allow any conclusions to be drawn. With some caution, one might interpret the almost overflowing flow of text and the methodical (dis)structuring of the Fencing Doctrine as a result of his university education. Once the student slips into Latin when he corrects an illustration with the remark: *Hic nihil ad implevit* ("This has nothing to do with it").[191] The Latin dedicatory letter of the Heidelberg manuscript is the eloquent example, and a flash of his erudition occurs there at

---

[185] I would like to thank Mr. Paul BECKER, Nordhausen (HEMA fencing school *In Motu*) for his constructive feedback and questions.

[186] Cf. ANDRESEN 2017, pp 91–152 (esp. 140–146).

[187] *Walpurgis Fechtbook* (F1).

[188] FORGENG 2021: p 8–9.

[189] ZINGERLE 1909: p 184.

[190] LM, ff 18r and 46r.

[191] LM, f 97r.

the end[192] through one of the most famous quotations from Aristotle's *Metaphysics*:

> *Omnis homo naturaliter scire desiderat*
>
> "Every man desires to know by nature"

Lecküchner uses a slightly modified version of the still common saying that one should not cast pearls before swine when he presents an elaborate technique that should be reserved for a small circle of initiates and is therefore also called ꝺer Ungenanꝺt:

> Wanne man soll ꝺy eꝺlen margarítten oꝺer rosen nícht fur ꝺy sweyn strewen, ꝺas sy ꝺurch ꝺy selbígen ních ungeertt werꝺen unꝺ getretten ín ꝺas kott
>
> "Thou shalt not cast margherites and roses before swine, lest they be defiled and trodden in dung"[193]

Perhaps a little of Lecküchner's rhetoric shimmers through here.

In any case, the use of catchy names for fencing pieces (e.g. the Notꞓstück, the Storchenschnabel, the Scorpion, or the Basilisk) is evidence of his didacticism, although this was not entirely Lecküchner's invention, and in part he took up names that had already been introduced (such as the Ungenannt or the Sonnenzeígen). Clever are his references to pieces to be kept secret, to particularly effective ones, sere gutt(e) unꝺ abenteꞏwrísch(e),[194] or even to techniques mocking the opponent (schímpflích zu treyꞏben).[195] Here Lecküchner shows himself to be an experienced rhetorician.

On 30 June 1478, Hans Lecküchner—by then a 'presbyter', i.e. priest, from the Bamberg diocese—enrolled at the Electoral Palatinate University of Heidelberg, which had been founded in 1386. The register lists him as "*Johannes Leckurchner de Nurenperga, presbiter Babenperg. Dyoc*".[196] He was there in notable company from his homeland: the following year, the Cistercian monk Sebald Bamberger also enrolled in Heidelberg, who was later to become abbot of the monastery of Heilsbronn between Nuremberg and Ansbach and wrote a contemporary chronicle. Sons of Nuremberg patricians also received their university education in Heidelberg at this time, most notably the Tuchers: Hieronymus and Nikolaus Tucher studied here from 1476 to 1479; Hans Tucher from Nuremberg also began his studies in Heidelberg in 1480.[197]

---

[192] *Hans Lecküchner-Heidelberg* (LH), f 116v.
[193] LM, f 91r.
[194] LM, f 135r.

[195] LM, f 183r.
[196] TOEPKE 1884: p 357.
[197] KIST 2014: pp 41–49.

Lecküchner apparently did not earn a doctorate at the University of Heidelberg.

The dedication to Elector Philipp I of the Palatinate

Hans Lecküchner dedicated his fencing treatise to the Elector Palatine Philipp I (1448–1508, r. since 1476) from the House of Wittelsbach. The oldest of his two manuscripts, which has been preserved until today in the Heidelberg University Library as the successor to the Electoral Palatinate Library, is dated 1478 on the basis of an indication in the later text (see below).[198] It consists of 116 sheets (format 21.7 x 16 cm) of paper in a somewhat later vellum binding (17th century). On one end, the fencing lesson is framed by a colored miniature with the Electoral Palatinate coat of arms with helmet and crest consisting of the crowned lion, behind which stands a coat of arms holder in armor and banner, bearing Philip's motto 𝕹u / 𝕰t / cetera ("Now and so on") on red cloth.[199] On the other end, it is framed by the post-bound Latin dedicatory letter (of obviously different ink).[200] It is therefore possible that the finished text was only dedicated to the sovereign of the University of Heidelberg in a second step.

Fig. 4: The frontispiece of LH, featuring Philip I's heraldry, facing the first folio (ff Iv–1r, ca. 1478)

---

[198] LH. Cf. Bartsch 1887: p 136 (Nr. 236). — Mittler and Werner 1986: pp 68–70 (Nr. 7). — Miller and Zimmermann: p 396f.

[199] Cf. Frommberger-Weber 1973: pp 35–145 (esp. 120f).

[200] LH, ff 115r–116r.

In any case, Lecküchner must have been intensely occupied with his fencing theory for quite some time before he was able to work out this elaborate manuscript, produce a fair copy and finally hand it over. The amount of text exceeds all other German fencing treatises of the 15th century. Here Lecküchner proceeded in a scholarly (dis)structuring manner; sometimes the impression arises that the academic claim outweighs the practical value. Overall, he probably intended to transfer Liechtenauer's teachings in the long sword to the knife, which, however, comes up against technical limits due to fundamental differences between the one- and two-handed weapons and their blade lengths.[201]

In its basic conception, it follows the well-known templates in the Liechtenauer tradition of mnemonic verses (𝔷𝔢𝔱𝔱𝔢𝔩𝔫) and subsequent explanations (𝔊𝔩𝔬𝔰𝔰𝔢𝔫). The rhymes of the mnemonic verses are not particularly elaborate and sometimes rhymes seem somewhat impure, for example 𝔞𝔫 𝔳𝔬𝔯𝔠𝔥𝔱 / 𝔪𝔦𝔱 𝔰𝔠𝔥𝔩𝔞𝔤𝔢𝔫𝔡𝔢𝔪 𝔬𝔯𝔠𝔥𝔱 (instead of 𝔬𝔯𝔱).[202]

At this point, the content of Lecküchner's fencing cannot be presented and analyzed in more detail.[203] The question of models and references must also be left aside in the context of a biographical essay. Therefore, it should only be briefly mentioned that Lecküchner, who studied to be a fencing master, structured his work didactically in a very orderly manner. After a short general preface and introduction, he presents the basics of knife fencing: The concepts of the 𝔦𝔫𝔡𝔢𝔰, "before", and "after", followed by an explanation of the strength and weakness of the knife and the correct hand position. This is followed by the 23 main pieces, divided into the six "hidden hews" (𝔡𝔶 𝔰𝔢𝔠𝔥𝔰 𝔳𝔢𝔯𝔭𝔬𝔯𝔤𝔢𝔫 𝔥𝔢𝔴)[204] and 17 other main pieces. Lecküchner sensibly introduces important terms, principles, and techniques such as the 𝔷𝔦𝔫𝔫𝔢𝔫 (i.e., "battlements", as he calls the 𝔅𝔩ö𝔰𝔰𝔢𝔫 here), duplicating and mutating within the framework of the first main piece. General advice concludes the Fencing Doctrine, which is concluded by a repeated mention of the author and a prayer of blessing.[205]

Fig. 5: Philip I "the Upright", Count Palatine of the Rhine (stained glass window in the church of Neckarsteinach, 1483)

---

[201] Cf. LEISKE 2018: pp 166–176.

[202] LM, f 194r.

[203] Cf. MÜLLER 1994: pp 355–384.

[204] LM, f 2r.

[205] LH, f 114v/LM, f 216v.

Lecküchner's study in Heidelberg may alone have sufficiently motivated the dedication, but one must not forget that the imperial city of Nuremberg bordered on the Upper Palatinate to the east, which as an electoral precinct belonged to the domain of the Elector Palatine.[206] The Elector of the Rhineland was thus an immediate neighbor of the Nurembergers. Another neighbor of Nuremberg was Duke Otto II of Palatinate-Mosbach-Neumarkt, a close relative of the Elector, who was to inherit this territory in 1499. So it certainly made sense for a learned Nuremberg citizen to keep an eye on a career in Palatine service.

Philipp, born in 1448, had come under the guardianship of his uncle Count Palatine Friedrich after the early death of his father, Elector Ludwig IV, in 1449. The latter adopted his nephew in 1451 and assumed the dignity of Elector of the Rhineland in the legal form known as 'arrogation'. This was never accepted by Emperor Frederick III, but in view of the Wittelsbach's political and military strength he could not change anything.

Thus, it was only after the death of Elector Frederick I in 1476 that Philip succeeded to the rank of Elector to which he was entitled under hereditary law. Philip had been married to Margaret, the daughter of Duke Louis IX 'the Rich' of Bavaria-Landshut, since 1474, a dynastic union that both houses consolidated through the marriage of his third son Ruprecht to Elisabeth of Bavaria-Landshut—and which would eventually lead to the Landshut War of Succession with the Bavarian-Munich line in 1503. The marriage between Ruprecht and Elisabeth produced the dukes Ottheinrich and Philipp, for whom the principality of Palatinate-Neuburg was created in the 'Cologne Decree' of 1505. This is still relevant to the topic insofar as LM[207] bears a later owner or dedication note, which refers to the Palatinate-Neuburg superintendent Johann Tettelbach (1517–1598) in Burglengenfeld.[208]

The fencing master Paul Kal wrote a fencing lesson for Duke Ludwig IX 'the Rich' of Bavaria-Landshut († 1479) in or around 1470.[209] Whether his son-in-law

Fig. 6: Paul Kal assists his lord Ludwig IX "the rich", Duke of Bavaria-Landshut (PKM, f 4r, 1470)

[206] The Principality of Palatinate-Mosbach (with the capital Neumarkt in Upper Palatinate, Bavaria) became extinct with Count Palatine Otto II in 1499 and fell to the Electoral Palatinate.

[207] LM.

[208] For Tettelbach, see below.

[209] PKM. "Laut der Bibliothek wurde dieser Kodex für Ludwig IV. angelegt."
http://www.pragmatische-schriftlichkeit.de/transkription/edition_paulus_kal.pdf
(accessed 08.12.2015). However, there is a confusion with

Philipp was aware of this manuscript is unfortunately unknown, although the fencing master Paul Kal is unlikely to have remained unknown to him.

Lecküchner's treatise remains—at least according to the current state of research—the only such writing from the 15th century in the electoral library. At least a few other works of instruction should be mentioned, which were collected at the Heidelberg Court and which indicate that Lecküchner's work can be placed in a broad context of military writings and other works of 'pragmatic writing'.[210] Even today, for example, the Heidelberg University Library still houses several medical works such as a "Franconian Pharmacopoeia" from around 1440,[211] several copies of the "Fireworks Book of 1420",[212] the "War Book" of Philipp Mönch from 1496,[213] and Hans Schermer's writing on the construction of bulwarks.[214]

Parish priest in Herzogenaurach

Two years after his matriculation in Heidelberg, on 15 March 1480, the Nuremberg councilor or the keeper of the Heiliggeistspital ("New Hospital of the Holy Spirit"), Nikolaus I Groß (1420–1491), in a letter to the bishop of Würzburg, presented the "*dominus Johannes Lekuchner*" to his parish after the resignation of the Herzogenaurach parish priest Johann Berger (who moved to the altar of the Apostle in the Frauenkirche in Nuremberg).[215] Since 1337, the parish of Herzogenaurach had been incorporated into the New Hospital of the Holy Spirit in Nuremberg, i.e. the hospital had the right of presentation and could freely dispose of all income.[216] The sovereignty over the New Hospital lay with the Imperial City Council, which exercised the actual right of presentation (until it relinquished it in 1601).[217] It was one of the largest hospitals in southern Germany with enormous influence and prestige; since 1423, the imperial regalia and imperial sanctuaries had been kept in the Nuremberg

---

Ludwig IX of Bavaria-Landshut, cf. LENG 2008: 65–66 (Nr. 38.5.1).

[210] Due to libraries currently (December 2020–February 2021) closed due to pandemic, could no longer be consulted: BACKES 1992.

[211] Cod. Pal. germ. 213

[212] CPG 585; Swabia, around 1440; CPG 502, Southern Germany around 1470; CPG 562, Northern Bavaria around 1490

[213] CPG 126

[214] CPG 562

[215] Staatsarchiv Nürnberg, Reichsstadt Nürnberg, Briefbücher des Inneren Rats 36, ff 255v–256r; Briefbücher des Inneren Rats 38, f 110v. About

Herzogenaurach, cf. SODER VON GÜLDENSTUBBE 1978: pp 117–154 (esp. 125). I would like to thank Dr. Manfred WELKER, Herzogenaurach, for his advice and help, as well as the staff of the Landeskirchliches Archiv der Ev.-Luth. Kirche in Bayern, Nuremberg, for their help in obtaining the literature.

[216] Cf. KNEFELKAMP 1989: pp 65–68.
In 1601, the now Protestant imperial city of Nuremberg renounced the right of presentation in favour of the Bishop of Bamberg.

[217] Staatsarchiv Bamberg, Hochstift Bamberg, Geistliche Regierung, Akten und Bände 5903. I would like to thank Dr. Johannes STAUDENMAIER, Bamberg, for his kind advice.

hospital church (and only moved from there for a short time to Aachen or Frankfurt am Main for the imperial coronation).

Herzogenaurach, only 19 kilometers from Nuremberg, was a town of the prince-bishop of Bamberg, but in spiritual terms the parish was subject to the Würzburg bishopric, an eloquent example of the complicated ruling relationships in Franconia. In the first half of the 15th century, however, the office of Herzogenaurach had been transferred to Eberhard Zollner († 1456) of Nuremberg. The Zollners vom Brand were an important family, considered noble, with branches in both Bamberg and Nuremberg (where they also sat on the Inner Council in the 15th century), so that the imperial city of Nuremberg undoubtedly had great influence in Herzogenaurach in several respects.[218]

Hans Lecküchner's brief work as a pastor in Herzogenaurach cannot be assessed due to a lack of sources.[219] It is said that between 1481 and 1484 a

Fig. 7: The gothic church St. Maria Magdalena in Herzogenaurach, where Hans Lecküchner served as priest until his death in 1482

---

[218] Cf. BÄTZ 2002: pp 1170–1172.

[219] No references to Lecküchner could be identified in the relevant holdings of the Staatsarchiv Bamberg, Hochstift Bamberg, Geistliche Regierung, Akten und Bände as well as Urkundenbestand A 95 (Bamberg parish documents, under Herzogenaurach). I would like to thank Dr Klaus RUPPRECHT, Würzburg, and Dr

Johannes STAUDENMAIER, Bamberg, for their information and efforts. Also lacking references to Lecküchner is the Herzogenaurach parish archive, which is now in the archives of the Archdiocese of Bamberg and was reorganized and indexed in 2013. My sincere thanks go to Dr. Andreas HÖLSCHER, Bamberg, for his information.

certain Franz Lebküchner (Lettküchner; perhaps a relative) worked there as a benefice.[220] Parish pastor Hans Lecküchner died at the end of 1482.[221] On the basis of the Nuremberg presentation of 31 December 1482, a certain Johann Pottensteiner succeeded Johannes Lecküchner in the Herzogenaurach parish, and he was soon succeeded by the priests Nicolaus Glob and Johann Troßler († 1503).[222] His successor, the priest Johannes Wydhössel, who also came from Nuremberg, has preserved a Herzogenaurach parish order and description of his office, which he wrote when he took office and thus provides an insight into the state of affairs at the end of the Middle Ages—he indignantly complained about the lack of documents on the property of the parish, which does not give his predecessors a good report card, at least in secular administrative matters.[223] The Gothic parish church of St. Magdalena is still preserved as a structural testimony from Lecküchner's time in Herzogenaurach, but the vicarage was replaced by a new building in 1846.[224]

Fig. 8: Typical page layout of LM (f 3r)

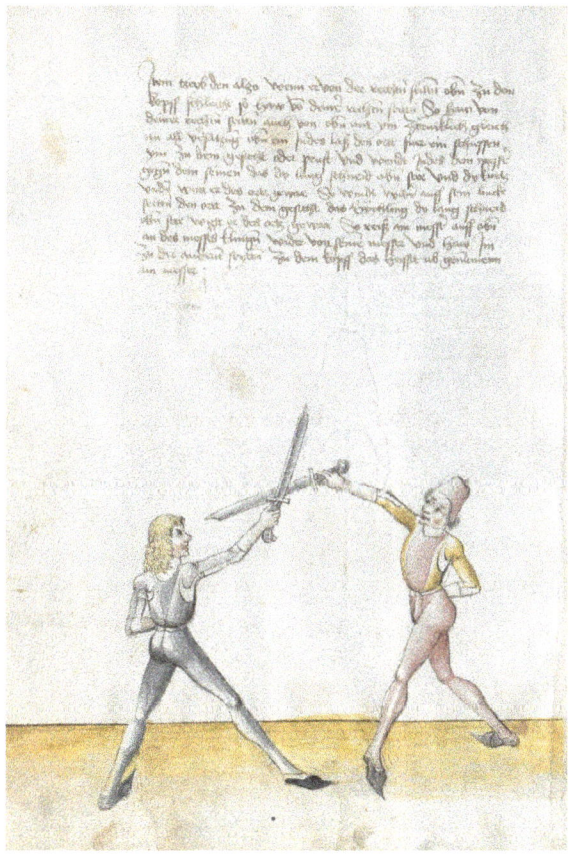

The 'last hand' manuscript from 1482

One year before his death, Lecküchner edited his manuscript of 1478 once again and put it into a particularly representative, richly illustrated form. The consistent illustration places Lecküchner's work in the top group of German fencing manuscripts of the 15th century: until this time, there were extensive text manuscripts of Liechtenauer's teachings with little or no illustration—and, even predominantly, illuminated manuscripts with extremely sparse text (for example, by Kal or Talhoffer); a consistent combination of text and image is known, for example, from the manuscripts of the Gladiatoria group.[225] Illustrated manuscripts were particularly important within the 'Nuremberg Group'. For this new illustration, however, the text of 1478 was no longer

[220] Cf. SODER VON GÜLDENSTUBBE 1978, p 125. According-ing to the *Nuremberg Briefbuch* no. 38, f 110v, he could not have acted as pastor and successor to Hans Leck-üchner. Possibly the quote by SODER VON GÜLDEN-STUBBE was the result of a mix-up.

[221] KIST 1965: p 254 (Nr. 3879).

[222] Staatsarchiv Nürnberg, Reichsstadt Nürnberg Briefbücher des Inneren Rats 38, f 110 (there the deceased priest is called "Johannes Lebkuchner"). The letter is already dated "M° CCC° octagesimo tercio" due to the usage at the turn of the year at that time.

[223] MEYER 1949; MEYER 1981: pp 532–558, 559–561.

[224] Realschematismus des Erzbistums Bamberg, hrsg. vom Erzbischöflichen Ordinariat Bamberg, Bd. 1, Bamberg 1960, pp 492–500 (esp. 499).

[225] Cf. LENG 2017: pp 211–234.

specially adapted, so that one can speak of a 'secondarily' illustrated version.[226]

The rich illustrations exclusively show civilians dressed in the typical South German fashion of the time. The artist is thought to be in the circle of Michael Wolgemut, and it is worth noting that there is a close relationship to fencing manuscripts of about the same period between about 1480 and 1500 by Master Peter Falkner.[227] Here we refer to the contribution by Falko Fritz in this volume. Alongside the *Baumanns fight book* (B) perhaps Lecküchner's way of combining text and illustration served as an inspiration for the similarly richly illustrated fencing manuscript by the Nurembergian Anthoni Rasch/Rast († 1549), which is preserved only in a later but obviously quite faithful copy.[228]

Lecküchner again dedicated his second version to Elector Philip I of the Palatinate. The text concludes with the dating note:

> *Composita est materia illa per domine Johannes Lecküchner tunc tempore plebanus in Herzogaurach, anno Domini M° CCC° septuagesimo octavo, sed iste librum scriptum est et completum Anno 8° secundo in vigilia sancti sebastiani etc.*
>
> "This material was put together by Mr. Johannes Lecküchner, currently a priest in Herzogenaurach, in the year of our Lord 1478, but this book was written and completed in the year (14)82 on the eve of the day of St. Sebastian, i.e. 19 January."[229]

In his second text, Lecküchner also explicitly names himself as the author at the beginning and end: this is the kůnst und zedel ym messer, dy er selbs gemacht und getickt hatt den text und dy auslegung darüber. This second oldest Lecküchner manuscript is now in the Bayerische Staatsbibliothek in Munich (LM).[230] Due to the extensive illustrations, it comprises 100 leaves more than the manuscript of 1478, i.e. 216 leaves (format 30.0 cm × 20.7 cm) in a modern half-leather binding. The paper has various watermarks in the form of an ox's head.[231] The script is by one hand throughout, a practiced pulpit script in Gothic bastarda—and apparently identical to the hand of the Heidelberg manuscript. The manuscripts in Heidelberg and Munich are considered to be Lecküchner's autographs.

---

[226] LENG 2017: p 219.

[227] *Peter Falkner* (PF) and *Paris fight book* (P) Cf. LENG 2008 (Nr. 38.1.5): http://kdih.badw.de/datenbank/handschrift/38/1/5; last changed 19.03.2018. — LENG 2008 (Nr. 38.2.3):http://kdih.badw.de/datenbank/handschrift/38/2/3; last changed 08.01.2020.

[228] AR. For dependence from B, see also WELLE 2014.

[229] LM, f 216r. The feast day of St. Sebastian (Fabiani et Sebastiani) is January 20th, the eve therefore January 19th.

[230] LM. Cf. SCHNEIDER 1978: p 177f. — LENG 2008: pp 74–76 (Nr. 38.6.1), Tafel VIII and Abb. 35.

[231] Piccard X, 656, Piccard XII, 807, continue to be similar to Piccard XVI, 132.

Due to the dedication to the Palatine Elector Philip and the careful text and image design of LM, it can be assumed that this was also a dedication copy intended for the prince. Perhaps Lecküchner had already hoped with the dedication of his first manuscript in 1478 to be accepted into Electoral Palatine service after his studies in Heidelberg. The position of pastor in Herzogenaurach, which he obtained instead due to Nuremberg's patronage, undoubtedly provided him with a good income, but this did not stop him from dedicating the manuscript of 1482 to the Elector as well.

Whether this luxurious, illustrated manuscript was commissioned by the prince must remain unknown. It obviously never reached Elector Philip I, for although the manuscript was once in the possession of the Elector Palatine and reached Munich via Düsseldorf, an owner's note, or rather a more recent dedication, speaks against the uninterrupted lineage in princely hands: on the flyleaf is named Magister Johann Tettelbach,[232] Superintendent at Burglengenfeld, who in 1579 gave the manuscript to Duke Philipp Ludwig of Palatinate-Neuburg (b. 1547 in Zweibrücken, r. 1569–1614). Obviously, it initially passed from Lecküchner's estate into private hands before coming into Wittelsbach possession late in 1579.

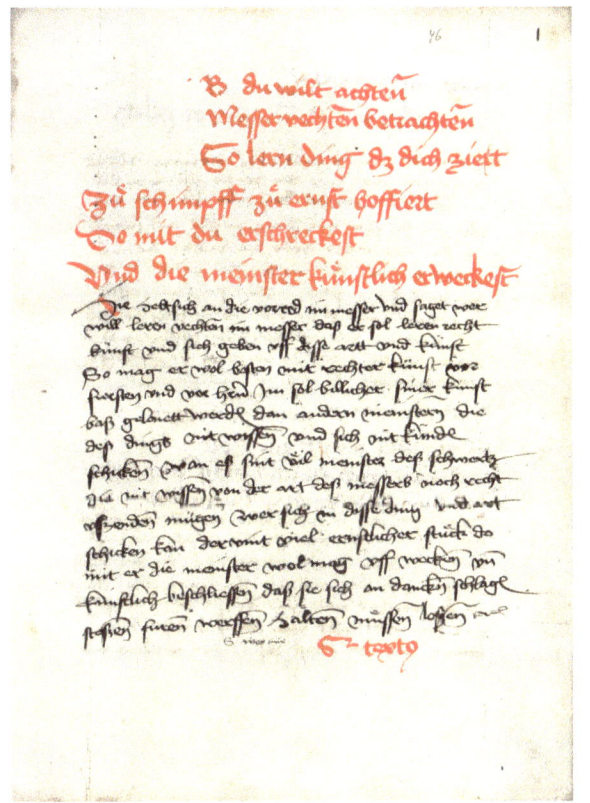

Fig. 9: First page of Lecküchner's treatise in HS (f 46r, 1491)

A short view on the impact in the region around Nuremberg

Hans Lecküchner's work found early adaptors. One of Lecküchner's younger contemporaries, the fencing master and 𝔐𝔞𝔯𝔵𝔟𝔯ü𝔡𝔢𝔯 Peter Falkner (born around 1460, died after 1506), is to be mentioned here, and several of his manuscripts are attested.[233] The earliest datable copy of Lecküchner's text corpus dates from 1491 by Hans von Speyer.[234] In Nuremberg, too, there must have been at least one text witness at the

---

[232] Johann Tettelbach was born in 1517 in the imperial city of Dinkelsbühl (Ansbach district, today Middle Franconia, but historically Swabia) and died on 25 March 1598 in Burglengenfeld (Schwandorf district, Upper Palatinate), which belonged to the principality of Palatinate-Neuburg. He was first rector of the Kreuzschule in Dresden (Saxonia); pastor of St. Afra in Meissen (Saxonia); superintendent in Chemnitz; then pastor in Schwandorf (Upper Palatinate) and since 1580 superintendent in Burglengenfeld. He had two sons, Johann the Younger (b. 1546 in Dresden, d. after 1586) and Dr. Heinrich Tettelbach (b. c. 1560 in Neuburg a.d. Donau, d. 1644 in Frankfurt a. Main).

[233] PF and P.

[234] *Hans von Speyer* (HS).

beginning of the 16th century, known to Albrecht Dürer around 1512.[235] Whether this was the illustrated manuscript of 1482 (later in the possession of Tettelbach), or another, lost text witness, is unknown. The Augsburg scribe and city treasurer Paul Hektor Mair (1517–1579) also had a copy in his extensive collection of fencing manuscripts, the source of which he may well have obtained from the Nuremberg region, as he did with the work of Anthoni Rast. One must assume the existence of at least one further, now lost source from Lecküchner's direct possession, because since his manuscript of 1478 had remained in Heidelberg, he needed at least one copy (or the original text) for himself in order to create the 'last hand' version (LM) from it. Further work on the textual evidence promises exciting new insights for the future.

---

[235] *Albrecht Dürer* (AD). Cf. WASSMANNSDORFF 1888: 138–145, and WELLE 2021.

FALKO FRITZ

The Messer and Its Use in Lecküchner's Teachings

If you want to deceive him	Wiltu In betrygen
As you have lied under	So du pist vnterligen
Wind to the left, go through in winding	Linck windt durchgee In winden
Learn to find strokes, thrusts and cuts	Hew stich schnytt ler vinden
To the head, to the body	Zu kopff zu leyb
Perform what you desire	Was du pegerst das teyb
In every action	In allem gefertt
Test the soft or the hard	Prüff weych oder hertt

~ Hans Lecküchner, 1482

ecküchner's treatise on Messer fencing is one of the most extensive descriptions of 15th century fencing. It contains poetic verses, detailed instructions, and pictures on almost all of its 216 folios. In contrast to most other fencing sources, it solely deals with the Messer.

The verses and chapters bear a strong resemblance to Liechtenauer's verses and general tradition, but Liechtenauer's name is not mentioned in the book. Many descriptions are very similar to the long sword fencing texts in, e.g., the manuscripts known under the names of *Peter von Danzig fight book* (PD) or *Lew fight book* (L). This raises the question of how much the teachings were adapted from the long sword to the single-handed Messer and its characteristics. The adaptations may influence the types of Messers used for fencing, the instructions given in the text, and also the pictures illustrating the techniques. An analysis of three particular aspects, 1) the influence of Messer shapes on fencing, 2) the instructions Lecküchner gives to end a technique, and 3) the stances depicted in *Hans Lecküchner-München* (LM), are presented in the following sections.

The construction of a Messer

The Messer is a single-edged sword type that was common in late medieval and early modern Germany. Messers come in various sizes and shapes. Based on two examples, this article looks at the characteristics of Messers and how their shape influences the fencing techniques.

The first example (A) is an original from the author's private collection. It was presumably made around 1500, has a relatively short, straight blade and a one-handed grip (fig. 1). It falls somewhat between the types M3do and M5bo in the Elmslie Typology of single-edged medieval swords, with a

(A)

(B)

Messer example	(A)	(B)
Total length [mm]	712	1025
Blade length [mm]	582	810
Grip length [mm]	130	215
Distance from crossguard to balance point [mm]	100	130
Total weight [g]	880	1150

Fig. 1: Two original Messers from ca. 1500 (top photo by author, bottom photo by Peter Johnsson)

Table 1: General dimensions of examples (A) and (B)

back edge that has neither a pronounced clip point nor a sharpened section near the point.

The second example (B) is a larger weapon from the European Hanseatic Museum, Lübeck, also from around 1500, with a long, curved blade and two-handed grip. Fig. 1 includes a probable outline of the missing point. Blades of this size and shape are usually referred to as 𝕶𝖗𝖎𝖊𝖌𝖘𝖒𝖊𝖘𝖘𝖊𝖗, and this one is an Elmslie type M4c+.

Looking at these two examples and other surviving Messers in museums, we see that the blades are often very similar to other single-edged swords, like falchions or sabers. The feature that characterizes them as Messers is not the blade shape, but the grip construction with its (mostly) wooden scales riveted to both sides of a full-width tang and a 𝖂𝖊𝖍𝖗𝖓𝖆𝖌𝖊𝖑 (guard spike), or simply 𝕹𝖆𝖌𝖊𝖑, protruding from one side of the crossguard. This 𝕹𝖆𝖌𝖊𝖑 may serve as a protection for the hand on the hilt and support the use of some techniques in Messer fencing, but it is also a byproduct of the way many Messers are constructed. This becomes most evident when compared with other designs.

So, before looking at the Messer, this is a brief recapitulation of the standard construction of European blades, be it the sword, the rapier, or the falchion. The last is very similar to the Messer in many ways but

Fig. 2: Construction of a falchion

1. crossguard, grip and pommel are mounted onto the tang

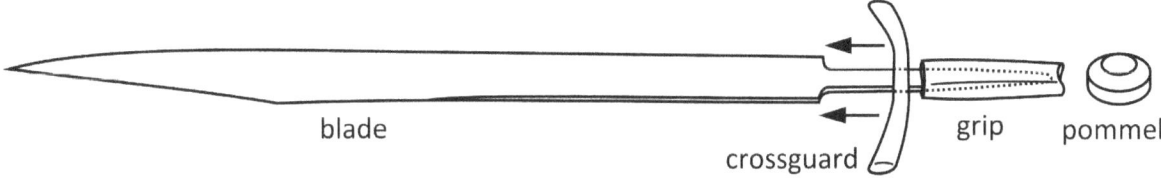

blade   crossguard   grip   pommel

2. the end of the tang is hammered down

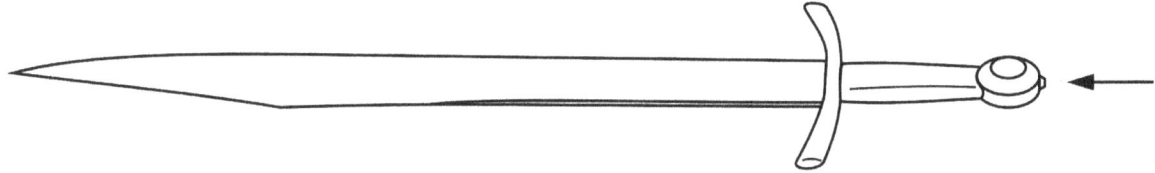

differs in the way the grip is designed. On the falchion, the cross-guard, grip, and pommel are all stacked onto the tang from the back and then held in place by a rivet at the very end (fig. 2).

Using this design, the crossguard rests firmly against the base of the blade while grip, pommel, and rivet keep it safely in place.

The Messer on the other hand is constructed differently. The characteristic scales riveted to the full-size tang are the most obvious parts of the alternative design. However, on some surviving originals including examples (A) and (B) the blade, tang, and pommel are either forged from a single piece of steel or forge welded together before the Messer is assembled. Thus, the crossguard cannot be threaded onto the tang from the back, it simply doesn't fit over the pommel. Consequently, it must be mounted from the front, over the blade, which in turn means that it cannot be set against the base of the blade as on a falchion or sword. It requires another rivet to fix the crossguard firmly to the blade. This rivet is the **Wehrnagel** (fig. 3). Therefore, in addition to providing protection for the hand, the **Nagel**'s first and foremost function in the Messers constructed this way is to prevent the crossguard from sliding back off over the blade.

As a consequence of this design, many crossguards are wider at their center to offer enough room for the hole that will hold the base of a solid **Nagel** as a rivet. Both examples (A) and (B) have this feature, and in both cases the center of the crossguard is drawn out into the grip, forming its base. Especially on example (A), the metal is shaped into a nice, smooth curve between grip and crossguard at this particular spot. It gives its wielder superb control with the index finger when the blade is stretched forward in a thrust or in winding (fig. 4, see also fig. 6). A similar curve is

Fig. 3: Construction of Messer example (A)

1. the crossguard is mounted from the front, spike and scales from either side

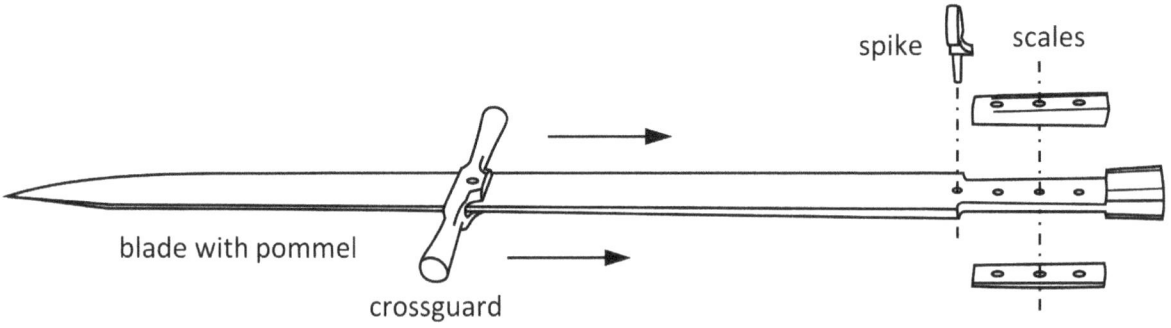

spike   scales

blade with pommel

crossguard

2. the spike end is hammered down, additional rivets fix the scales

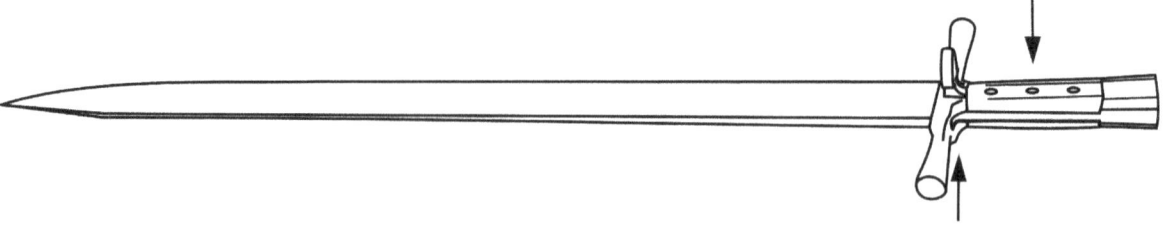

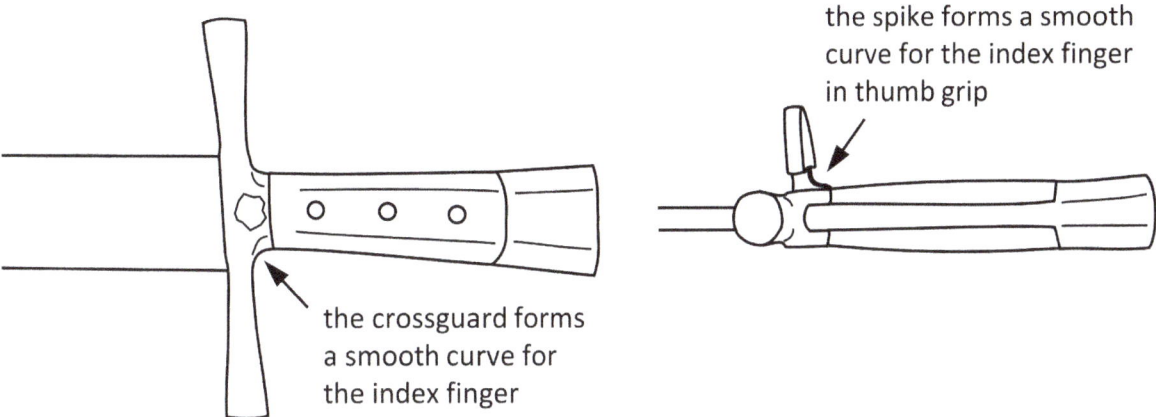

the spike forms a smooth curve for the index finger in thumb grip

the crossguard forms a smooth curve for the index finger

Fig. 4: Grip details resulting from the riveted construction.

Fig. 5: A typical Messer illustration from LM (detail from f 82r)

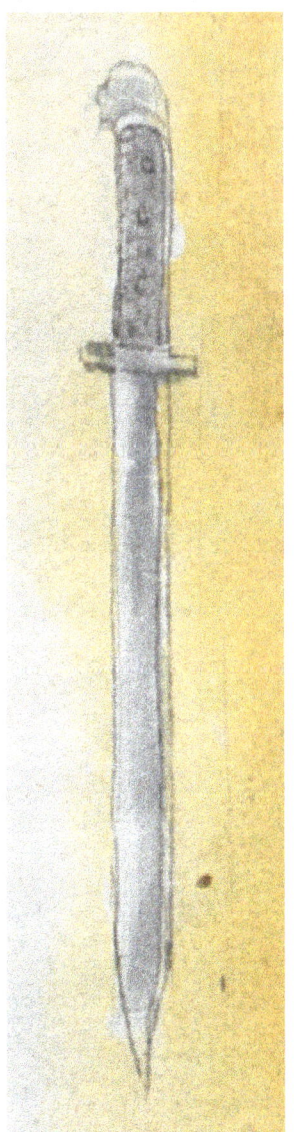

often found on the spike and comes into play when the weapon is held in thumb grip.

Influences of curvature on winding and thrusting

The examples (A) and (B) were chosen for this article because they nicely illustrate the wide range of blade shapes that can be found in Messers. While example (A) is a relatively short, light, and straight Messer, (B) is longer, heavier, and curved. The Messers shown in Lecküchner's treatise are very similar to example (A), albeit with a much longer grip and a shorter crossguard (fig. 5).

Before looking at the instructions in Lecküchner's text and the fencers pictured in LM in more detail, some thoughts on using straight vs. curved blades in fencing shall be discussed here.

Holding a blade, we feel the grip between our fingers and we perceive the weight of the blade in front of us (fig. 6). We are used to working with all sorts of tools, and so were our ancestors in the late Middle Ages. The smaller the tools, the more we tend to hold them with thumb and index finger only, sometimes aided by the middle, ring, and little fingers as required. But essentially, when it comes to fine motor skills, the thumb and index finger give us the most valuable tactile information. These two fingers are also closest to the blade, too, thus experiencing the most direct feedback.

The point of the blade we actually feel is the balance point, meaning the center of its mass and its leverage resulting from the position. The greater the distance between grip and balance point, the more top heavy a weapon feels. But disregarding the distance from the grip, the location of the balance point also determines the line of a straight thrust. Pushing a blade forward means pushing its mass along the center line shown in fig. 6 and fig. 7.

With example (A) and other straight blades, the point of the blade is very close to this natural line of thrust (fig. 7). Furthermore, when the

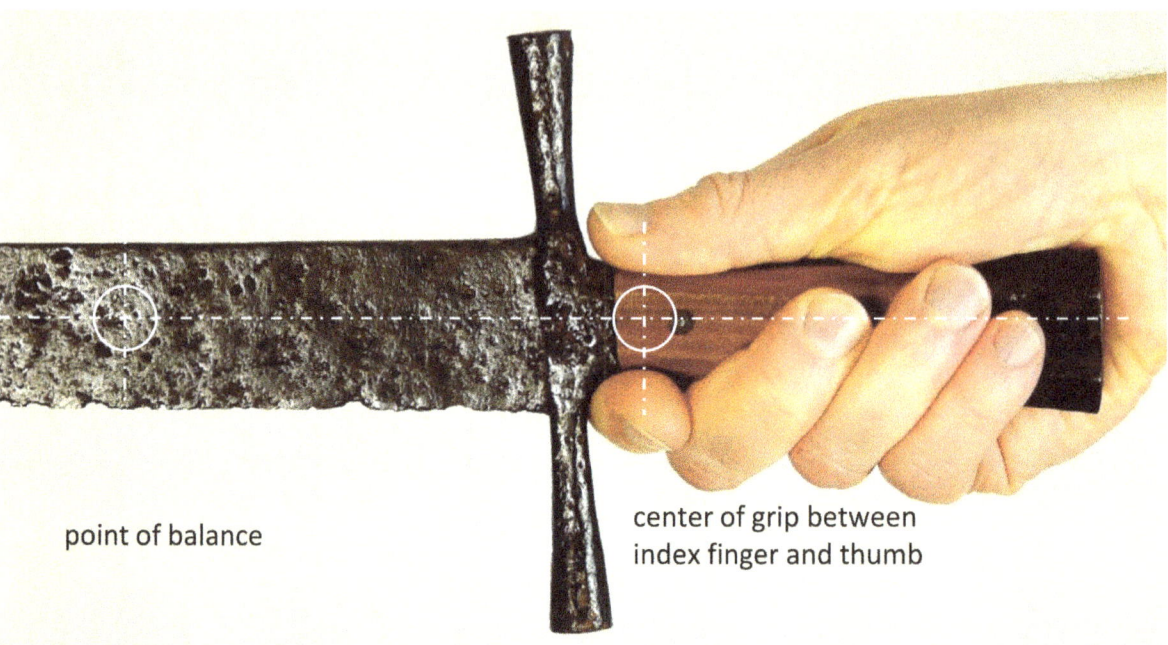

point of balance

center of grip between
index finger and thumb

blade is turned around its longitudinal axis, e.g. when winding from longpoint to an upper hanging, the point keeps its orientation almoſt perfectly. No matter if the true or false edge is making contaċt with the opponent's blade in the bind, thruſting the balance point forward moves the point precisely towards the same target point, without requiring any additional modifications by the fencer.

Fig. 6: Grip on example (A) during thrusting or winding, shown here with reconstructed, detachable wooden scales

Fig. 7: Orientation of the point when thrusting and winding with straight or curved blades

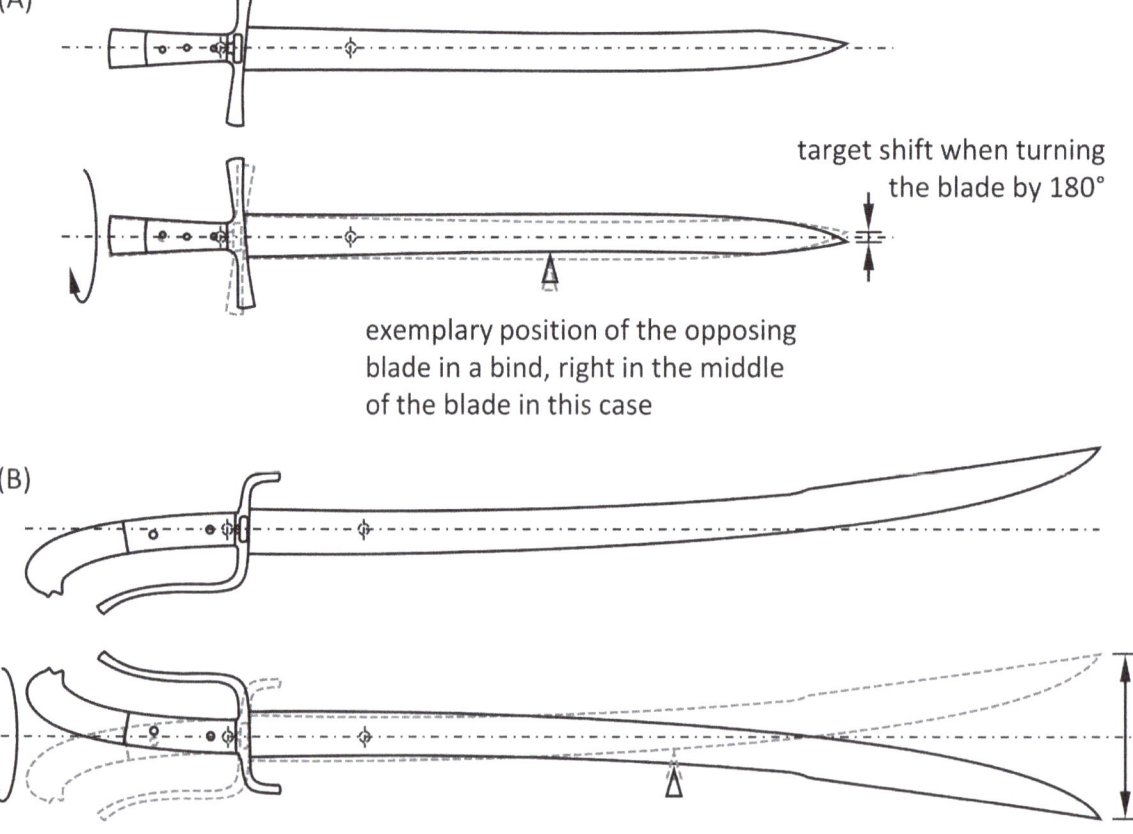

(A)

target shift when turning
the blade by 180°

exemplary position of the opposing
blade in a bind, right in the middle
of the blade in this case

(B)

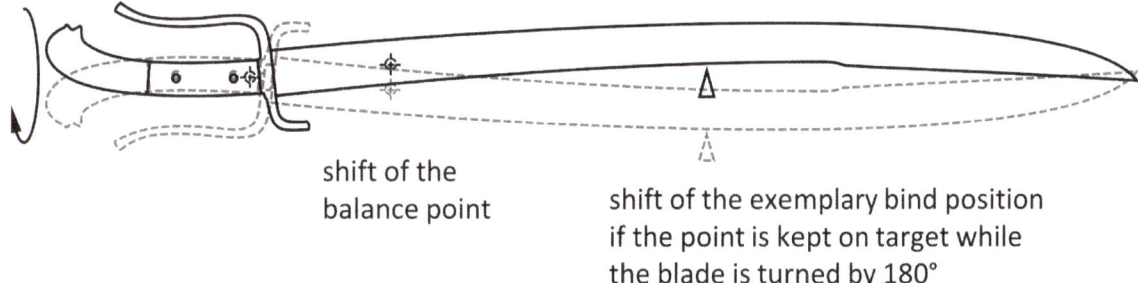

shift of the
balance point

shift of the exemplary bind position
if the point is kept on target while
the blade is turned by 180°

Fig. 8: Shift of balance
point and bind position
if example (B) is turned
by 180° while keeping its
point on target

With a curved blade on the other hand, like example (B) shown in fig. 7, turning the Messer from one edge to the other will affect the target area of a thrust significantly. This can be compensated by the fencer by learning to control the point so that it stays on target, but there is a tradeoff. By keeping the point centered, the curve of the blade shifts the contact point with the opposing blade drastically, as illustrated in fig. 8. This gives the opponent noticeable feedback and a much better chance to react quickly. In addition, this action requires a repositioning of either the grip or the balance point, which makes the motion more complicated.

When fencing in the bind, turning the blade while keeping the point on target and also keeping the information to the opponent as subtle as possible calls for a straight blade. Or even better, a slender and straight blade. When we look at the 𝔉𝔢𝔡𝔢𝔯-type swords shown in the Liechtenauer tradition manuscripts, this is exactly what we see. Of course, there are other fencing styles that use curved blades for thrusting and even use the special characteristics of these blades to their advantage. But when using a Messer in a system originally conceived for a double-edged, two-handed sword, a blade where the center of the grip, the balance point, and the blade tip are all aligned on a straight line makes thrusting and fencing from the bind easier. The clip point we see in the majority of illustrated sources on Messer fencing may just be the result of this, shaping the blade so that the point drops back closer to the center line.

If there is some truth to this hypothesis, larger 𝔎𝔯𝔦𝔢𝔤𝔰𝔪𝔢𝔰𝔰𝔢𝔯𝔰 like example (B) would not be designed for fencing in the bind and could consequently be made with more curvature and an off-center point.

Statistical analysis of instructions in Lecküchner's text

As discussed in the previous section, different shapes of Messers are optimized for different uses. Liechtenauer's method of long sword fencing prominently mentions three ways of harming or threatening the opponent with the blade: the strike, the thrust, and the cut. In the texts, these are called the three 𝔚𝔲𝔫𝔡𝔢𝔯, a presumably deliberate ambiguity meaning

either "wonders" or "wound-inflictors". Being firmly rooted in the Liechtenauer tradition, Lecküchner's text uses the same concept, e.g. in the verse where he says that you shall learn to find the strike, the thrust, and the cut from the winding.[236]

A statistical approach to evaluate the relevance of these actions in Lecküchner's fencing is presented in this chapter.

Methodology

This analysis categorizes the instructions Lecküchner gives on how to win a technique or how to overbear the opponent. It is not an account of all strikes, thrusts, or wrestling actions mentioned in the book, but only of those at the end on a technique, intended to harm, threaten, or lock the opponent to force him to surrender. It is a deliberately narrowed subset of all fencing actions that helps to take a structured approach based on Liechtenauer's concept of three Wunder.

The direct attacks on the opponent's body were sorted in the following four categories:

- Thrusts and threatening with the point, including setting the point to an opening and pushing the opponent backwards.
- Strikes with the true or false edge, including a few strikes with the flat of the blade.
- Cuts with an edge, where the cutting action is explicitly mentioned.
- All actions not focusing primarily on the blade, including all types of wrestling, arm locks or breaks, disarms, throws, pommel strikes, etc. (These often involve pressing the opponent with parts of the Messer which may result in a cut, too, but as long as the objective is to manipulate the opponent's balance or disarm him rather than cutting him, these were counted in the wrestling category.)

In cases where the text lists a number of alternative options for what the fencer could do, all of them were counted, but if the text ends with the instruction "then do whatever you like", this was not attributed to any category.

In essence, these first three categories represent the three Wunder, while the fourth sums up everything else. Since the Messer is usually shorter than a two-handed sword, and its blade is shorter than an arm in most cases, a fencer can reach the opponent's hand with his own hand whenever the distance is close enough for an attack to the head or chest. Consequently, grappling and wrestling are likely to play a much bigger role in Messer fencing

---

[236] LM, f 211v

		first folio	last folio	total pages	total instructions	thrusts	strikes	cuts	non-blade (wrestling, disarms, pommel hits)
Zornhau	*Zornhau*	3r	14r	23	29	17	12	0	0
Wecker	*Wecker*	14v	24v	21	32	15	16	1	0
Entrüßthau	*Entrüßthau*	25r	28r	7	15	0	14	1	0
Zwinger	*Zwinger*	28v	28v	1	1	0	1	0	0
Geferhau	*Geferhau*	29r	29v	2	2	2	0	0	0
Wincker	*Wincker*	30r	32v	6	11	4	7	0	0
Vier Läger	four guards	33r	34r	3	0	0	0	0	0
Versatzungen	displacing	34v	38v	9	8	8	0	0	0
Nachreisen	drawing after	39r	45v	14	17	8	6	2	1
Überlaufen	**overrunning**	**46r**	**61v**	**32**	**38**	**6**	**1**	**1**	**30**
Absetzen	setting off	62r	62v	2	2	2	0	0	0
Durchwechseln	changing through	63r	66v	8	10	8	2	0	0
Zucken	twitching	67r	72r	11	13	6	7	0	0
Durchlaufen	**running through**	**72v**	**102r**	**60**	**64**	**1**	**5**	**0**	**58**
Schnitte	cutting off	102v	110v	17	18	7	3	7	1
Drücken	**pressing the hands**	**111r**	**117r**	**13**	**14**	**1**	**0**	**0**	**13**
Ablaufen	running off	117v	118r	2	2	0	2	0	0
Pnehmen	taking over	118v	126r	16	19	9	6	0	4
Durchgehen	going through	126v	128r	4	7	3	3	1	0
Pogen	bow	128v	130v	5	5	0	5	0	0
Messernehmen	**taking the Messer**	**131r**	**197v**	**134**	**145**	**30**	**25**	**6**	**84**
Verhängen	hanging	198r	211r	27	34	15	11	0	8
Winden	winding	211v	216r	10	16	6	9	0	1
all chapters combined				427	502	148	135	19	200
						29%	27%	4%	40%
chapters prioritizing bladework				188	241	110	104	12	15
						46%	43%	5%	6%
chapters prioritizing non-blade results (wrestling, disarms and pommel strikes)				**239**	**261**	**38**	**31**	**7**	**185**
						15%	12%	3%	71%

Table 2: Number of instructions by category, sorted by chapter

than in the long sword. The following analysis of the text in Munich 582 may shed some light on the priorities presented in the book.

Distribution of instructions per chapter

Table 2 shows the results per chapter. All in all, 502 instructions on 427 pages were counted. 40% of the results are wrestling or other non-blade

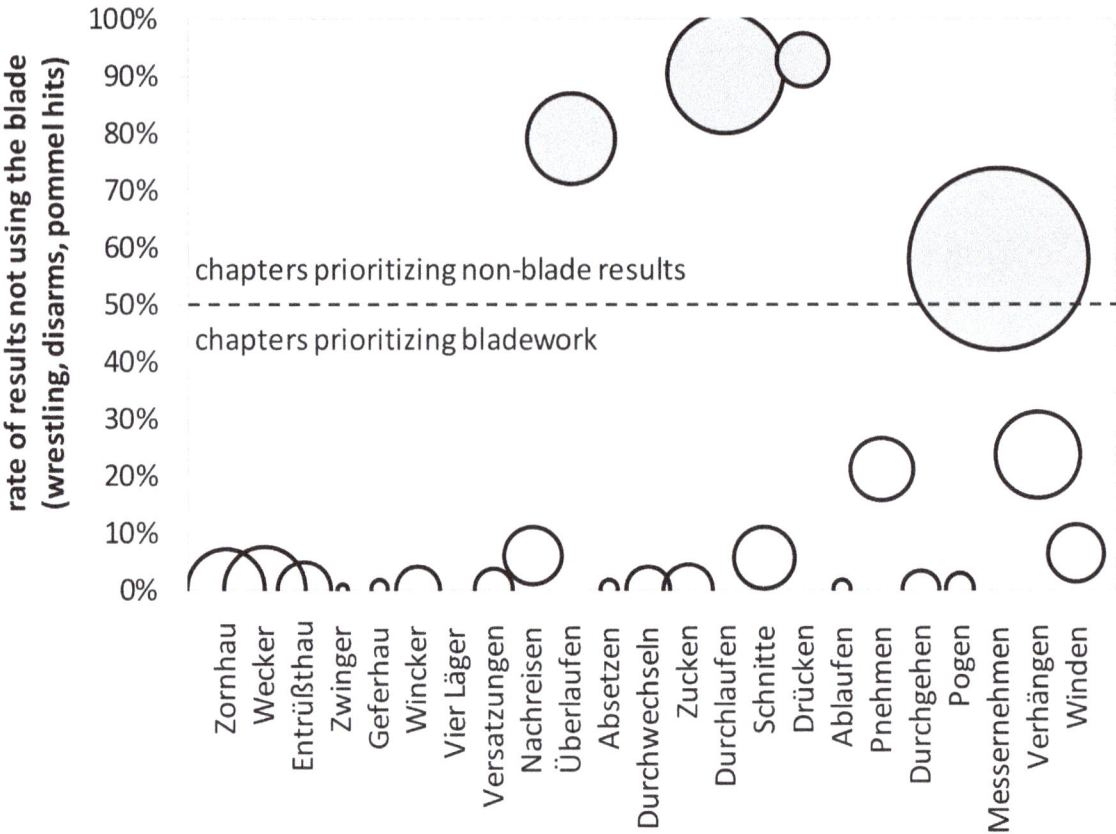

actions. The great majority of these, however, is found in only four of the 23 chapters, namely the Überlaufen, Durchlaufen, Drücken, and Messer nehmen. These are set in **bold text** in Table 2. In these chapters, 71% of the results fall into the wrestling category, while it is only a rate of 6% in all other chapters combined. The chapters prioritizing non-blade results also represent more than half of the book's volume, 239 pages with 261 instructions, vs. 188 pages and 241 instructions in chapters focusing on bladework.

This distribution is also shown in fig. 9, using the same scheme to highlight the chapters prioritizing wrestling in grey. The bubble sizes in the diagram represent the number of counted instructions in each chapter.

In the chapters prioritizing bladework, there are 46% thrusts, 43% strikes, 5% cuts, and 6% and wrestling. So, there is a slight preference for the thrust over the strike, but the difference is not significant. Among the three Wunder, the cuts have by far the least importance in Lecküchner's text. For all detailed results, see table 2. More research is needed to compare these results with distributions in, e.g., the long sword sources.

Fig. 9: Distribution of instructions per chapter: the bubble size represents the total number of instructions found in the respective chapter, the rate of blade vs. non-blade results within this number is shown in percent

Statistical analysis of stances depicted in LM and other sources

A second statistical analysis was done regarding the stances and body alignment shown in Lecküchner's pictures.

When fencers accustomed to the two-handed sword pick up a Messer today, they can use a lot of their existing repertoire. Since many of the plays in LM are copied from Liechtenauer's teachings, this is not surprising. But very often you see them fencing with the right foot leading most of the time, which isn't always agreeable with the text.

In the long sword, standing with the left foot forward makes perfect sense. If you try to stand in a position like the 𝔓flug or Ochs on the right-hand side, but with the right foot forward instead of the left, it requires an unnaturally twisted body. Stepping forward with the left foot makes the stance a lot easier and more stable. With a Messer, though, this is not the case. Having the left hand tucked behind the back, you can move the Messer into any guard while keeping the right foot forward without feeling obviously uncomfortable. In some techniques it even seems quicker and easier.

So, if Lecküchner's text instructs us to stand with the left foot leading, is that because it was directly copied from the long sword, or were the fencers in the 15th century used to a different set of body mechanics and advancing the left foot came perfectly naturally to them, even with a Messer? After all, they did not grow up with the movements of modern Olympic fencing which modern people envision when hearing the term 'fencing'. In Olympic fencing, having the same foot forward all the time is the absolute norm, or even a regulatory requirement. People who grew up with wrestling, dagger, and two-handed sword would be primed differently than we are today.

In order to see if there are any noticeable differences in the stances in long sword and the Messer, the following chapter presents a foot position tally of all pictures in Lecküchner's treatise and compares the result with the stances in other illustrated manuscripts from the 1420s to the 1590s.

Methodology

In each picture, both fencers were evaluated for whether the left foot or the right foot is closer to the opponent's center of gravity, or neither of them. The latter may happen if, e.g., the stance is directly perpendicular to the line connecting the two body positions—or if one fencer is flat on the ground and the other sits on top playing backgammon.[237] All in all, 2872 positions were counted in this fashion, 830 in LM and 2042 in 14 other manuscripts.

---

[237] LM, f 91v

The following four fencing styles were compared:

- Messer
- Dussack
- sword without armor (𝕭𝖑𝖔𝖋𝖋𝖊𝖈𝖍𝖙𝖊𝖓)
- sword in harness

The fourth group is clearly different to the previous three. It was added because it is an important aspect of Liechtenauer's tradition, and it is also a way to test the methodology. If the stances depicted in 𝕭𝖑𝖔𝖋𝖋𝖊𝖈𝖍𝖙𝖊𝖓 and harnessed fencing don't show significantly different results in the analysis, there is very good reason to question the results altogether.

The list of the 15 evaluated manuscripts is by no means exhaustive. It is rather a selection of sources that show illustrated techniques page by page and that were available to the author at the time of taking the tally. It must also be noted that the pictures show a wide variety of situations either at the start, during or at the end of the described technique. So, if the picture shows two fencers having the right foot forward, of course that does not mean they had the right foot leading all the time. However, it was assumed that this effect would influence all fencing styles in all manuscripts equally, so that a dominance of right feet in one section vs. left feet in another would actually mean that the fencer in the former style would have the right foot leading more often than those in the latter.

Distribution of foot alignment in LM

Before comparing Lecküchner's Messer with other manuscripts, table 3 and fig. 10 show the results per chapter in LM. As before, the chapters prioritizing wrestling are highlighted in grey.

All in all, 66% of the 830 depicted fencers have the right foot closer to the opponent's center of gravity than the left foot. The rate of right feet forward is higher in the chapters prioritizing bladework than in the four chapters that contain mostly wrestling (74% vs. 59%). Fig. 10 puts the numbers in perspective, again using the bubble size as an indicator for the number of pictures included in each chapter.

The results from the analysis of the pictures concur to a great extent with the previous statistical evaluation of the text. In the chapters that focus on fencing with the blade and where the techniques end with a thrust, strike, or cut, almost ¾ of the pictures show the fencers with the right foot closer to the opponent, while it's less than 60% in the chapters 𝖀𝖇𝖊𝖗𝖑𝖆𝖚𝖋𝖊𝖓, 𝕯𝖚𝖗𝖈𝖍𝖑𝖆𝖚𝖋𝖊𝖓, 𝕯𝖗𝖚𝖈𝖐𝖊𝖓, and 𝕸𝖊𝖘𝖘𝖊𝖗𝖓𝖊𝖍𝖒𝖊𝖓. The text and the pictures both show a comparable grouping of chapters.

Looking at Lecküchner's fencing, we actually find two books in one.

- 19 chapters arranged on 188 pages teach Messer fencing with the blade, mostly with the right foot forward and either a thrust or a strike at the end.
- 4 chapters on 239 pages contain mainly wrestling instructions where the Messer is used as a tool to get into reach and to support locks, levers, throws, or disarms.

However, it seems unlikely that the fencing masters in the late 15th century would have made this distinction themselves. Wrestling was a natural part of Messer fencing, and the chapters prioritizing non-blade results are spread out between others in the book rather than being treated in one particular section.

Table 3: Rate of foot alignments, sorted by chapter

			foot facing opponent		
		fencers	right	left	equal
Zornhau	Zornhau	44	70%	30%	0%
Wecker	Wecker	42	81%	19%	0%
Entrüßthau	Entrüßthau	12	58%	33%	8%
Zwinger	Zwinger	2	50%	50%	0%
Geferhau	Geferhau	4	50%	50%	0%
Wincker	Wincker	12	58%	42%	0%
Vier Läger	four guards	6	67%	33%	0%
Versatzungen	displacing	16	88%	13%	0%
Nachreisen	drawing after	28	79%	21%	0%
Überlaufen	**overrunning**	**64**	**44%**	**44%**	**13%**
Absetzen	setting off	4	50%	50%	0%
Durchwechseln	changing through	14	79%	14%	7%
Zucken	twitching	22	82%	18%	0%
Durchlaufen	**running through**	**116**	**59%**	**30%**	**11%**
Schnitte	cutting off	34	82%	18%	0%
Drücken	**pressing the hands**	**26**	**58%**	**42%**	**0%**
Ablaufen	running off	2	50%	50%	0%
Pnehmen	taking over	34	62%	38%	0%
Durchgehen	going through	8	63%	38%	0%
Pogen	bow	10	60%	30%	10%
Messernehmen	**taking the Messer**	**260**	**63%**	**35%**	**2%**
Verhängen	hanging	52	77%	23%	0%
Winden	winding	18	89%	11%	0%
all chapters combined		830	66%	31%	3%
chapters prioritizing bladework		364	74%	25%	1%
chapters prioritizing non-blade results		**466**	**59%**	**36%**	**5%**

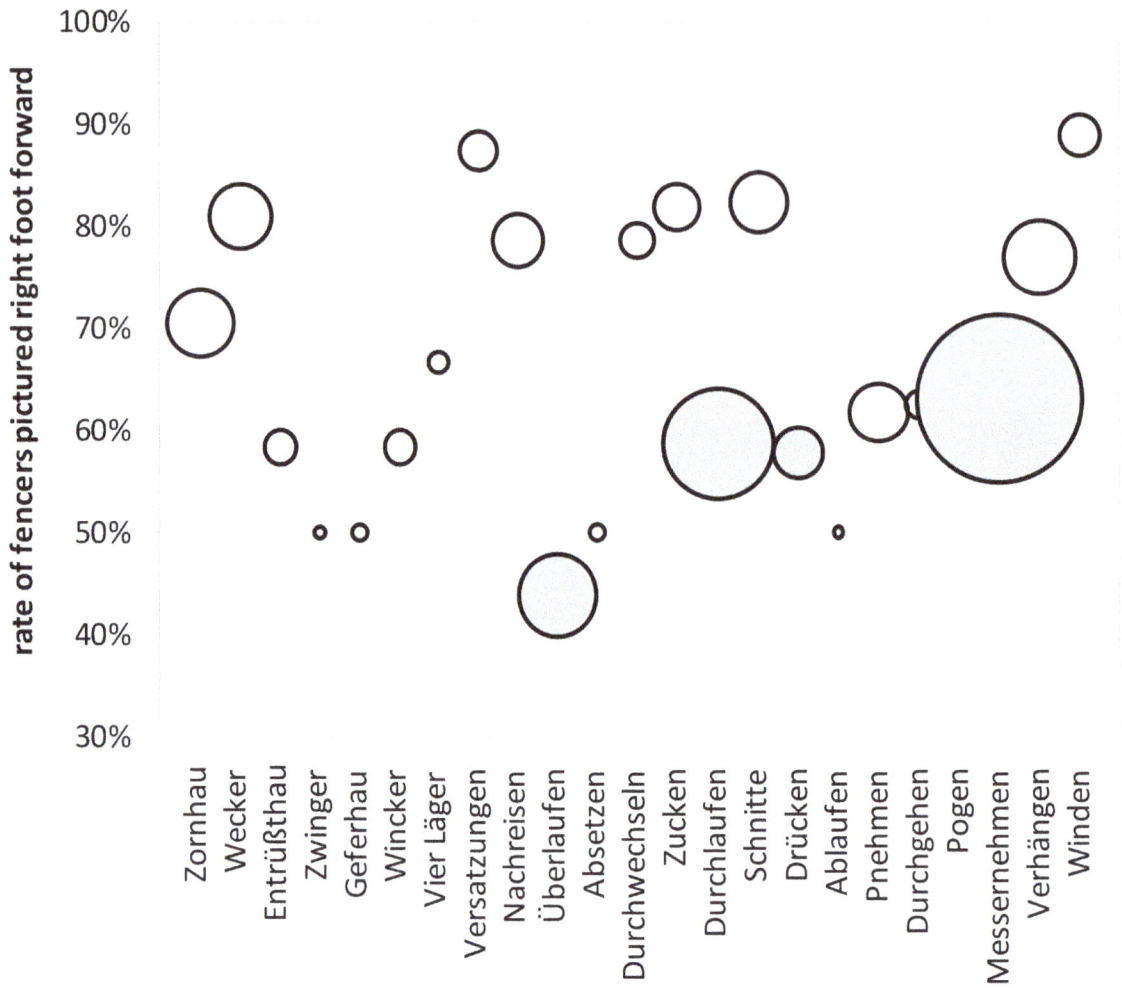

Comparison with other sources

In order to see if the rate of right foot forward ſtances in LM differs from the fencers' positions shown in other treatises, a similar count of feet was also done in 14 other manuscripts. Table 4 liſts the results sorted by weapon, while fig. 11 shows the diſtributions in chronological order.

As expected, there is a clear difference between armored fighting and fencing without. Only 28% of the fighters in harness are depicted with the right foot forward, while it's 54% in sword without armor (𝔅𝔩𝔬ß𝔣𝔢𝔠𝔥𝔱𝔢𝔫). Since the sword blade is held in the left hand in harnessed fighting, it makes sense to advance the left foot when engaging with the opponent.

Similarly, the higher rate of right feet forward is an obvious advantage when having both hands on the sword grip in 𝔅𝔩𝔬ß𝔣𝔢𝔠𝔥𝔱𝔢𝔫 and extending the right arm to maximize the reach. However, 42% of the fencers are shown with the left foot leading, so there is a relatively even diſtribution of ſtances and foot positions vary depending on the technique. There is also a wide spread of results between manuscripts. While only 39% of the long

Fig. 10: Distribution of foot positions per chapter: the bubble size represents the total number of depicted fencers in the respective chapter

manuscript	author / common name	year	fencers	foot facing opponent		
				right	left	equal
Messer						
HTM	*Hans Talhoffer-München*	1467	16	63%	31%	6%
PKM	*Paul Kal-München*	1470	14	71%	29%	0%
B	*Baumann's fight book* (part A)	1470s	16	44%	56%	0%
LM	*Hans Lecküchner-München*	1482	830	66%	31%	3%
PF	*Peter Falkner*	1495	100	63%	36%	1%
AD	*Albrecht Dürer*	1512	72	56%	44%	0%
	all manuscripts combined		**1048**	**64%**	**33%**	**3%**
Dussack						
PMD	*Paul Hektor Mair-Dresden*	1542	80	58%	35%	8%
JWA1	*Lienhart Sollinger*	1564	16	31%	56%	13%
W01	*Wolfenbüttel fight book #1*	1591	38	66%	29%	5%
	all manuscripts combined		**134**	**57%**	**36%**	**7%**
Long sword (Bloßfechten)						
B	*Baumann's fight book* (part C)	1420s	32	56%	44%	0%
HTM	*Hans Talhoffer-München*	1467	136	39%	58%	3%
PKM	*Paul Kal-München*	1470	52	65%	27%	8%
B	Codex Wallerstein (part A)	1470s	52	67%	31%	2%
PK	*Peter Falkner*	1495	64	48%	41%	11%
PKS	*Solothurner Fechtbuch*	1500s	32	63%	31%	6%
AD	*Albrecht Dürer*	1512	26	54%	46%	0%
JWA1	*Jörg Wilhalm-Augsburg #1*	1523	158	63%	37%	1%
G	*Goliath*	1535	74	58%	38%	4%
PMD	*Paul Hektor Mair-Dresden*	1542	240	50%	47%	3%
JWA1	*Lienhart Sollinger*	1564	54	44%	50%	6%
W01	*Wolfenbüttel fight book #1*	1591	72	67%	28%	6%
	all manuscripts combined		**992**	**54%**	**42%**	**4%**
Sword in harness						
B	*Baumann's fight book* (part C)	1420s	80	34%	64%	3%
GN	*Gladiatoria-New Haven*	1430s	116	21%	73%	6%
GK	*Gladiatoria-Kraków*	1440s	122	25%	73%	2%
HTK	*Hans Talhoffer-København*	1459	34	24%	65%	12%
HTM	*Hans Talhoffer-München*	1467	22	36%	64%	0%
PKM	*Paul Kal-München*	1470	56	32%	68%	0%
JWA2	*Jörg Wilhalm-Augsburg #2*	1522	96	29%	54%	17%
PMD	*Paul Hektor Mair-Dresden*	1542	172	29%	59%	12%
	all manuscripts combined		**698**	**28%**	**65%**	**7%**

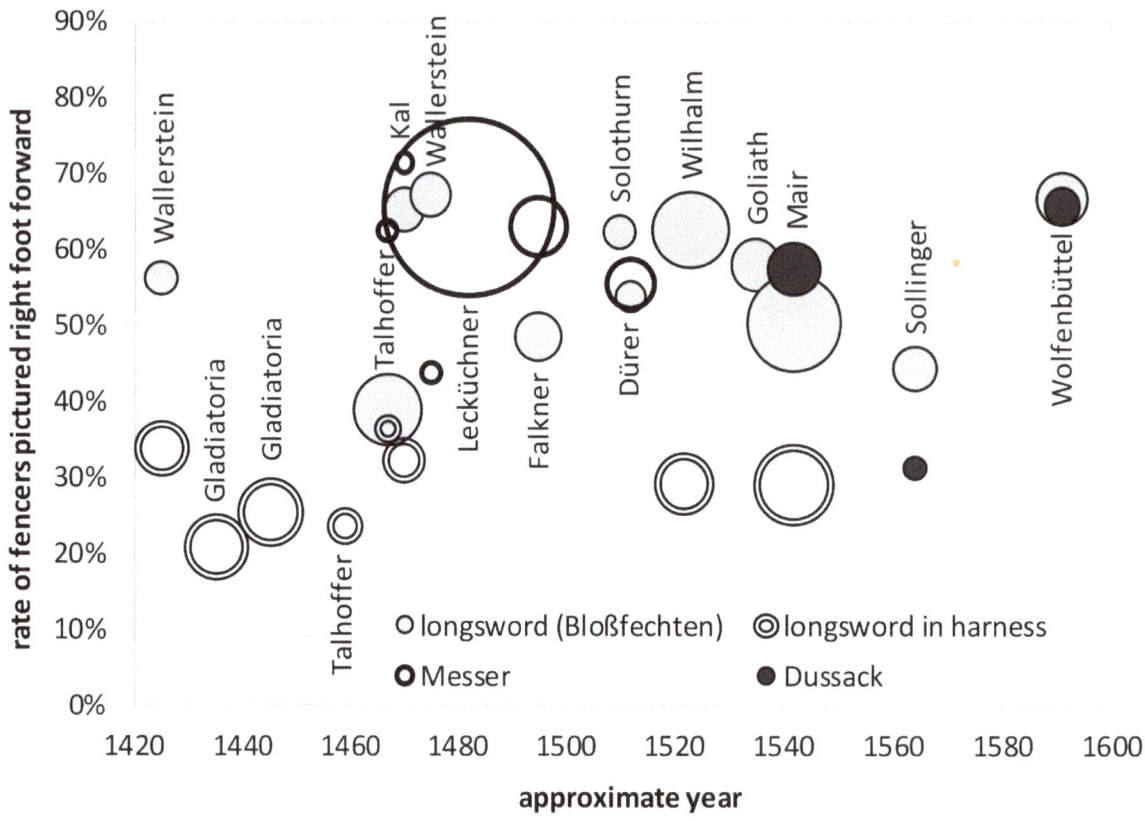

Fig. 11: Rate of right feet forward stances in chronological order: the bubble size represents the total number of depicted fencers in the respective source

Table 4 (op.): Rate of foot alignment in 15 late medieval/early modern fencing manuscripts

sword (Bloßfechten) fencers in HTM have the right foot forward, it's 67% in the commented section of B.

The Messer and the Dussack are first and foremost separated by time rather than foot position. After 1520, the Messer disappears from the treatises in favor of the Dussack. Interestingly, the right foot forward rate decreases with this development. Both weapons are held with the right hand only, but the distribution of stances in the Dussack is more similar to the long sword than to the Messer, at least in the evaluated pictures.

In Messer fencing, Lecküchner's treatise contains the bulk of the analyzed material. 830 out of the total 1048 depicted Messer fencers are shown in LM. At 66% in total, it has a very high rate of right feet forward compared to the other sources. Only the short Messer section in PKM, the commented long sword in B, and W01 from 1594 have an even higher rate. Ignoring Lecküchner's chapters that focus on wrestling rather than bladework, the rate would be even higher. But since the distinction between wrestling and bladework was not done in the other sources, further work is needed for a more detailed analysis.

Conclusion

LM contains almost as much wrestling as fencing with the blade. Even without a statistical analysis of the Liechtenauer tradition's long sword treatises, it is safe to say that wrestling has a higher importance in Lecküchner's

Messer fencing than in comparable long sword sources. The Messers shown in LM have a straight blade very similar to example (A) and don't appear to be overly long, which favors closing in and getting into wrestling distance. More than half of the book's pages are part of four chapters that focus on wrestling.

When used for bladework, the straight blade of the Messers depicted in LM enables winding techniques without giving the opponent an excess of feedback he could benefit from. Thrusts and strikes are the predominant intended results of the techniques, with a slight preference for the thrust or other techniques that threaten the opponent with the point in the end. All three evaluated aspects, the statistical evaluations of the text, the stance of fencers in the illustrations, and the properties of the depicted Messers unanimously support this result.

Particularly when disregarding the four chapters prioritizing wrestling, the depicted fencers have a strong tendency to stand with the right foot forward, more so than in comparable sword or Dussack sources. It seems like the fencers in the late 15th century already preferred this stance in practice due to the characteristics of the Messer and its one-handed use. Even though much of the text is very similar to the instructions in the Liechtenauer tradition's long sword teachings, the significant differences in the depicted stances support the Messer's role as a weapon for a distinctive, independent fencing style. So, when modern long sword fencers pick up a Messer and keep their right foot in front most of the time, there is an indication that this is true to the original way of Messer fencing rather than being a product of a different priming by modern fencing and the media.

With its sheer volume and coherence in text and illustrations, Lecküchner's manuscript is an extraordinary resource for the recreation of a fencing style that links the Middle Ages to the early modern era. There is more material on the Messer in LM than in all other sources before 1500 combined, and when the Messer is mentioned in later treatises, many of them are based on Lecküchner's work.[238]

As a weapon that is handled with one hand only and keeps the other hand free for grappling, the Messer is not only the Dussack's predecessor, but with its preference for thrusting and keeping the right foot forward, it also anticipates key elements of rapier fencing and other more modern styles.

---

[238] See Dierk HAGEDORN's contribution to this book for reference.

JESSICA FINLEY

"What's in a Name?" A Comparative Analysis of the Nomenclature of Johannes Lecküchner and Johannes Liechtenauer

Anger Cut, Waker,	Zoren |haw |vecker
Anger and Constrainer,	Entrust: hat zwinger /
Danger with the Winker	gefer mit |vincker
The Bastion, parry,	Dy pasteyn |versetz
chase, overrun, and set off,	Nachrayß |vberlauff |vnd absetz
the change pull through,	Den wechsel |durch / zuck
run through, the slicing off, press,	Lauff durch |dy abschneydt |druck
run off, snare,	Ablauff |benym
go through, Bow, take the weapon,	Durchge |pogen |dy wer nymb
hang the windings against the openings;	Heng dy winden |gen plössen
strike the strokes, learn to vex him.	Schlag dy straych: ler verdrossen

~ Hans Lecküchner, 1482

y the end of the fifteenth century, the tradition of fencing according to the teachings attributed to Johannes Liechtenauer was well established in German speaking lands, as attested by the number of surviving manuscripts which elaborate on his Zettel.[239] In the creation of this poem, Liechtenauer used a number of techniques to aid in memorization common to the medieval scholastic tradition, including rhyme, allegory, opposition, and similarity.[240] Lecküchner, as a priest trained in the university system of the latter Middle Ages,[241] would have been familiar with these techniques of memory aid. Rather than regurgitating the Zettel by rote, he built upon it, adding his own fresh ideas as he applied it to the Messer, and at the same time created his own "ornaments" and images to allow for the memorization and internalization of his contribution to the Liechtenauer tradition.

---

[239] A mnemonic poem, with rhyming couplets which allow for the memorization and internalization of the whole of the art of fencing with the sword. It includes fencing on foot out of armor (*Blossfechten*), and techniques for use in the duel both on horseback (*Rossfechten*) and on foot in armor (*Harnischfechten*).

[240] CARRUTHERS AND ZIOLKOWSKI 2004, p 117, lists the features of "ornaments", that is the images used to aid in memory, as elementary principles of mnemonics: "surprise and strangeness (for example, *metaphora*, metonymy, *allegoria*, oxymoron, and in art, grotesquery),

exaggeration (hyperbole and litotes), orderliness and pattern (chiasmus, tropes of repetition, various rhythmic and rhyming patterns), brevity (ellipsis, epitome, synecdoche, and other tropes of abbreviation) and copiousness (all tropes of amplification), similarity (similitude), opposition (paradox and antithesis) and contrast (tropes of irony). All of these characteristics are essential for making mnemonically powerful associations."

[241] FORGENG 2015, pp x–xi.

To the modern audience, this use of source material is unusual in two different ways. On one hand, a modern reader may feel that this approach smacks of plagiarism, that Lecküchner is taking another person's work and presenting it largely as his own.[242] This use of earlier texts from authoritative sources with one's own fresh additions is, however, expected by the medieval audience. In contrast to our approach, to create a work divorced from authority would be the greater offense. On the other hand, a modern reader may find the alteration of the 𝕵ettel to be offensive. However, yet again, this is in line with medieval practices, where one is expected to take *memoria* and build upon it to create new "original" contributions.[243] "Indeed, those who practiced the crafts of memory used them—as all crafts are used—to *make* new things: prayers, meditations, sermons, pictures, hymns, stories, and poems."[244] In this case, Lecküchner took the 'old craft' of the long sword to create a 'new craft' of the Messer.

As we endeavor to unravel the changes made by Lecküchner as he built upon Liechtenauer's imagery,[245] we must ground ourselves first in that inspirational *memoria*, and from there seek to understand the symbolism of the changes and additions, and lastly make hypotheses regarding the reasons these changes were made. In short, to approach an understanding we are required to analyze both masters' use of symbolic imagery, allegory, comparison, and contrast. I will be using figures to assist in understanding these concepts for those who are new to this style of memory aid. This is no small task and as such, my efforts below should be considered preliminary exploration into these topics, and not authoritative conclusions.

Sechs Verporgen Hew—Six Secret Strikes[246]

We begin in each text with a listing of strikes, which Lew and Lecküchner refer to as "secret" strikes. What is meant by "secret" is unclear, but when the anonymous glossator of Liechtenauer's Rossfechten describes the verporgen ringen ("secret wrestling"), he says that "you should not let it come to light or let any man see it".[247] Lew suggests that by performing it in front of other people you must be doing so "for the sake of

---

[242] FORGENG 2015, p xi.

[243] CARRUTHERS AND ZIOLKOWSKI 2004, p 22.

[244] CARRUTHERS AND ZIOLKOWSKI 2004, p 3.

[245] For the purposes of this article, all references to Liechtenauer's art, *Stücke*, and *Zettel* will use the version found in *Lew fight book* (L), as transcribed and translated in HAGEDORN 2017. Folio and page numbers

will be from the same. Likewise, page numbers referencing to Lecküchner's art will come from FORGENG 2015 and folio numbers and transcription will be from *Hans Lecküchner-München* (LM), which can be found on Wiktenauer.com (LORBEER 2006).

[246] LM, f 2r; FORGENG 2015, p 5.

[247] TOBLER 2010, p 141.

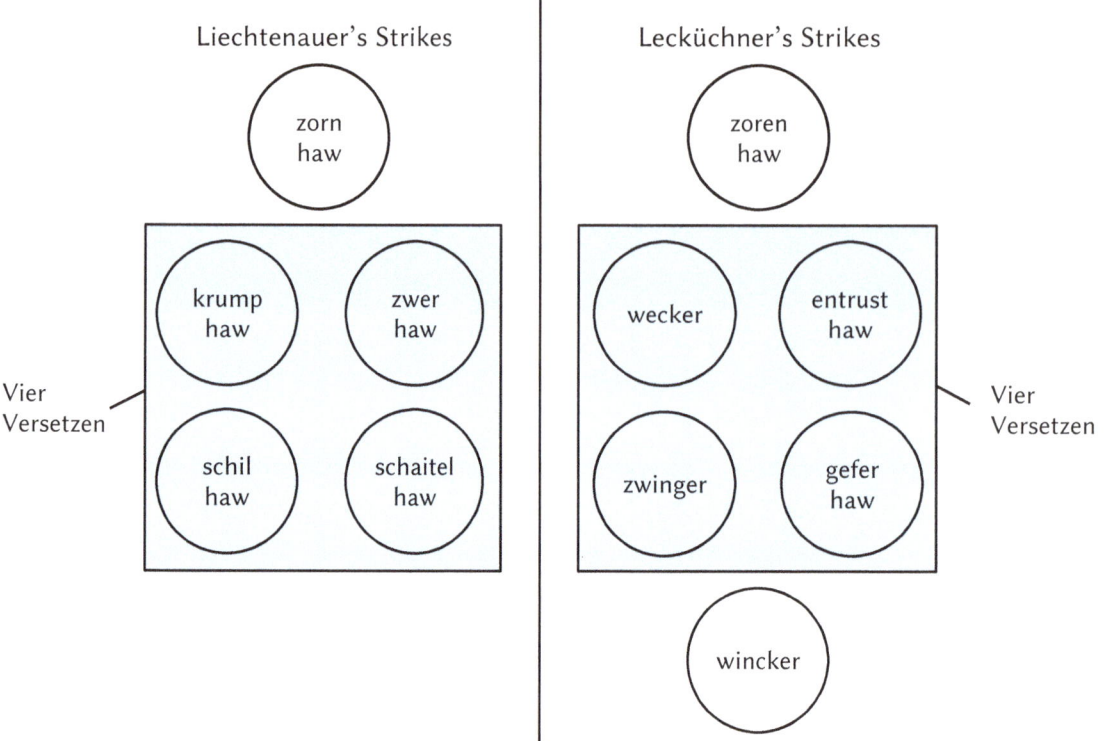

Fig. 1: A visualization
highlighting the
names and relation-
ships of the strikes

bragging" and that it "disrespects their art".[248] Surely this is not implied by Lecküchner when he uses the word "secret", as he gives examples of these in use in 𝔉𝔢𝔠𝔥𝔱𝔰𝔠𝔥𝔲𝔩𝔢𝔫, which are typically public affairs. Still, he likely wanted to keep their precise use in-house and for the study of his own students, to have a tactical advantage when fencing both in play and in earnest.

While Liechtenauer lists five strikes, Lecküchner gives us six, so we must consider the ways that numbers were perceived and used mnemonically in the Middle Ages. In the case of Liechtenauer, the strikes should be broken down into 1 + 4 (𝔷𝔬𝔯𝔫𝔥𝔞𝔴 plus 𝔙𝔦𝔢𝔯 𝔙𝔢𝔯𝔰𝔢𝔱𝔷𝔢𝔫).[249] Lecküchner's six, then, can be broken into 1 + 4 + 1 (𝔷𝔬𝔯𝔫𝔥𝔞𝔴 plus 𝔙𝔦𝔢𝔯 𝔙𝔢𝔯𝔰𝔢𝔱𝔷𝔢𝔫 plus 𝔚𝔦𝔫𝔠𝔨𝔢𝔯). He kept the 𝔷𝔬𝔯𝔫𝔥𝔞𝔴—and maintained the name—so that we understand where these come from, but then also manages to maintain a reference to 5 strikes by renaming the 𝔙𝔢𝔯𝔰𝔢𝔱𝔷𝔢𝔫 and adding the 𝔴𝔦𝔫𝔠𝔨𝔢𝔯 as a new strike. In this way he creates for himself a foundation, grounding his work solidly in both Liechtenauer's concepts (how the strikes are performed and how they counter guards) but also in the numeration. We will see that this is not the only time he adds new material but maintains a reference to the important numbers from Liechtenauer's art.

---

[248] HAGEDORN 2017, p 351.

[249] *Zornhaw* stands apart from the other four strikes as it is used solely against an incoming cut, whereas the other four each can be used against specific kinds of strikes and also against a specific guard adopted by the opponent. We will also see how the *Vier Versetzen* can be further broken down and analyzed as pairs.

As we work to understand what the intention is behind the naming conventions found in either Liechtenauer or Lecküchner it is important to understand these words not as unconnected names, but rather as a part of a mnemonic device. As such, we should expect to see connections between each name, and that these should highlight contrasts between the elements, or similarities between them, or even to provide opportunities for wordplay and punning.[250] We shall see this both internally within the meanings found within a single technique, as well as without and between techniques.

The first of the specialized strikes which are used out of armor, whether with long sword or Messer, is called the 𝔃𝔬𝔯𝔫𝔥𝔞𝔴[251] ("Wrathful Strike"). The initial strike is described with relative uniformity[252] between the two texts, and the name is indicative of the way we are supposed to strike onto the incoming sword of our opponent: wrathfully—and without any parrying. It is interesting as it is in some ways set apart from the other strikes: it is not called out as one of the 𝔙𝔦𝔢𝔯 𝔙𝔢𝔯𝔰𝔢𝔱𝔷𝔢𝔫.[253] Instead, it is making a play on words which prepares us to understand the later use of 𝔳𝔢𝔯𝔰𝔢𝔱𝔷𝔢𝔫 to be something which simultaneously is, and is not, a parrying action. This circular and self-referential pedagogical style is used extensively by Liechtenauer.

The second strike, the 𝔎𝔯𝔲𝔪𝔭 𝔥𝔞𝔴 or 𝔚𝔢𝔠𝔨𝔢𝔯, is a versatile strike against an incoming blow from the opponent which arcs over and across the top of their sword or hand.[254] 𝔎𝔯𝔲𝔪𝔭𝔥𝔞𝔲 means "crooked" in the sense of a twisting path and appears to be descriptive of the arc of the strike in space. 𝔚𝔢𝔠𝔨𝔢𝔯, however, is more figurative than physical, and means something like "something that awakens", or in modern German, "alarm clock". For Lecküchner, however, this could be referencing the town watch, whose job it would be to ring the alarm bells to wake up

---

[250] It is in this context that I provide possibilities for understanding the meaning of these words in English. I do not intend to imply that these are the only possible reads of these words. The study of medieval memory devices has shown that there are many possible reads of symbols, and that these reads are sometimes even contradictory with each other. The medieval person would not have been frustrated about this, but would simply note it as a way to encourage more connections, and thus, easier recollection of the element in question. To quote CARRUTHERS AND ZIOLKOWSKI regarding medieval mnemonic images, "Their authors did not consider their efforts to be truthful so much as helpful." Consider my efforts to be in this vein. (CARRUTHERS AND ZIOLKOWSKI 2004, p 30.)

[251] Spelling in the Middle Ages is notoriously variable. In these manuscripts this is rendered as either *zorn haw* or *zoren haw*.

[252] The initial strike is performed "wrathfully and without any parrying" whether with the long sword or with the Messer, but at this point the plays diverge as Lew tells us to shoot the point "straight toward his face or chest" while Lecküchner tells us to shoot to the face or breast while turning the Messer "toward his, so that the long edge stands upward". This "inverted hand" position is an important tactical position in many Messer techniques.

[253] *Versetzen* will be treated more extensively below.

[254] While there are many differences between the performance of the blows with long sword versus the Messer, it is beyond the scope of this article to discuss extensively the performance of the techniques in detail, and only a brief description will be included for context and clarity.

the town when threatened by an incoming army.[255] The physical mnemonic could be maintained in the arcing across and back in of the sword during the strike and its successive hitting blow, much as the striker on a bell moves back and forth.[256]

The **3wer Haw** or **Entrust Haw** is similar to the **Wecker** in that the **3wer** is used against the incoming blow from above, but comes across in a loosely horizontal direction, intercepting the opponent's strike and threatening to hit the head. **3wer** is, again, a physical description, meaning "across" in the sense of movement in opposition. In fact, the phrase **krump und 3wer** is used to mean something like "back and forth", "zig-zag", or "to and fro". This tendency of these words to be used in concert in MHD sources creates connective tissue between these two strikes, and indeed, a **Krump haw** is frequently followed by an action similar to the **3wer**

haw. Lecküchner again departs from the physical description, and instead uses **Entrust**, meaning to remove armor or to disarm. It can also mean to be 'out of sorts', to bring something in anger, or to get out of position. This implies its use against a blow thrown in anger,[257] that it removes a threat by taking it out of position, or that it can take away an opponent's protection.

The fourth strike, called alternately the **Schil haw** or the **3winger**, is an interesting strike as it comes down with the short edge of the Messer[258] after a turning motion of the hand brings it into position. **Schil** is an interesting word in its application here, as it means both "slanted" and "squint" (in the sense of 'cross-eyed'). Liechtenauer appears to play with these dual meanings by both telling us to "squint at the point" of the opponent's sword in order to mislead our opponent, implying that we will strike the sword away when we instead drive a thrust into their throat with the **Schiel**. Likewise when we are told "squint at the head" as we throw this "slanted strike" to the hands. The word **3winger** means "something that constrains" and is related physically to the space between a moat and a city wall, or to a specially-built area in a castle that would funnel an army into a smaller space

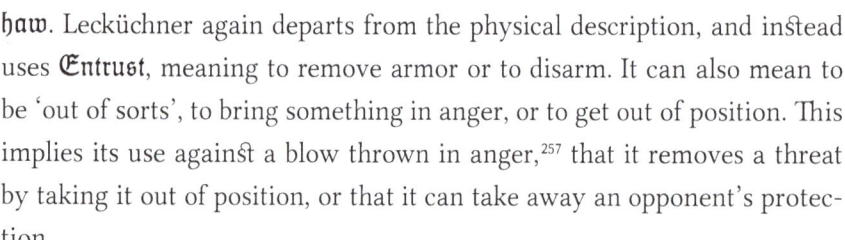

---

[255] GRIMM AND GRIMM 2004

[256] One could even use a modern English mnemonic for this, in that we can use the *Wecker* to 'ring someone's bell'.

[257] Perhaps it is a reference to the *Zornhaw*, however that would require a broader reading of the term *Zornhaw* than is implied in Lecküchner's texts.

[258] The "short edge" is sometimes referred to as the "blunt edge" in Lecküchner's text and is more commonly known in modern English as the "false edge". This is, of course, in opposition to "long edge" or "true edge".

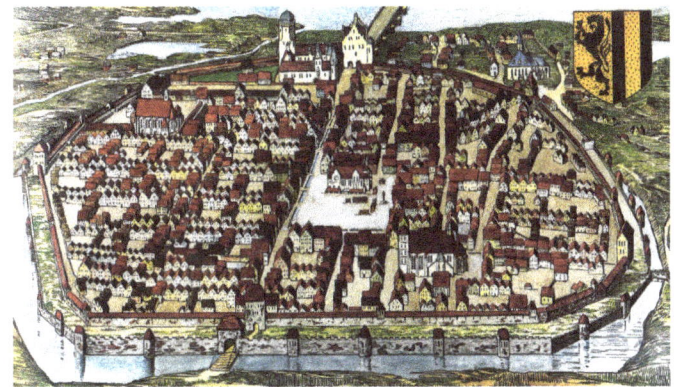

Fig. 3: The city of Dresden in 1521, surrounded by a visible *Zwinger*

where their movements can be predicted and more easily dealt with. It is also used to refer to a taskmaster or a tax collector, and in this way is more related to the idea of "something that creates pressure". 𝔷𝔴𝔦𝔫𝔤𝔢𝔫 is a concept that is core to Liechtenauer's art, and is especially glossed in the 𝔯𝔬𝔰𝔰𝔣𝔢𝔠𝔥𝔱𝔢𝔫 section where it is clear that every strike should "compel and constrain with the point".

The 𝔰𝔠𝔥𝔞𝔦𝔱𝔱𝔩𝔢𝔯 𝔥𝔞𝔴 or 𝔤𝔢𝔣𝔢𝔯 𝔥𝔞𝔴 is a strike from above to the top of the head, and is the least versatile strike, relying upon the maximization of reach afforded by the relative positions of the opponents rather than creating opposition against the sword. 𝔰𝔠𝔥𝔞𝔦𝔱𝔢𝔩 refers to a central part of the hair at the uppermost point of the head. This strike could therefore mean "something that divides vertically" and provide a mnemonic contrast to the slanting nature of the 𝔰𝔠𝔥𝔦𝔩 𝔥𝔞𝔴. 𝔤𝔢𝔣𝔢𝔯 probably refers to "danger" or "threat" and is likely a callback to Liechtenauer's verse where he says it is a "danger to the face, with its turn, a danger to the breast". This use of the 𝔷𝔢𝔱𝔱𝔢𝔩 to inspire his name creates a bookend in our mind, parenthetically breaking out 𝔷𝔬𝔯𝔢𝔫 𝔥𝔞𝔴 to 𝔤𝔢𝔣𝔢𝔯 𝔥𝔞𝔴 as Liechtenauer's.

The sixth and final strike is the 𝔚𝔦𝔫𝔠𝔨𝔢𝔯, and while it is not found in Liechtenauer's long sword 𝔷𝔢𝔱𝔱𝔢𝔩, it is performed similarly to the 𝔰𝔱𝔲𝔯𝔱𝔷 𝔥𝔞𝔲[259] from earlier treatises by masters within and without the Liechtenauer tradition. This strike, like the 𝔷𝔴𝔦𝔫𝔤𝔢𝔯, is a short edge strike from the dominant side; however, to perform this strike, the sword hand turns outward to the side from which one is striking. When Lecküchner uses this word there are two possible meanings we can use to understand the imagery he may be calling to our mind's eye. The first is that 𝔚𝔦𝔫𝔠𝔨𝔢𝔯 can have the sense of "something that moves side to side" and in that way be similar to a wave with the hand. This certainly evokes the gross physical movement of the strike as it is typically done twice, one from each side. Additionally, 𝔚𝔦𝔫𝔠𝔨𝔢𝔯 also has the sense of "to give a sign toward something by moving your eyes or hands" or "a provocation". So this, then, encodes a dual meaning: both of moving the hands to direct the opponent's attention and the tactical meaning and use of the 𝔚𝔦𝔫𝔠𝔨𝔢𝔯, which is as a provocation in one direction, presuming a strike to the other.

---

[259] *Sturtzhau* or "Plunging strike" is frequently seen used in sword and buckler, including in Andres Lignitzer's treatise found in the *Peter von Danzig fight book* (PD), f. 80v. (TOBLER 2010, p166.)

Focusing on Lecküchner's mnemonics between techniques, we can immediately see that we have three strikes named "X ḥaw" and three strikes that end in "er"[260] indicating the root verb has become a noun.

Considering the **Zoren**, **Entrust**, and **Gefer**, each of these are strikes that work in a simple trajectory: **Zoren** downward against the incoming strike, **Entrust** across against the incoming strike, and **Gefer** downward to the head. On the other hand, **Wecker**, **Zwinger**, and **Wincker** all have compound elements: **Wecker** crossing to strike the blade and follow up with a secondary technique to the opponent, **Zwinger** turning to cover the far side with the short edge and strike or thrust, and **Wincker** turning to provoke with the short edge, dropping short, and striking from a second **Winker** to the other side. In this way, the order in which these are presented can be seen as a contrast of simple and complex, and this transition back and forth between straight and crooked can even be extrapolated out into a play between 'preventing' and 'provoking' or 'inviting'. Of course, many more imaginative connections can be made by each individual fencer, but this alone highlights the potential usefulness of these mnemonic figures.

Sibenzehen Haubtstuck—Seventeen Chief Techniques[261]

The remaining techniques largely require less explanation, many of them being words which are specifically fencing jargon.[262] However, there are some interesting additions by Lecküchner when compared to Liechtenauer. As with the strikes, what stands out here is the numerical play at work. Liechtenauer has 5 strikes and 12 **Stücke**, totaling 17 **Hauptstücke**. Lecküchner, however, has 6 strikes and 17 **stuck** for a total of 23 **Haubtstuck**.

When he added a strike, Lecküchner's number of techniques would no longer be in alignment with Liechtenauer, creating a problem. However, it is one that he could solve by adding five new **Stücke** to the list, so that he could make the statement that there are seventeen, and thus find himself again in agreement with the earlier tradition. We cannot know if he first wished to add to the list of strikes, necessitating addition to the rest of the **Stücke** to create balance, or if he chose to add **Stücke** and thus needed the new strike. Either way, considering he easily could have added or deleted any number of items, the fact that he stuck with numbers known to be important to the textual tradition seems intentional.

The first of the **Stücke** are the **Vier Leger**. **Leger** implies a number of compelling ideas, all of which have a sense of a "place to cease movement and which provides protection". This could be the lair of a prey animal, such as

---

[260] This is not consistent across later treatises and so should be treated with caution as a data point.

[261] LM, f 2r; FORGENG 2015, p 5.

[262] This is indicated in many MHD dictionaries, including LEXER 1992, HENNING 2014, and GRIMM AND GRIMM 2004.

Fig. 4: The encampment of an army besieging a city (*Luzerner Chronik* by Diebold Schilling, Korporation Luzern ms. S 23 fol., p 218)

a rabbit's warren, a fox's den, or the tall grass within which a deer lies down for shelter.[263] It also can refer to an army's camp, or even to the army while it is at camp.[264] It is possible that this word can refer to very temporary positions of defense, such as when wagons are circled around to provide temporary shelter, as well as a longer-term camp.[265]

Of course, in the context of fencing, a Leger is a temporary guard from which one enters the fight. Both of these 'modes' of thinking are useful, because they tell the story of temporary protection and shelter and of the need for preparation for fight, flight, or both.

Liechtenauer lists four Leger which he names Ochs, Pflug, Alber, and Vom Tag. Likewise, Lecküchner has four main guards, called Stier, Eber, Bastei, and Luginsland.[266]

Ochs means the same as its English counterpart Ox, and when one assumes the guard, their point hangs toward the face from either side of the body. Importantly, this domesticated animal was always worked and trained as a pair, often 4 pairs hitched to a single plow at a time. This 'paired nature' helps create an immediate expectation that a guard named Ochs should have two that work together.

Lecküchner's Stier, however, isn't an animal used for its work, but rather, is used for meat. As the citizenry of fifteenth century Nuremberg became more prosperous, the demand for large quantities of meat grew, and perhaps his use of Stier indicates its importance.[267] Of course, the Steer has two horns, and this image is evocative, for when these animals attack, after they hang their horns down, they tend to throw either one or the other forward.

[263] DALBY 1965, p 132.

[264] GRIMM AND GRIMM 2004.

[265] See OMAN 1885 regarding Hussite tactics called *"laager"* or *"laeger"*. Particular thanks to Sigmund WERNDORF for alerting me to this usage of a related word.

[266] Special thanks must be given to Oskar TER MORS for first making me understand the connections that these words could have, not just as general terms, but terms that were specific and relevant to Nuremberg. His work was a major source of inspiration for my current thoughts on this nomenclature. (TER MORS 2021)

[267] BLANCHARD 1986, pp 427–460.

Liechtenauer's 𝕻flug, or "Plow", is an obvious companion to the Oxen above. The plowshare cuts the earth, while its moldboard turns the soil over to the right, and so as it tills a field, it works by setting aside the earth. There is a pun in this idea, as it is from one 𝕻flug that we 𝕬bseʒ ("set off") an opponent's thrust, moving into 𝕻flug on the other side and threatening our own thrust. Lecküchner's 𝕰ber, or "Boar", appears to refer specifically to the domesticated male meat pig[268] and provides a connection to the 𝕾tier as they are the same class of animal. Of course, the boar has two tusks, one on each side, and these point upward.[269]

𝕬lber is probably best thought of as indicating the quality of "cunning". 𝕬lber is a word that changes meaning multiple times from the 13th century to present, transforming from meaning "simple-minded" in its earliest renditions, to "cunning" during the latter Middle Ages, and finally to "silly" today. This follows the many facets of the "Fool" in medieval German literature.[270] Lecküchner's 𝕭astei refers to a bastion, which is a protruding block from which a defender can have an unobstructed view to attack the invaders. This word has connotations of a wide range of defense, and of a strong foundation from which to defend oneself.

Finally, Liechtenauer's last guard is 𝖁om 𝕿ag, which translates, frustratingly, to "From the Day" in English, but carries the meanings of something that is understandable, true, or proven. A similar English idiom might be 'above board', and anything 𝖁om 𝕿ag is the opposite of something which is buried or concealed. Another way to read this name is 𝖁om 𝕯ach, meaning "from the Roof". Both "roof" and "day" physically indicate "above the ground" or "highest point" and as such would be related to the heavens in medieval thought. 𝕿ag might indicate the sun shining its rays of light down upon those below, much as a strike coming from this guard will rain down upon the opponent.

Lecküchner's 𝕷uginsland means literally "Look out over the land" and is a watchtower from which you can oversee all of the area you need dominion over. There still stands in Nuremberg the 𝕷uginsland which was built

---

[268] The wild boar in MHD is, interestingly, called *Hauer,* and while it is still possible that *Eber* in this case refers to the wild animal, that would seem at odds with the domesticated *Stier.* (LEXER 1992, p 47.)

[269] This also can lead to a dirty joke, as *Eberzahn* was a name given to a medieval fool, teeth were always considered to be references to "horniness", and this is the "rising thrust".

[270] The stories of the *Fastnachtspiel* tradition are a fascinating look into the variety of ways the Fool could present in the German Middle Ages. Sometimes they were a 'natural fool', meaning someone with a learning disability who was expected to tell the truth because of their nature. Other times they were a 'witty fool', meaning that they were someone who was smarter than everyone else and told the truth through pun and innuendo. Either way, fools had a tendency to turn the world upside-down to their forthright counterparts. See one of the many versions of "The Dialogue of Solomon and Marcolf" for a particularly striking example of the opposition of the 'Upright and truthful Solomon' against the 'Base and honest Marcolf' in the form of a battle of wits.

Fig. 5: The Nürnberg *Luginsland* (photograph taken between 1890 and 1905, LC)

by the city in 1377, which overlooks not only the city itself but also the noble courtyard in the 𝕭urg.[271]

Connecting the idea of the 𝕷uginsland with the 𝕭aſtei, we begin to see the importance of imagery around city defense,[272] and this is further supported by Lecküchner's description of the areas where you can hit your opponent. Liechtenauer calls these 𝕭loßen, which carries the connotation of an exposed area.[273] This could be nakedness of the human body, a tree without leaves, or even a clearing in a forest where an animal could be driven.[274] In this way, we see loosely connected imagery about 𝕷eger, where prey animals are protected rather than being in a clearing where they are not.

Lecküchner calls these exposed areas 𝕴ynnen,[275] which are "battlements": the top of the walls which are surrounding the town. This word carries the sense of alternating peaks and valleys, which the word "crenulation" in English probably fits best. He says that there are four 𝕴ynnen, two above the belt and two below, and describes them both in his section on the 𝕴oren ḥaw and in the 𝕰ntruſt ḥaw. Liechtenauer names these openings 𝕺chs and 𝕻flug after the 𝕷eger that protects each of them, but Lecküchner declines to do the same, though there is no obvious reason why he should not. Instead, he simply calls them "den ʒynnen". Of course, it is over these battlements that the enemy will attempt to enter the town during a siege, using all of the specialized equipment which can be applied to such a problem.

The 𝕍ier 𝕍erſetzen are the four specialized strikes mentioned previously and are used to "break" the 𝕍ier 𝕷eger. Liechtenauer clarifies that 𝕶rump ḥaw breaks 𝕺chs, 𝕴wer ḥaw breaks 𝕍om 𝕿ag, 𝕾chil ḥaw breaks 𝕻flug, and 𝕾chaittler ḥaw breaks 𝕬lber. Likewise, Lecküchner maintains this structure, with the corresponding strike countering the corresponding guard.[276]

---

[271] Nuremberg's *Luginsland* can be seen in context here:
https://www.kaiserburg-nuernberg.de/bilder/burg/gesamtplan.pdf

[272] Thanks to Oskar TER MORS.

[273] When we fight out of armor, Liechtenauer calls it *Blossfechten*, or "fighting while exposed". In armor, he describes the *Bloßen* as being any exposed area such as the eyeslots of the helmet or the palms of the hands.

[274] LEXER 1992, p 24.

[275] LEXER 1992, p 336.

[276] A fun way to play with the imagery is to create a story from these meanings: *Wecker* ("town guard") breaks the *Stier* ("steer") and has a nice dinner.

Fig. 6: A hare is unharbored and driven into a trap (BL MS Yates Thompson 13, f 70v, 2nd quarter 14th c.)

Verſetzen, however, is its own intereſting word, particularly when it comes to creating imagery and internal puns. If we are thinking of hunting as an allegory for the fight, then we can call up **Verſetzen** as driving an animal from its den, that is out of its **Leger**, by sending a weasel in a rabbit's warren or sending dogs to where the deer are laying down. Once the animal is "unharbored" then it can be pursued, driven to a clearing where it has no cover, and dispatched.[277] **Verſetzen** also could mean to make an opposing army leave camp, to displace them, again, with the goal of attacking the newly exposed troops. These ideas share a commonality in that what is "broken" is the protection, but not the opposition itself.

Of course, **Verſetzen** also means "displace" in the sense of a fencing parry with a sword. And so it is that Liechtenauer plays on this pun and discusses how to parry correctly (with the sword point before the opponent), and how not to parry (by focusing on the opponent's sword and letting your own sword go wide to the side or with the point high). Likewise, Lecküchner admonishes us to "Guard yourself againſt parrying", a beautifully pointed pun, as the word **Hütten** is a synonym for, and often used interchangeably with, **Leger**, and has a much ſtronger implication of being hidden or under cover. He goes on to say "…you shall not parry too much, if you do not wish to be shamed and hit…"[278]

Nachreyſen holds the meaning of chasing or pursuit, but particularly with the idea of catching up as you chase. This is important for your mental map of this element of the art, because it is easy to think of moving at the same time or in concert with the opponent when you think of chasing, but this is not the whole idea of **Nachreyſen**. If you have unharbored your opponent with a **Verſetzen** then you want to pursue them to their **Bloßen** or **Zynnen**.

---

[277] See *Nachreyſen.*   [278] FORGENG 2015, p 70.

An interesting deviation that Lecküchner makes in this section is that he does not place Jndes here in the middle of Nachreysen as was done by Liechtenauer, but instead places it at the very end of his treatise. Of course, the places of importance in any work are the beginning, the middle, and the end.[279] In some ways, the move of this gloss to the end of his art highlights its significance; however it does eliminate a useful pun that Liechtenauer seems to make that Jndes is central to the art.

Uberlauffen's primary meaning is that of overtopping an obstacle, such as water running over a dam or pouring from a pitcher. This certainly has a relationship to the fencing actions of Liechtenauer's Uber-lauffen; however, it also carries the meaning of launching an attack, particularly with surprise.[280] This is relevant to Liechtenauer's use of the word, which is primarily about attacking a higher opening than the opponent to create a geometric advantage and his glossers even suggest that one should ignore low attacks and simply hit by shooting the point at an upper opening. Lecküchner includes this older idea, but then includes a winding motion with the hilt flowing around and over the opponent's sword, picking up on the primary meaning.

After this point, we shall see that many of the words used in both masters' verses are less evocative symbolically, and instead are straightforward. Whether this is due to the fact that more allegorical meanings have yet to be understood, or if this is an intentional shift by the authors to indicate that these techniques are also more narrow in scope rather than being conceptual is yet to be determined. However simple, a brief overview of the meanings of the next four techniques where the two masters share language is still a useful inclusion.

When working with the word Absetzen,[281] one finds it is typically related to removing something and putting it down, such as a hat from one's head, or a rider being thrown from the back of a horse. It also is noted as a fencing term for preventing or stopping an incoming blow from the opponent, and as such we can use the imagery of shaking off their attack and sending their blow elsewhere. It is interesting to observe that the techniques of Absetzen instruct the fencer to "create an exposed area" which invites the opponent to launch an attack, which you can then set aside.[282]

Durch wechseln carries a connotation of giving up one thing to get another. It is noted in GRIMM as being a word particular to fencing, and

---

[279] For more on the importance of these points, see DE MACHAUT 1994, pp 45–46.

[280] GRIMM AND GRIMM 2004.

[281] Ibid.

[282] Special thanks to Nathan WESTON for sharing his observation of this use of exposure to invite attack as the common thread in the use of the term *Absetzen* in both the *Absetzen* chapter as well as in the *Krumphau* chapter of the glosses in the *Sigmund Ainringck fight book* (SR).

this meaning is sensible as it is described as threatening an attack on one side, and when the opponent moves to intercept it, acting by going around and through their parry to another opening. In other words, the precipitating action which allows for performing 𝔇urᴄh weᴄhseln is provoking a parry, as compared to the 𝔄bseᴛen which is preceded by an inviting exposure.

𝔷uᴄken is a word which means "pulling", but in particular a strong and fast pull, rather than a long and steady pull.[283] It is also used when describing snatching or stealing something from someone else, and both medieval masters indicated that this fencing technique is to be used against opponents who want to parry the sword and that it will trick or fool them. This physical action is an abrupt retraction of the sword backward and is followed by a blow returning inward, but unlike 𝔇urᴄh weᴄhseln, it does not assume or require changing sides of the opponent's sword. Lecküchner expands the use of the word to refer to pulling the leg back to slip out of an attack to the lower half of the body, as well as pulling the body aside to avoid a blow.

𝔇urᴄh laufen, or "running through", is again indicated by GRIMM to be a fencing term, but it can also refer to crossing a room or passing through a space, such as when celestial bodies cross the heavens. 𝔇urᴄhlaufen is to be used in fencing against opponents who run in at you and intend to crowd you with strength and seems to refer to running through that pressure as they come close in order to have access to grip their body or arms (or even their sword, for a disarm). Whereas Liechtenauer's gloss of this element tends to express a tight system of eight throws and two disarms, Lecküchner greatly expands upon this and pulls in all of the wrestling techniques shown in the ℜossfeᴄhten and 𝔥arnisᴄhfeᴄhten glosses,[284] as well as adding a few unique techniques. It is in this section that we can see most clearly that Lecküchner was not limiting his scope to only Liechtenauer's 𝔅lossfeᴄhten, but was pulling inspiration for his Messer techniques from the whole 𝔷ettel.[285] In use, then, both 𝔷uᴄken and 𝔇urᴄhlauffen involve getting free from pressure that the opponent gives, but one involves withdrawal and the other advances.

---

[283] The related word *ziehen* is found in many medieval wrestling manuscripts and would seem to imply a more steady pulling motion rather than the abrupt nature of *zucken*.

[284] Of course, when these techniques are applied to the Messer, Lecküchner modifies them where necessary to make them relevant to work on foot or work out of armor.

[285] In particular, we should consider *der Ungenannt* ("the unnamed") (LM, f 90v; FORGENG 2015, p 181) which

is taken from the previously mentioned *Verporgen Ringen* (HAGEDORN 2017, pp 350–51). The *sonnenzaigen* or "sun pointer" (Munich 582, f 111r; FORGENG 2015, p 222; HAGEDORN 2017, pp 344–49) in another example, which has a number of counters in Lew that are recounted and added to by Lecküchner. It will require future research to create a full accounting of the connections between the techniques in Lecküchner's treatise and other treatises in the Liechtenauer tradition.

Abschneiden means to "slice off" and has the connotation of separating or severing something. It can also mean to harvest grain, which is evocative as it immediately reminds one of Liechtenauer's plowing imagery.[286] Lecküchner tells us that **Abschneiden** (slices against the arms or hands of the opponent) should be used against opponents who tend to withdraw from a bind and strike around to another side,[287] and this should immediately remind us of **Zucken**. Lew glosses an additional idea which appears to be punning on the idea of "slicing off" but instead directs us to "slice away" from the sword when an opponent binds on top of our sword, but as we have seen before, Lecküchner does not include these techniques in this section.

The last technique in this section of shared ideas is **Hent Drucken** which means "pressing the hands"—a particularly benign sounding name for a gruesome technique. We are told to use it against opponents that free themselves from the bind and run in at you, and as such is likely recalling our thoughts to the setup of **Durchlauffen**. It is a technique that uses both the **Unterschnitt** and the **Oberschnitt** (referenced in **Abschneiden**) as a way to maintain control over the opponent and their weapon despite their pressure. This is a limited section in Lew, but Lecküchner adds a number of counters and related techniques such as the **Sunnen Zaigen**.[288] The proximity of **Abschneiden** and **Hent Drücken**, as well as the overlapping nature of their performance, is indicative of a mnemonic technique of "logical progression". While the triggers to perform these (pulling free from the bind to come to the other side) and the initial slice are initially the same, the choice to abandon a singular technique and move to a more complex one depends on the development of the engagement in the moment.[289]

Ablaufen is the first of the additional **Haubtstuck** added by Lecküchner and means "run out" and tends to be applied to the idea of thread running off of a spool.[290] Indeed, Lecküchner expands on this and says that you perform it by "...roll[ing] it like the women when they reel off yarn on both sides..." and this instruction is evocative of the use of a 'niddy-noddy', which is moved in a figure-8 pattern to wind thread into

---

[286] LEXER 199r, p 201.

[287] As the opponent returns in, they will tend to do so in either by cutting downward with their hands, in which case we should take a position like *Ochs/Stier* to slice the underside of their arms, or they may be cutting across with a *Zwer haw/Entrusthaw*, in which case adopting a position like *Pflug/Eber* allows us to slice over the top of their arms.

[288] The *Sunnen Zaigen*, or "sun pointer" is an important technique in *Rossfechten*. Lecküchner adapts it

to Messer on foot, and after its inclusion into his treatise, it begins to show up on foot with long sword in later treatises, and even is applied to dueling with long shields (see TOBLER 2011).

[289] This is a common theme in Liechtenauer's art, and is the idea of using *Fühlen* ("feeling"), that is as Lecküchner says "working... based on the property of the things that are executed against you..." (FORGENG 2015, p 432.)

[290] GRIMM AND GRIMM 2004.

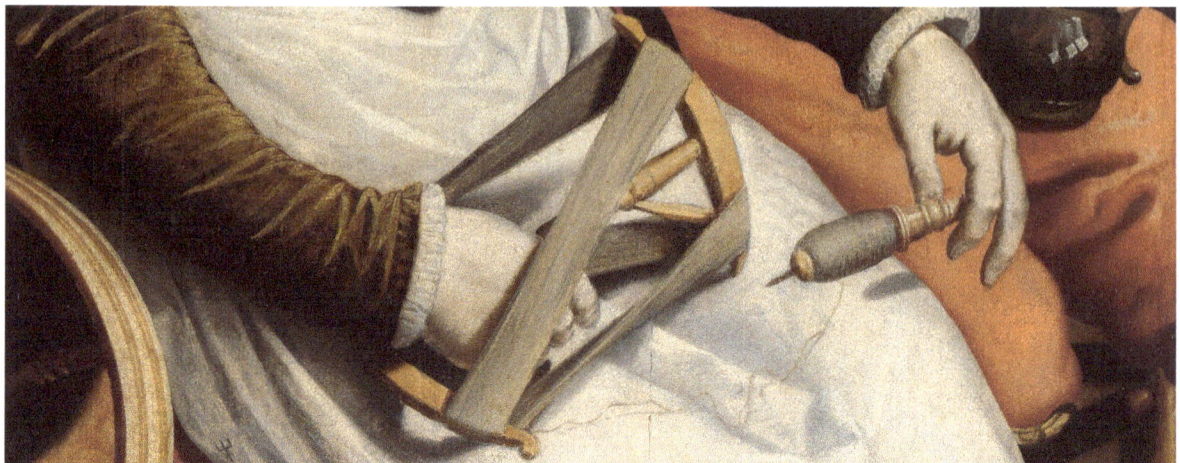

Fig. 7: A niddy-noddy (*Man en vrouw bij een spinnewiel* (detail) by Pieter Pietersz, Rijksmuseum SK-A-3962, 1560s)

a skein.[291] The fencing action is a way to free oneself from the bind without using 𝕯urch wechselen or 𝖅ucken, instead turning the hand and rolling the wrist to be able to get free and strike to the other side.

𝕻nemen appears in HENNIG[292] as a synonym to 𝖅ucken in some ways, carrying a meaning of snatching or robbing, but also means taking something away,[293] driving out, shutting out, or preventing.[294] I expect that this is again a punny meaning, in that we should think of both the fact that we are robbing them of their hanging point, driving them out and preventing their threatened attack, but then afterward pulling away and striking them out of a bind.

While 𝕯urchgen means most simply "go through", in this case it is used differently than it is seen elsewhere in the earlier Liechtenauer tradition where this is a wrestling technique similar to a fireman's carry.[295] Lecküchner uses this word to refer to an action of turning the hand so that the blade can drop down and go through with a cut to the other side. If 𝕬balufen is a wheeling movement of the sword hilt, 𝕯urchgen can be thought of as a wheeling movement of the sword blade around to hit to the other side.

𝕭ogen or "Bowing" is described by Lecküchner as a setup for disarms and other techniques. It involves rising up into a position similar to a hanging parry, with the point below the hand and the hand held high. 𝕭ogen as a verb means to form or move in an arch, but also to spring away from blood and wounds, or to make a bow. In the earlier German sword traditions, the vidilbogen[296] is a position in which the sword is held over the arm

---

[291] FORGENG also notes this meaning in a footnote. It is interesting to note this inclusion of cottage industry symbolism, perhaps more suited to the city life of a man in Nuremberg than the pastoral symbols of Liechtenauer. (LM, f 117v; FORGENG 2015, p 235.)

[292] HENNIG 2014, p 26.

[293] SAYCE 1967, p 282.

[294] FORGENG translated this word as "snaring", a choice he does not footnote but could derive from a more modern meaning of "encompassing", though this seems to be related to mental work of coming to a conclusion rather than a physical capture.

[295] In this case you are "going through" the arms of the opponent, rather than through the sword. *Hans Talhoffer-Gotha* (HTG), f 129v.

[296] *Walpurgis Fechtbuch* (F1), f 22r.

Fig. 8: The fiddle-bow position and a disarm (Γ1, † 22r)

in the manner of a fiddle bow, and many have assumed connections between these techniques. This may be so; however, it seems also to be referring to both forming an arch, and the large leap that is always described as accompanying the returning blow.

𝕸𝖊𝖘𝖘𝖊𝖗 𝕹𝖊𝖍𝖒𝖊𝖓 is "taking the Messer" and is the simple-to-understand concept of taking the opponent's sword from them. Lecküchner not only pulls from Liechtenauer's disarming techniques,[297] but also from unarmed techniques against a dagger attack, and then again expands into techniques at the half-sword[298] which appear to be inspired by 𝕳𝖆𝖗𝖓𝖎𝖘𝖈𝖍𝖋𝖊𝖈𝖍𝖙𝖊𝖓 techniques. He includes not only disarms here but also serpentine wraps around and through the arms that are similar to ones found in armored sword and poleaxe treatises.

He seems to use this entry into these other techniques as a segue into a large section which appears to be unconnected to 𝕸𝖊𝖘𝖘𝖊𝖗 𝖓𝖊𝖍𝖒𝖊𝖓, and with this he deviates quite wildly from earlier treatises whose inclusion of techniques within the heading of a particular 𝕳𝖆𝖚𝖇𝖙𝖘𝖙𝖚𝖈𝖐 appears to always indicate that the techniques are related to it in some way. However, Lecküchner simply appears to have a lot of material he wanted to adapt to the Messer (or perhaps to create out of whole cloth) and did not have anywhere else to put it but at the end of his five extra 𝕳𝖆𝖚𝖇𝖙𝖘𝖙ü𝖈𝖐𝖊. In some ways, this is similar to other fight books that simply lump extra material at the end with a heading which reads something like "other good techniques". At this point, however, after his deviation, he ends the

---

[297] These techniques are found in the *Durchlauffen* sections of various glosses.

[298] This refers to a specialized gripping of the sword with one hand on the hilt, and another gripping the blade. This allows for use of the sword in closer range, with the ability to parry powerfully, and, when necessary, as a lever for throws or the breaking of limbs.

book by returning to the final two elements of Liechtenauer's 𝔷ettel, which we are told work together.

𝔥engen or "hanging" has diverse usage in MHD sources and is a versatile word, making it hard to know precisely which meaning we should take from it. It means, of course, to hang, but is more specific in that it can mean to let free of a horse's reins, or to let a hunting dog slip free from its leash.[299] It can also refer to allowing the tracking-hound to freely follow the trail of a prey animal. It carries connotations of extension, of rushing forward, and of driving behind something.

Fig. 9: Hunting dogs, leashed and unleashed (HTG, f 30v, 1448)

Liechtenauer describes the 𝔥engen as being 𝔒c𝔥𝔰 and 𝔓flug on both sides and tells us we should use them with winding, but gives very little further detail. Lecküchner also describes these as 𝔖tier and 𝔈ber on both sides but goes on further to outline techniques from 𝔅a𝔰tei and 𝔏ugin𝔰land where you threaten the opponent with the point, despite initiating these thrusts from point-retracted guards.

The final element of both Liechtenauer's and Lecküchner's art is 𝔚inden or "winding". Winding, of course, brings to mind a "winding trail" in English, but in MHD seems to more frequently be about spiraling actions such as winding up a spool, or a falcon circling above you in the sky,[300] rotating or turning, and most importantly for our punning purposes, it means "finding the end", or as we would say in English "wrapping up".[301] At the sword, winding is described as a movement turning from one 𝔒c𝔥𝔰 or 𝔖tier to another, and back. Lecküchner says "If you wish to deceive him when you lie underneath, wind left, go through in winding, learn to find cuts, thrusts, slices."[302] While Lew's gloss reads as though there is a 'back and forth' component to winding, Lecküchner leaves those windings in the 𝔥engen chapter and creates entirely new windings using 𝔡urc𝔥gen in combination with winding movements and creating spiraling actions. So with this we see that there are two (or really three) meanings that provide us with our puns: to turn, to wind around, and to reach the end. It makes sense that Lew doesn't

---

[299] Lexer 1992, p 86.

[300] Thiebaux 1974, p 170.

[301] As Liechtenauer uses *Ochs* and *Pflug* to wind, and frequently refers to winding as their *Arbeit* ("work", "effort"), which brings again a mental association with plowing a field to the idea of winding. Interestingly, the agricultural technology of the latter Middle Ages

resulted in plows spiraling outward in a clockwise direction along a strip of land, and forming a 'ridge' in the center and deep furrows on each side. This is quite different to the modern expectation of plows that move back and forth. (Alcántara 2017.)

[302] LM, f 211v; Forgeng 2015, p 422.

title the chapter head "winding" but instead 𝕯𝖎𝖊 𝖇𝖊𝖘𝖑𝖎𝖊𝖘𝖘𝖚𝖓𝖌 𝖉𝖊𝖗 𝖓𝖊𝖜𝖊𝖓 𝖟𝖊𝖙𝖙𝖊𝖑𝖓, or "The Conclusion of the New Notes".[303]

Conclusion

Lecküchner, in his 1482 manuscript, maintains continuity with the earlier traditions of Johannes Liechtenauer, while also building his own nomenclature and adding new 𝕳𝖆𝖚𝖇𝖙𝖘𝖙𝖚𝖈𝖐𝖊, and thereby establishing a new direction for the material. While he relies on the basic structure of the 𝕭𝖑𝖔𝖘𝖘𝖋𝖊𝖈𝖍𝖙𝖊𝖓 𝖅𝖊𝖙𝖙𝖊𝖑, he delves into the whole art of fencing to pull his techniques for the Messer, applying these principles to this different weapon in a novel way.

Despite these forays into different areas of the art, Lecküchner grounds his fresh conceptualization in Liechtenauer's ideas. While creating new techniques, he maintains important numbers within the treatise, including 17 𝕳𝖆𝖚𝖇𝖙𝖘𝖙𝖚𝖈𝖐 referencing Liechtenauer's 17 𝕳𝖆𝖚𝖕𝖙𝖘𝖙𝖚̈𝖈𝖐𝖊. He begins and ends his treatise as Liechtenauer did, starting with 𝖅𝖔𝖗𝖓𝖍𝖆𝖚 and ending with 𝖂𝖎𝖓𝖉𝖊𝖓.

The major shifts in theme with his naming conventions is seen in his six strikes and four guards, where he seems to shift from Liechtenauer's allegories of hunting, which would have great appeal to the nobility for whom those treatises were written, and instead applies language which brings to mind city defense and the interests of citizens rather than courtly concerns. (A matter for further research is whether or not this supposition of a shift in audience is accurate.)

While it is clear that Lecküchner is grounded in the scholastic tradition of the latter Middle Ages, examination of the specific nature of his work's relationship to this tradition has only begun. This preliminary work into the mnemonic and pedagogical thrust of Lecküchner's treatise will need future research to confirm these findings, to compare his application across his various treatises, and to determine whether other areas of his treatise not discussed here follow similar logic.

---

[303] HAGEDORN 2017, pp 192–193.

Dierk Hagedorn

Many Magnificent Messer Manuscripts (Plus Plentiful Picturesque Prints)

The distinguished art of fencing was not created by mediocre nor insignificant men, and not without purpose; on the contrary, it was created by great men of sharp intellect, as a means to teach martial exercises.

~ Hans Lecküchner, 1478

quom presignis dimicandi ars non a mediocribus nec parui momenti hominibus, haudquoque ab re, sed per magnificos acutissimi ingenii viros, et ut militaria per eam exercitia comprobentur inuenta sit

edieval and Renaissance fighting techniques have been codified in a special kind of books known as **Fechtbücher** (literally: "fight" or "fencing books"). The German name is due to the circumstance that the majority of these works was written in the German language of the time, namely Early New High German. This brief article is therefore centered around the German manuscripts and early prints from ca. 1400 to the 17th century, and even more specifically around the teachings for one single and peculiar weapon—the Messer (literally a knife).

While the German term **fechten** (as in **Fechtbuch**) would translate in today's English as "to fence", the meaning encompassed a far greater variety of disciplines in the past—not merely the Olympic sport with foil, epée, and saber. The etymological similarity between **fechten** and 'fight' is obvious and telling. The medieval fight consisted of unarmed wrestling as well as combat in armor with a considerable array of weapons such as pollaxes, daggers, and, of course, swords. The sword is the most iconic and characteristic weapon and for most people the one that comes to mind first when they think of medieval combat. It features prominently in popular literature, movies, and TV series; what is Conan without his sword, what a Jedi knight without her lightsaber? (It's actually not a saber, but more on that later.)

The Messer, however, has so far received comparably little attention, and the lack thereof in pop culture is hardly understandable considering it was an almost omnipresent weapon in its time.

The weapon

The Messer is essentially a single-edged weapon with only a short back edge. The blade can be straight or slightly curved. Characteristic are the riveted handle scales as well as a riveted cross-guard, where the rivet is prolonged on the outside to form a so-called **Wehrnagel** (guard spike) for

Fig. 1: Messers and "Hungarian" shields (GK, f 55r, 1440s)

added hand protection. The Messer was popular roughly up to the middle of the 16th century; after that it fell out of fashion but was followed by the Dussack, also a one-edged weapon that lacked a regular hilt and usually only had a grip hole. Later incarnations of the weapon, from the 16th or 17th c., featured more conventional hilts and often a shell guard instead of a **Wehrnagel**. Some of the early Dussacks, on the other hand, retained the guard spike, despite the simplicity of their hilt designs. Dussacks for training purposes were frequently manufactured out of wood, though there were also specimens made of metal; in any case, the Dussack existed in a variety of forms, particularly outside Germany. The origin of the weapon is not entirely clear, the name, however, is most likely of Czech origin. Many techniques that had been devised for the Messer were adapted to the Dussack.[304]

Neither the Messer nor the Dussack seem to have had a significant fan base outside of Germany; Britain, France or Italy for instance preferred the falchion, *fauchon* or *falcione*—similarly one-edged weapons but rather with a sword-like hilt.

Other similar weapons (removed in time or place of origin) are likewise not the subject of this paper, including the backsword, the scimitar, or the saber, whose signature characteristic is a curved blade—as with many Messers. The curved blade distinguishes it from the straight-bladed sword, and therefore the aforementioned 'lightsaber' is *per definitionem* a 'light-sword'.

Concerning the usage of the word "Messer" in its period, there's an interesting detail in fight books that treat the teachings of Master Liechtenauer, his mounted combat in particular. In a very exhaustive codex, the so-called *Peter von Danzig fight book* (PD), there are some noteworthy passages:

---

[304] For more on dussacks, see Adam Franti's chapter in this book.

> Druck vast stoss von tzawm · sueche sein Messer
>
> "Press hard, push from the bridle; look for his Messer."[305]

> Der Mezzer nemenn · vnd behalden gedenck
>
> "Remember to take the Messer and to keep it."[306]

These are two verses that, among others, serve as an introduction to the mounted combat section and are called 'figures'; these brief, rhymed couplets appear again, later in the glosses—which is the text that explains and comments on the verses. One of these glosses reads as follows:

> zeuch wieder swert noch Messer vnd wendt dich mit deiner lincken seitten an sein rechte.
>
> "Draw neither sword nor Messer and turn yourself with your left side to his right."[307]

Yet another passage repeats one of the aforementioned verses and reads as follows:

> Der Messer nemen behalten lere an schemen [...] peug dein lincken armm auswertz an dem gepint seins swertz So mues er das swert lassen vallen [...] das lert die xv figur Die spricht also Der Messer nemen vnd behalden gedenck.
>
> "Learn to take and keep the Messer without shame [...] bend your left arm on the outside to the hilt of his sword, so he must drop the sword [...] The 15th figure teaches this: Remember to take the Messer and to keep it."[308]

Another manuscript from the Glasgow Museums (SE) has a similar but a bit more clarifying version of that passage:

> Der Messer nemen behalten ler an schemen. Glosa Hie merck wie ainem sein swert nemen solt oder sein Messer oder den tegen [...] Darumb sprich die xxv figur der Messer nemen etc.
>
> "Learn to take and keep the Messer without shame. Gloss: Notice here how to take someone's sword, or his Messer, or the dagger. That is why the 25th figure says: Take the Messer etc."[309]

---

[305] PD, f 7v.
[306] PD, f 8r.
[307] PD, f 40ar.

[308] PD, f 49r.
[309] ES, f 80v.

A final example from the manuscript PD features the term Messer in another discipline, in armored combat:

> Wil er aber zucken swert oder degen vnd wil das sper fallen lassen
> So lug auf den degen oder Messer oder swert zyehen.

> "But if he wants to draw his sword or dagger and drop the spear, then observe his drawing the dagger, Messer or sword."[310]

It is curious that in some cases the verses of the figures exclusively mention the Messer, while in other places it is the knightly sword that is being handled. We are left with a certain ambiguity here. But while in some instances it appears as if both sword and Messer were used synonymously, the majority of the passages distinguishes clearly between the two weapon types and so we can assume that the Messer in Master Liechtenauer's terminology alludes to a short weapon, ranging in length between a dagger and a sword.[311]

The sources

One cannot deny that a huge and possibly the most popular part of fighting instructions from the Middle Ages and the Renaissance revolves around the sword. There is hardly a fight book in which it does not appear. The Messer, on the other hand, features not as prominently. Nevertheless, among the roughly 100 books that serve as the foundation of this examination, there is a surprising amount of 41 sources in which the Messer, or its successor, the Dussack, is dealt with, either in writing or picture or both—more than a third. Similarly surprising is the amount of variation in the individual treatises. While the major part of sword fighting techniques go back to a certain Master Johannes Liechtenauer— for whom we possess no biographical data whatsoever but who likely lived at the turn of the 14th to the 15th century—there is no such predominant figure for Messer fighting. An aspiring candidate, however, is Master Johannes Lecküchner (ca. 1430–1482) who was a South German clergyman and who has spawned at least ten renditions of his combat lore.

The following overview lists the manuscripts in order of repository and the prints in order of author. Each source is followed by a more evocative nickname than the often-unwieldy shelf mark designation, plus an additional abbreviation, a short siglum.

---

[310] PD, f 111r.

[311] I am indebted to Christian TOBLER for directing my attention to some of the quotes mentioned above and who came up with the conclusion that we have to speak of Master Liechtenauer's Messer as a rather short weapon in the first place.

Folio 75r of the Biblioteka Jagiellonska's *Goliath* (G) proclaims the beginning of a Messer section which, however, was never realized. Therefore, this manuscript is not included in the following overview. Two more sources have not been incorporated in this investigation: One is MS Fol.U. 423.792, a Dutch manuscript in the Newberry Library in Chicago which features Turkish sabers on two illustrated pages (fig. 2). The weapon is referred to as a **tashack naer teurcksche wyse** ("Dussack in the Turkish manner"), but its appearance

Fig. 2: "Dussack in the Turkish manner" (SK, f 12r, 1595)

with the strongly curved blade and the pommel-less hilt deviates too much from the average Dussack to include this specific manuscript. The second excluded source is Sebastian Heußler's *Neu Kunstlich Fechtbuch* from 1615. One image on p 215 portrays two Eastern fighters with possibly Turkish sabers, and another image on p 229 shows two fighters with rather straight-bladed weapons that resemble Messers with complex hilts.[312] Since the accompanying text explains how to fight with the rapier, the images probably only serve as a means to showcase a variety of bladed weapons.

Manuscripts

1. Augsburg, Stadtarchiv, *Schätze 82 Reichsstadt.* Anton Rast, AR.
2. Augsburg, Universitätsbibliothek, *Cod. I.6.4°.2.* Baumann's fight book, B.
3. Augsburg, Universitätsbibliothek, *Cod. I.6.2°.4.* Paulus Hector Mair, Augsburg, PMA.
4. Berlin, Kuperstichkabinett der Stiftung Preußischer Kulturbesitz, *78 A 15.* Hans Talhoffer, Berlin, HTB.
5. Bologna, Universitätsbibliothek, *Ms. 1825.* Paulus Kal, Bologna, PKB.
6. Dresden, Sächsische Landesbibliothek, *Mscr. Dresd. C.93/94.* Paulus Hector Mair, Dresden, PMD.
7. Erlangen, Universitätsbibliothek, *Ms. B 200.* Liber Quodlibetarius, LQ.
8. Glasgow, R. L. Scott Collection, *E.1939.65.341.* Sigmund Emring, SE.
9. Glasgow, R. L. Scott Collection, *E.1939.65.354.* Gregor Erhart, GE.
10. Gotha, Forschungsbibliothek Schloss Friedenstein, *Ms. Chart. B1021.* Paulus Kal, Gotha, PKG.
11. Heidelberg, Universitätsbibliothek, *Cod. Pal. Germ. 430.* Johannes Lecküchner, Heidelberg, LH.

---

[312] See fig. 4 on p 233.

12. Kassel, Landesbibliothek, *2° Ms. iurid. 29*. Hans Talhoffer, Kassel, HTKa.
13. København | Copenhagen, Det Kongelige Bibliotek, *Thott 290 2°*. Hans Talhoffer, København, HTK.
14. Köln | Cologne, Historisches Archiv, *Best. 7020*. Köln fight book, K.
15. Kraków | Cracow, Biblioteka Jagiellonska, *Ms. Germ. Qu. 16*. Gladiatoria, Kraków, GK.
16. London, British Library, *Sloane MS No. 5229*. Albrecht Dürer, London, ADL.
17. Lund, Universitetsbibliotek, *MS A.4°.2*. Joachim Meyer, Lund, JML.
18. München | Munich, Bayerisches Nationalmuseum, *MS Bibl. 2465*. Joachim Meyer, München, JMM.
19. München | Munich, Bayerische Staatsbibliothek, *Cgm 558*. Hugo Wittenwiler, HW.
20. München | Munich, Bayerische Staatsbibliothek, *Cgm 582*. Johannes Lecküchner, München, LM.
21. München | Munich, Bayerische Staatsbibliothek, *Cgm 1507*. Paulus Kal, München, PKM.
22. München | Munich, Bayerische Staatsbibliothek, *Cgm 3712*, Jörg Wilhalm, München #2, JWM2.
23. München | Munich, Bayerische Staatsbibliothek, *Cod. icon. 393*. Paulus Hector Mair, München, PMM.
24. München | Munich, Bayerische Staatsbibliothek, *Cod. icon. 394a*. Hans Talhoffer, München, HTM.
25. München | Munich, Bayerische Staatsbibliothek, *Cod. icon. 394*. Hans Talhoffer, München #2, HTM2.
26. Nürnberg | Nuremberg, Germanisches Nationalmuseum, *Cod. Hs. 3227a*. Nürnberger Hausbuch, Nicolas Pol fight book, N.
27. Paris, Musée National du Moyen Age, *CL23842*. Paris fight book, P.
28. Salzburg, Universitätsbibliothek, *M.I.29*. Hans von Speyer, HS.
29. Towson, Sammlung Amberger. *Codex Amberger*, A.
30. Wien | Vienna, Albertina (Graphische Sammlung), *Hs. 26-232*. Albrecht Dürer, AD.
31. Wien | Vienna, Kunsthistorisches Museum, *KK 5012*. Peter Falkner, PF.
32. Wien | Vienna, Kunsthistorisches Museum, *KK 5126*. Paulus Kal, Wien, PKW.
33. Wien | Vienna, Österreichische Nationalbibliothek, *Cod. Vindob. 10825/6*. Paulus Hector Mair, Wien, PMW.
34. Wolfenbüttel, Herzog August-Bibliothek, *Cod Guelf. 38.21 Aug. 2°*. Lienhart Sollinger/Jörg Wilhalm, Wolfenbüttel, JWW.
35. Wolfenbüttel, Herzog August-Bibliothek, *Cod. Guelf. 83.4 Aug. 8°*. Wolfenbüttel fight book #1, W01.

36. Wolfenbüttel, Herzog August-Bibliothek, *Cod. Guelf. 1074 Novi.* Fechtbüchlein, FB.

Prints

37. Anon.: *Der Allten Fechter gründtliche Kunst.* Christian Egenolff, Frankfurt am Main, ca. 1530. Egenolff, E.

38. Joachim Meyer: *Gründliche Beschreibung der freyen Ritterlichen vnnd Adelichen kunst des Fechtens in allerley gebreuchlichen Wehren mit vil schönen vnd nützlichen Figuren gezieret vnd fürgestellet.* Thiebolt Berger, Strassburg 1570. Joachim Meyer, JM.

39. Andre Paurenfeindt: *Ergrundung Ritterlicher Kunst Der Fechterey.* Wien | Vienna, Hieronimus Vetor, 1516. Andre Paurenfeindt, AP.

40. Jacob Sutor: *New Künstliches Fechtbuch.* Wilhem Hoffman, Frankfurt am Main, 1612. Jacob Sutor, JS.

41. Theodor Verolinus: *Der Künstliche Fechter—Anderer Theil.* Joann Bencard, Würzburg, 1679. Theodor Verolinus, TV

Description

The extent to which the Messer is explained in the sources varies considerably from just a few lines of text to splendid, illustrated manuscripts with as many as 400 illustrations or more and accompanying text. Some sources even contain more than one Messer section. The following list offers a brief description of each Messer passage in the examined sources.

Fig. 3: Messer against sword (AR, f 44v, 1553)

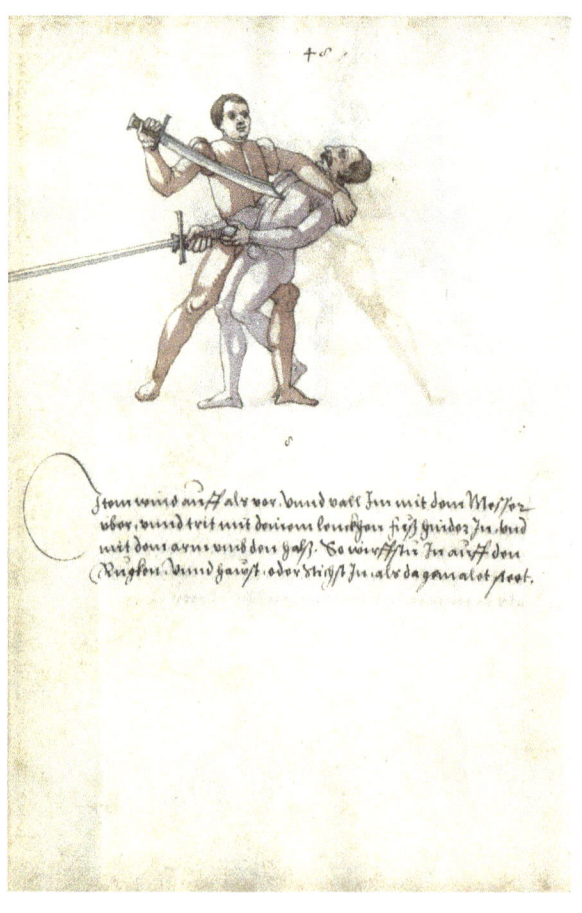

1. Anton Rast, AR (43r–47v)

Illustrated version similar to and possibly a (partial) copy of B. The manuscript omits some images but also adds additional ones next to a completely different text (fig. 3).

2. Baumann's fight book, B (29r–32v)

Archetypal illustrated version with a series of eight images that has spawned a number of copies or related teachings, such as AR or AD.

3. Paulus Hector Mair, Augsburg, PMA (34r–36r)

Known as Paulus Hector Mair's sketch book for his huge double volumes PMD, PMM and PMW,

this slim manuscript shows five pairs of Dussack fighters in sepia-colored outlines. One of the images (although mirrored) goes back to E which in turn is based on P (fig. 4 & 8). It is noticeable that the images show lefthanded fighters throughout while the manuscripts PMD, PMM and PMW show the same situations but reversed so that we see righthanded fencers. Possibly the images in PMA served as templates for woodcuts.

Fig 4: Mirrored illustrations in PMA (top; f 35v, 1540s) and PMD (bottom; f 123v, ca. 1542)

4. Hans Talhoffer, Berlin, HTB (54r/v)

Only two pairs of fighters are shown that fight with Messers and bucklers (small shields). Apart from that, folio 62v shows two examples of Messers with straight blades.

5. Paulus Kal, Bologna, PKB (25r–26v)

A brief illustrated section without captions in an incomplete, fragmentary manuscript.

6. Paulus Hector Mair, Dresden, PMD (C93, 114r–135v & 136r–180v)

Two Dussack sections appear in the first part of this two-volume manuscript: The first section consists of teachings in word and image that are partially based on E (fig. 4); the second is a text-only abridged version of Lecküchner's fight books, adapted to the Dussack.

7. Liber Quodlibetarius, LQ (118r/v & 120v–121r)

The fight book section in this larger volume is considerably small and the Dussack part is even smaller; it consists of only eight images—that in part are similar to P but mostly mirrored—with short captions. The Dussacks are peculiar since they lack the characteristic grip hole and look more like sticks with a pommel-like end.

8. Sigmund Emring, SE (25v–26v)

Several singular techniques for the Messer without images.

9. Gregor Erhart, GE (111r–144v & 172v–178r)

This manuscript contains two sections of the Messer; the first one goes back to Johannes Lecküchner although here it is attributed to Johannes Liechtenauer. No images accompany the text which is a heavily

abridged version of the original. The second section offers several independent Messer lessons that also contain only text.

10. Paulus Kal, Gotha, PKG (47r–48v)

This copy of PMM or PMW shows only four out of seven images. No text.

11. Johannes Lecküchner, Heidelberg, LH (1v–116r)

The first archetypal version of Master Johannes Lecküchner's fight books that contains only text except one dedicatory illumination on fol. 1v that shows the coat of arms of Philip 'the Upright' of Wittelsbach, Elector Palatine of the Rhine.

12. Hans Talhoffer, Kassel, HTKa (302r–309r)

Among several law texts, this massive volume also contains copies of Hans Talhoffer's manuscripts, such as HTM. The same eight images from this source are included.

13. Hans Talhoffer, København, HTK (119v–123v)

A series of nine illustrations with brief captions. The images are similar but only in part identical to Talhoffer's later work from Munich, HTM, where comparable stances are assumed, not only by the Messer fighters but also by those who are armed with swords and bucklers. Here, in HTK, the fighters are equally equipped not only with their Messers, they also wield both spiky and round bucklers (fig. 5).

14. Köln fight book, K (13r–17v)

Unillustrated instructions for the Messer that borrow some nomenclature from other sources without any recognizable relationship.

Fig. 5: Spiked and round bucklers (HTK, f 122r, 1459)

15. Gladiatoria, Kraków, GK (55r)

This manuscript that mostly treats armored combat in full harness, contains also 13 images of unarmored fighting techniques, and a single one of these portrays Messer fighters who also carry so-called Hungarian shields (fig. 1). A brief text explains that the techniques with this combination can be performed for pleasure and in earnest.

Fig. 6: Two Messer plays (ADL, f 69r, ca. 1512)

16. Albrecht Dürer, London, ADL (67v–69r)

On two pages, four pairs of roughly-sketched fencers, drawn by Albrecht Dürer, show four guards and their counters, accompanied by the respective names of the positions (fig. 6).

17. Joachim Meyer, Lund, JML (44r–65v)

Joachim Meyers's handwritten version of his Dussack exercises that would later become part of his printed edition JM. It is embellished by a diagram and seven colorful illustrations. Like in the printed book, he draws in part on Lecküchner's nomenclature without borrowing from his technical advice.

18. Hugo Wittenwiler, HW (131r–132v & 133r/v)

A Swiss manuscript with several fighting disciplines, including techniques for the Basselet (baselard), a weapon equally similar to the Messer and the dagger. There are also a handful of additional techniques for the 'short knife' whose length is not specified.

19. Joachim Meyer, München, JMM (22r–48r)

An illustrated Dussack treatise within an extensive collectanea of various disciplines, authored by Joachim Meyer.

20. Johannes Lecküchner, München, LM (1r–216v)

The most extensive fight book about the Messer. Unlike most other sources which are compendia of various fighting disciplines, Johannes Lecküchner deals exclusively with the Messer. A stunning 416 pairs of fighters populate the pages of this volume.

21. Paulus Kal, München, PKM (71v–74v)

Seven illustrations, accompanied by only brief captions. No recognizable relation to other sources.

22. Jörg Wilhalm, München #2, JWM2 (53r–65r & 66v–68v)

This codex, comprised of various sources from various dates, contains the same two Messer teachings as GE. Due to the ambiguous

dating, it has so far not been ascertained what manuscript served as the template for the other.

23. Paulus Hector Mair, München, PMM (393, 100r–123r & 124r–152v)

A Latin adaptation of PMD; the artwork is similar but four additional techniques are shown.

24. Hans Talhoffer, München, HTM (113r–116v)

The most famous and lavishly illustrated manuscript authored by Master Hans Talhoffer. On eight pages he shows pairs of Messer fighters, accompanied by brief explanatory captions. It is remarkable that unlike most sources this manuscript shows bloody wounds and even severed hands.

25. Hans Talhoffer, München #2, HTM2 (pp. 221–228)

A traced copy of HTM with the identical images and texts.

26. Nürnberger Hausbuch, Nicolas Pol fight book, N (82r/v & 84r–85r)

An introduction to fighting with the long Messer, presumably based on Johannes Liechtenauer's teachings. After only a few short paragraphs the section is discontinued.

The Messer is mentioned another time in the same manuscript, in a section that speaks of techniques for 'the dagger or the short Messer'. This passage uses both weapons synonymously.

27. Paris fight book, P (60r–69v)

This giant illustrated compendium contains 20 Messer techniques, six of which are identical or very similar to those in Peter Falkner's fight book (PF). There's no text so it cannot be ascertained whether the same text as in Falkner—that goes back to Johannes Lecküchner—would have appeared here too. Consequently, a relationship to Lecküchner's teachings is probable, although it cannot be proven.

28. Hans von Speyer, HS (5r–7r)

The manuscript contains two Messer sections. The first one is authored by a certain Master Andreas who appears among the corpus of the German fight books exclusively in this codex with several techniques that alternate between the sword and the Messer. The sword passages quote Master Liechtenauer's text while the Messer is based on Master Lecküchner's teachings.

The second part is similar to LH, although several techniques are missing, while on the other hand other unique ones appear that have no counterpart in other versions of Lecküchner's teachings.

Both versions have no images.

29. Codex Amberger, A (11r–13r)

This small fragment possesses three Dussack images without text that can also be found in E.

30. Albrecht Dürer, AD (74r–91v, 96r–97r & 97r–100v)

This outstanding manuscript contains three versions of Messer teachings. The first illustrated part, a series of 55 images without any accompanying text, is attributed to Albrecht Dürer, although it is possible that not the master himself but rather members of his workshop are responsible for the fine line drawings with subtle watercolor washes (fig. 7). The illustrations from B reappear here plus many additional ones, based on a so far unknown or unidentified origin.

The second section presents an abridged version of Lecküchner's verses, which is followed by section three, another version based on Lecküchner, but this time verses and explanatory glosses are joined again, in this manner resuming Lecküchner's original arrangement. Having the verses placed in front of the verse-and-gloss version is unprecedented in the realm of Messer sources but mirrors an identical procedure that was frequently applied in fight books that feature Johannes Liechtenauer's teachings.

This manuscript's version is similar to LH but contains only about half of the text.

Fig. 7: Messer against Messer and dagger (AD, f 85v, 1512)

31. Peter Falkner, PF (18v–43v)

The Messer forms the largest part of this fight book and consists of a series of 50 illustrations, accompanied by verses that are based on Lecküchner's originals. There's an overlap of six images with the Paris fight book (P) which otherwise is quite similar in the style of the artwork.

32. Paulus Kal, Wien, PKW (76v–79v)

The same images as in PKM sans text.

33. Paulus Hector Mair, Wien, PMW (10825, 97r–118v, 120r–147r & 149r–153v)

A Latin-German version of PMD; the images and their number equal PMM. The illustrated section is followed by a German and a Latin version of Lecküchner's text in an abridged form.

34. Lienhart Sollinger/Jörg Wilhalm, Wolfenbüttel, JWW (1r–7v & 9r–14v)

Two next-to-identical sections of Messer techniques by Lienhart Sollinger. Techniques 46–49 are missing in the first version (fol. 1r–7v), which may hint at a loss of material. The section ends with the 53rd technique. The second version (9r–14v), however, has 54 techniques. Nevertheless, techniques 50–53 from the first version correspond to techniques 51–54 in the second.

35. Wolfenbüttel fight book #1, Wo1 (68r–73r & 75r–85r)

This fight book features an illustrated sequence of ten burly fencers, each taking a Dussack position on each page, named below the image. After a couple of blank pages, there is a second illustrated Dussack section with more detailed descriptions below the images.

36. Fechtbüchlein, FB (7r–12v)

Exercises for the Dussack with no identified connection to other sources; no images.

37. Egenolff, E (XVIv–XXXIIIv)

This book, printed in at least four editions between 1530 and 1558, offers a short rendition of Lecküchner's teachings, although the master's name has been corrupted to 'Hans Lebkommer' (fig. 8). It contains three illustrations of dussack fencing, some of which are repeated a number of times. All of them show fighters in a mirrored stance in comparison to the images from AP.

38. Joachim Meyer, JM (2.Ir–2.LXIXv)

Joachim Meyer has dedicated the entire second part of his four-part *magnum opus* to the Dussack. While he is influenced by Master Johannes Liechtenauer and draws on portions of Lecküchner's nomenclature, his lessons are a unique creation of his own. Fourteen woodcuts enrich his edition, some of which appear two or even three times.

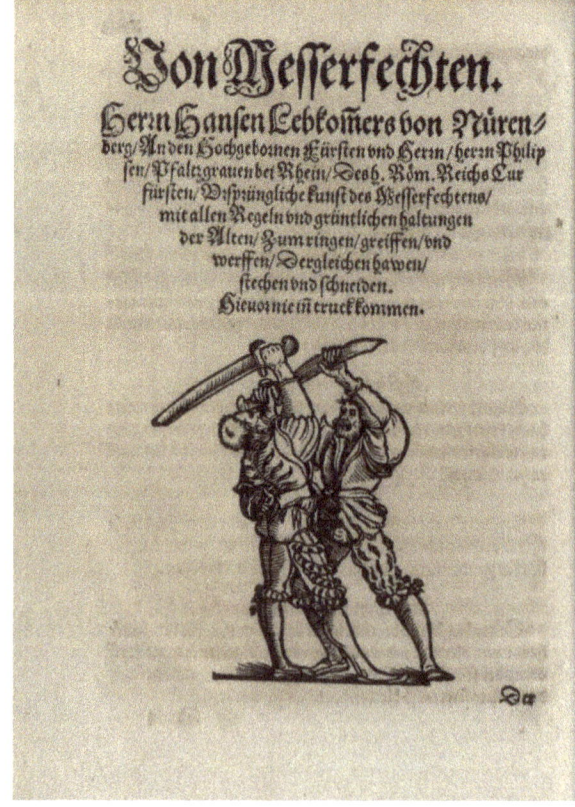

Fig. 8: Beginning of "Lebkommer's" chapter (E, p XIVv, 1530s)

39. Andre Paurenfeindt, AP (Gv–Ir)

The earliest printed fight book in German language has eight wood-cut illustrations, depicting Dussack techniques. The accompanying text, however, speaks of the Messer, and it is a solitaire as far as we know.

40. Jacob Sutor, JS (pp. 20–35)

Sutor's book is a greatly abridged adaptation of Meyer's original (JM) and the illustrations are also vastly simplified and coarsened.

41. Theodor Verolinus, TV (Second part)

The second volume of Verolinus' four-part edition treats the Dussack which is yet another simplified version of JM.

Timeline & dependencies

The following table attempts to sort the sources in a tentative timeline. Not every date can be ascertained unambiguously since some sources are not internally marked and can therefore only be dated rather approximately based on watermarks or paleography. The table also shows whether a treatise contains images and/or text.

Note that in some cases we need to consider compendia that were composed of multiple individual elements; one manuscript may contain both text and image, but they need not belong to each other necessarily.

Three additional columns inform whether a certain source belongs to one of the most widespread tradition lines, one that goes back to Master Johannes Lecküchner, another one to *Baumann's fight book* (B) and the third to *Egenolff* (E) or *Paurenfeindt* (AP) respectively.

Several unique teachings stand next to other fighting lores that have generated a number of descendants.

Eight sources contain two versions of Messer fighting techniques (PMD, GE, JWM2, PMM, HS, PMW, JWW, and W01), and one single source, AD, even offers three. This adds up to a total of 50 Messer sources in 41 volumes.

Note that the last entry in the table, HTM2, has been added due to its appearance in the specialized literature; it is a connoisseur's copy from the early 19th century of an original from 1467.

Conclusion

An astonishing amount of fight books transfers teachings for the one-edged Messer, although not in a uniform but on the contrary in a rather heterogenous way. Many masters and teachers of the Middle Ages and

Table 1: Timeline and dependencies

Siglum	Colloquial name	Date	Image	Text	Lecküchner	Baumann	Egenolff/Paurenfeindt
N	Nürnberger Hausbuch/Nicolas Pol fight book	15th c.		X			
GK	Gladiatoria, Kraków	ca. 1440–50	X	X			
HTB	Talhoffer, Berlin	before 1459	X	X			
HTK	Talhoffer, København	1459	X	X			
PKB	Paulus Kal, Bologna	1458–1467	X				
PKM	Paulus Kal, München	ca. 1460	X	X			
PKW	Paulus Kal, Wien	late 15th c.	X				
PKG	Paulus Kal, Gotha	late 15th c.	X				
HTM	Hans Talhoffer, München	1467	X	X			
B	Baumann's fight book	ca. 1470	X	X		X	
HW	Hugo Wittenwiler	ca. 1470		X			
LH	Johannes Lecküchner, Heidelberg	1478		X	X		
LM	Johannes Lecküchner, München	1482	X	X	X		
PF	Peter Falkner	1480–1500	X	X	X		
P	Paris fight book	1480–50	X		X?		
HS	Hans von Speyer	1491		X X	X X		
K	Köln fight book	ca. 1500		X			
SE	Sigmund Emring	1508		X			
AD	Albrecht Dürer	ca. 1512	X	X X	X X	X	
ADL	Albrecht Dürer, London	1512	X	X			
AP	Andre Paurenfeindt	1516	X	X			X
LQ	Liber Quodlibetarius	1524	X	X			X
E	Egenolff	ca. 1530, 1st ed.	X	X			X
GE	Gregor Erhart	1533		X X	X		
FB	Fechtbüchlein	16th c.		X			
PMA	Paulus Hector Mair, Augsburg	1545/46	X				X
AR	Anton Rast	1553	X	X		X	
A	*Codex Amberger*	1555–70	X				X
JWM2	Jörg Wilhalm, München #2	1556 ?		X X			
JML	Joachim Meyer, Lund	ca. 1560	X	X			
JMM	Joachim Meyer, München	1561	X	X			
PMD	Paulus Hector Mair, Dresden	after 1566	X	X X	X	X	X
PMM	Paulus Hector Mair, München	after 1566	X	X X	X	X	X
PMW	Paulus Hector Mair, Wien	after 1566	X	X X	X	X	X
JM	Joachim Meyer	1570, 1st ed.	X	X			
JWW	Lienhart Sollinger/Jörg Wilhalm, Wolfenbüttel	before 1588					
Wo1	Wolfenbüttel fight book #1	1591	X X	X			
JS	Jacob Sutor	1612	X	X			
HTKa	Hans Talhoffer, Kassel	17th c.	X	X			
TV	Theodor Verolinus	1679	X	X			
HTM2	Hans Talhoffer, München #2	1820	X	X			

the Renaissance found it worthwhile to devote themselves to this fascinating weapon, which, though not as charismatic as the sword, was nevertheless just as efficient.

To my knowledge, a complete overview of the entirety of the surviving Messer sources has never before been undertaken. This paper shall therefore serve as a starting point for further research and examination of the individual Messer passages in the fight books in order to better showcase this weapon, which has been underrepresented in the literature so far.

Olivier Dupuis

Fighting with Long Knife for Leisure or Self-Defense: Discussions Around the Production Contexts of Lecküchner's Fencing Manuscript

If you wish to consider	[O]𝖡 ðw wilt achten
And study Messer fencing,	Messer vechten betrachten
Then learn that which adorns you,	So leren ðing das ðich zirtt
And ennobles you in jest and in earnest,	Zu schimpff ze ernnst hofirt
With which you will intimidate	Do mit ðu erschreckest
And artfully surprise the masters.	Und ðy meyster künstenlich erbeckest

~ Hans Lecküchner, 1482

ans Lecküchner's manuscripts appear as a curiosity to a certain extent. In the first place, it seems surprising that a priest from the University of Leipzig composed a fencing treatise, taking inspiration from the glossators of Liechtenauer. The fact that he dedicates his treatise to a single weapon is also unusual, but the fact that it concerns a one-handed weapon, treated marginally until then, is even more so. To tell the truth, not until Henri de Sainct-Didier in 1573 may another treatise be found thus entirely focused on a form of fencing without a second weapon (such as a buckler or a dagger). But what imposes itself above all on the reader is the extraordinary dimension of his work, in the physical sense.

Indeed, the two Lecküchner manuscripts are both physically imposing even if the dimensions of the second (*Hans Lecküchner-Munich*, LM) appear more clearly out of the ordinary. If we measure the entire area, that is to say by multiplying the area of each page by their number, it is possible to have a system of comparison between the different manuscripts. However, it is necessary to keep in mind that we do not always have the manuscripts in their original format: the surface of the pages may have been reduced during subsequent binding operations, or entire quires may be missing inside a manuscript. These precautions being taken, it is possible to estimate the total surface area of paper required for the second manuscript of Lecküchner, to 13 m². In the 15th century, only one fencing treatise topped this area: *Paul Kal-München* (PKM) with an area of 15 m². No other fencing manuscript from this period exceeds 10 m².

However, we should mention *Hans Talhoffer-München* (HTM) from 1467 which has a global area of 9.5 m² of parchment, a material much more expensive than paper. And in fact, the development of the paper industry in Germany from the beginning of the 15th century, even accelerating in

the second half of the 15th century, is on the one hand an essential factor for the development of printing and on the other hand a paper price reduction factor thus facilitating the production of larger manuscripts.

This disproportion in the size of Lecküchner's manuscript, as well as its necessary cost of production, obviously pleads for a serious consideration of his work and in particular the context of production.

Is it really serious? Lecküchner's work in the light of schimpf and ernst

The first lines of Lecküchner's treatise paraphrase Liechtenauer's poem and apostrophize the reader by inviting him to learn what will allow him to stand out: "so lerne ding das dich zirtt / zu schimpff ze ernest hofiert". That is to say, to learn the martial knowledge described in his huge handbook.

Fig. 1: A duel in earnest (ÖNB Cod. 3062, f 27r, 1437)

In a way, Lecküchner responds to the invitation of this poem which does mention the Messer along with the spear and the sword among the weapons whose mastery makes it stand out. But for his part he completely dismisses the other two weapons and deals exclusively with the Messer. The second part of these two lines is ambiguous for a modern reader who naturally sees in these words a figure of strong attenuation or euphemism. Its wording differs slightly from that of the glossators, of which I take the two versions attested before the composition of LM:

> lere kunst die dich zyret vnd in kriegen zu eren hofieret[313]

> "Learn art that adorns you, and in wars brings honor"

> lern dinck das sich zieret vnd in kriegen ser hofieret[314]

> "Learn things that adorn and flatter you greatly in wars"

By the time Lecküchner is writing, this pair of words, **ernst** and **schimpf**, is already well established in the German language as two opposing terms, one conveying seriousness, the other derision. It is

[313] *Peter von Danzig fight book* (PD), f 10r. Translation by Cory WINSLOW.

[314] *Lew fight book* (L), f 1r. Translation by FLEISCHHAUER, FRITZ, HAGEDORN, and REHM.

possible to find for example from the 13th century the poet Ulrich von Liechtenſtein who wrote in his poem *Vrouwen Dieneſt*, "Dô fuor ich witen in díu lant, swâ iemen ritterſchaft dô vant, eʒ waer ʒe ſchimpf ode ernstlich", which can be translated as: "I journeyed far in the country to find any knightly activities, be they in earneſt or for leisure."[315]

These words are immortalized in German literature by Johannes Pauli, a Franciscan preacher. He began his preaching activities in southeaſtern Germany around 1480, and in 1522 published a collection of hundreds of *exempla* entitled **Schimpf und Ernst**. The success of this book was quick and very important, both in terms of spreading and of the impact on German literature. It consiſts of short moralizing ſtories, often amusing, grouped by categories of characters. Each ſtory, however, is directly associated with one of the two qualifiers, **ernst** and **ſchimpf**. For example, one of them, associated with the category **ſchimpf**, tells the ſtory of a fencing maſter who finds himself having to face one of his former ſtudents in a duel to the death. He kills the ungrateful ſtudent with a trick and concludes the farce with the moral that a teacher should not pass everything on to his ſtudents.[316]

During the 14th and 15th centuries, **ſchimpf** can also take a more positive meaning, at leaſt in literature, and refer to chivalrous games, as was probably the case in the aforementioned poem by Ulrich von Liechtenſtein. This will perhaps be even clearer in this excerpt: **wir ſůln turniren híe, mîn lîp in ʒehen jâren níe durch ſchimpf kom in kein wâpencleit**, which means "we muſt jouſt here, in ten years I didn't carry my equipment to enter a game".[317] This reference to chivalrous game muſt certainly have been on a contemporary reader's mind when he discovered the only mention of this couple of words in the glosses made on Liechtenauer verses:

> **Das ſie mit rechter kunst des swerʒ můgen besteen Jnn ſchimpff vnd Jn ernst**[318]
>
> "that they can persiſt with the rightful art of the sword in jeſt and in earneſt"

Fig. 2: A duel in jest (ÖNB Cod. 3062, f 27v, 1437)

---

[315] LACHMAN 1841, p 102. Unless otherwise stated, all translations given are by the author.
[316] Pauli 1522, *exemplum* 311.

[317] LACHMAN 1841, p 7.
[318] L, f 5r. Translation by FLEISCHHAUER, FRITZ, HAGEDORN, and REHM.

193

Lecküchner takes it up almost in full, removing the mention of the sword:

> das sy mit rechter kunst wol bestann yn schimpff und yn ernest[319]
>
> "so that they stand well with the rightful art in jest and in earnest"

One should note here that the other glossator renders the word, not schimpf, but Kampf:

> das sy mit rechter gunst des swertz wol mügen besten In kampff und in ernst[320]
>
> "that they can persist with the rightful art of the sword in combat and in earnest"

This is interesting because while Kampf may have the general meaning of combat, in the Middle Ages it also referred to a duel in a closed field. And ernst can also take the meaning of fight to the death. This last glossed sentence can thus reveal a pair of synonyms designating any combat situation where life is at stake, which is not reflected in the given translation which I have deliberately made by using the wording of the extract of Lew. The change from Kampf to schimpf modifies the meaning of couple of words. It swaps from any potentially deadly fights to any contexts where the art of combat is necessary.

In the context of fencing, whether it is with the long sword or the Messer, the words schimpf and ernst do not oppose as much as complement each other: the first refers to games where the goal is not to kill or seriously injure the opponent, the second in combat situations where life or physical integrity is at stake.

Lecküchner uses this pair of words on two occasions later in his text, which he uses again to generalize the use of techniques to all combat contexts:

> Item du hast manigerlay stuck und pruch Schimpfflich und ernest-
> lich gefellt dir eynes nicht so nym eyn anders wer dy ding recht
> verstett und ytlichs zu seyner zeytt treyben kan dem gefallen dyse
> ding
>
> "Then, you have various elements and breakings, for amusement and serious combat. If you do not like one, take another one.

> Who understands the things rightfully and can perform a partic-
> ular one in its time, such a man likes these things"[321]

> **al3o magstu dich eynes ydlichen weren In schimpff oder ernest etc.**

> "This way you can defend against anyone for amusement or in serious
> combat"[322]

Before studying the contexts of serious or leisure use of the Messer sep-
arately, I propose to leap forward twenty years from the writing of Lecküchner's text to Hettingen, a small village in the Swabian Alps.

In 1503, the priest of the parish of Saint-Martin, along with his chaplains
and the members of two other surrounding parishes, established a collegial
regulation governing essentially life rules of the ecclesiastics. The sixth paragraph is particularly interesting and deals mainly (but not only) with
clothing. Among the various restrictions of wearing colors, this text speci-
fies that neither layman nor religious people should wear any hat, sword,
or Messer in the church.[323]

This ban says a lot about the importance of carrying weapons on a daily
basis, since it is necessary to make the exclusion of weapons inside a place
of worship explicit in a regulation of rural parishes.

Ernst

It seems incongruous nowadays to imagine a parish priest in his 50s using
his free time to write a large fencing treatise. Yet Lecküchner's situation fits
rather well to this task. On the one hand, his intellectual training is solid,
and guarantees him not only to know how to read and write in German and
Latin, but also how to compose and argue. On the other hand, his profes-
sional activities could give him the free time needed for writing, but probably also the income making it possible to buy the large stock of paper and
ink, and possibly also to travel, read existing fencing treatises, and meet
other experts. Finally, this ecclesiastical status did not keep him so isolated
from the world; the previously-reported Hettingen dress code shows this.

The 15th century saw the Messer settle alongside the burghers of the
towns, a population much better documented than the rural areas by various archive documents. The city of Strasbourg, to name but one, produced
more than ten police regulations prohibiting all or part of its population
from carrying weapons in the streets of the city, whether during the day or
in night. As a general rule, the carrying of a weapon is forbidden to all,

---

[321] LM, f 171r. Translation by Żabiński, Mitchell,
Fritz.

[322] LM, f 186r. Translation by Żabiński, Mitchell,
Fritz.
[323] Kraus 1950, p 168.

foreigners or citizens, except for the watch, the executioner, and the personal guards of the city councilors. These weapon-carrying regulations always mention a list of weapons as an example, without aiming to be exhaustive, and among these weapons the large knife always finds its place, thus showing, if necessary, that it was commonly carried by urban and peri-urban populations. Visitors were invited to leave their weapons at the inn, and the innkeepers were then required to ensure that this was the case, at the risk of them being fined for letting their customers carry forbidden weapons out onto the streets.[324] Regarding the men in charge of the watch, the regulatory equipment in 1477 consists of a breastplate, a gorget, a helm, iron gauntlets and a long messer or a sword. This is not a specificity of this free city: a census in 1477 of the heritage of a burgher in the city of Nuremberg lists among many pieces of armor, a halberd, two swords and a long Messer.[325]

This mention of the long knife, lang Messer, immediately highlights the problem of size. If the weapon is short, the most spontaneous and obvious defensive action is to block the hit with the hand, arm outstretched, grabbing the wrist holding the weapon. This action only makes sense if, once the wrist has been grasped, the length of the blade does not allow the head or the neck to be reached. In summary, large knife fencing differs from dagger fencing by the fact that the weapon is too long to allow one to defend oneself with one hand.

Let's look at four 15th-century fight books that predate Lecküchner's work, all of them containing a repertoire of fencing techniques with a sword or a long knife held in one hand and without a shield.[326]

The first repertoire is made up of the manuscripts of Fiore dei Liberi, produced at the turn of the century, all of which present a series of commented and illustrated techniques on one-handed sword fencing. In terms of quantity, this section falls behind the wrestling, dagger, or two-handed sword for all relevant manuscripts, but is by no means negligible. For the sword used one-handed, Fiore dei Liberi develops the sole role of the defender to which he offers a fairly simple driving strategy depending on the type of attack, then develops the consequences and some options depending on the reaction of the attacker. All defensive actions start from the only defensive posture proposed, "because this guard is good and powerful".[327] This defense can be applied against all the offensive actions that can be carried out with the one-handed sword: cuts, thrusts as well as the projection of the weapon. This starting

Fig. 3 (op.): A guard position and three counters that follow from it by Fiore dei Liberi (Novati facsimile, cc 13b-14a (details), 1409)

---

[324] *"ein jeglicher Scharwachter der sin naht zuor huot duot, soll die naht stetes anhaben und tragen ein pantzer, ein kragen, ein iserin houbtgedeckte und zwen iserin hentschuo und sin hant gewer, ouch ein lang alter ods ein swert, al harkommen ist".* HATT 1929, pp 115–17.

[325] WEISS 1980, p 136.

[326] The following development borrows heavily from DUPUIS 2013.

[327] CHIDESTER 2016, pp 216 and 271.

posture is the following: to hold the sword as if it were in the scabbard, i.e. point down, along the left leg.

Paul Kal produced several fencing manuscripts, mostly illustrated, and is the author of the second repertoire. His manuscript conserved in Munich is almost contemporary with Lecküchner's treatise, but only presents a small set of Messer fighting techniques.[328] Comments associated with images, usually a terse sentence, provide an ambiguous interpretation of the techniques illustrated. Nevertheless, it is possible to see an overall logic in the seven techniques. In six of the seven illustrations, the action is carried out by the character dressed in blue. The seventh illustration presents a counter to the technique presented in the sixth, and it is logically performed against the character in blue. As with Fiore dei Liberi, the first figure is the starting point, the main actor placing himself in what Paul Kal calls the first guard, the weapon hanging at the level of the left leg set back, as if it were in the scabbard. The following two illustrations describe two possible defenses, then the consequences that can occur.

Hans Talhoffer is the author of various weapon books containing illustrations of excellent quality but unfortunately, like Kal, little commented on. Among the manuscripts attributed to him (dated 1448 to 1467), only the final one presents a significant number of long Messer fighting techniques.[329] But as with the previous two authors, Talhoffer begins the sequence from the same posture, the knife placed next to the left leg, point down. He proposes three ways of defending oneself from a top-to-bottom cut, and the consequences that can follow from these defenses.[330]

Finally, the last repertoire is represented by an anonymous manuscript, often named *Baumann's fight book* (B), containing a small set of eight defensive techniques, each time against a single cut given from top to bottom on the left side of the defender.[331]

This corpus represents the essential material of detailed technical repertoire existing for sword-only or long Messer fencing before Lecküchner's writing. Each of them presents

---

[328] PKM, ff 71v–74v. Manuscript 20 in the article from Dierk HAGEDORN, "Many magnificent Messer manuscripts", in this same volume. It will be referred as *MMMM* in the following footnotes.

[329] HTM, ff 113r–16r. *MMMM*, Ms. 24,

[330] For more on Paul Kal, Hans Talhoffer, and B, see the article from Oskar TER MORS and Casper J VAN DIJK, "The art that Hans Lecküchner made and devised himself", in this same volume.

[331] *MMMM* Ms. 2.

a subset of the possible defenses against a simple, uncompounded offensive action; it is possible to draw any parallel between the amount and details about Messer techniques and the variety of actions proposed by these same sources for the long sword, where many more aspects are discussed, including different ways of attacking, and not only to defend oneself.

We are very far from the program defined in the principles of fencing preserved in an anonymous manuscript of Nuremberg, written before the middle of the 15th century: "The techniques of fighting with the large knife are derived from those of the long sword, so that the fundamentals and principles that apply to the sword, also apply to the knife."[332] According to this document, nothing would prevent either the development or the writing of a synthesis on an art dedicated to long Messer fencing which can rival in richness what is presented for the long sword. However, most of the repertoire of one-handed sword or long Messer fighting is limited to defensive actions against a simple attack. In fact, only Lecküchner seems to have applied and developed this principle laid down long before.

These few examples at least demonstrate an interest in the art of fencing with the long Messer, and Peter Schwyzer's example confirms that it was taught by fencing masters during that time. This native of Bern was a working fencing master at the end of the 15th century and seems to have travelled from town to town looking for people interested in a teaching class of a few weeks. He came to Baden im Aargau, near Zurich in 1495, petitioned to organize a public 𝔉𝔢𝔠𝔥𝔱𝔰𝔠𝔥𝔲𝔩𝔢, and it seems offered his teaching to anyone for potentially any set of weapons: sword, dagger, staff, axe, halberd, as well as the preparation for the judicial duel (𝔎𝔞𝔪𝔭𝔣) both on foot and on horseback, but also for what interests us, the Messer.[333]

The art of knife fighting as a leisure activity

The example of Peter Schwyzer proves, if necessary, that the Messer was well taught among other weapons, but this archival document does not specify whether this weapon was used during the 𝔉𝔢𝔠𝔥𝔱𝔰𝔠𝔥𝔲𝔩𝔢 which he intended to organize.

This term 𝔉𝔢𝔠𝔥𝔱𝔰𝔠𝔥𝔲𝔩𝔢 also appears in the regulation of 1503 of Suab parishes already mentioned before: in addition to prohibiting Messer carrying to laity and religious, the regulation enjoins religious people to avoid inappropriate and dishonorable games, in particular with lay

---

[332] *Nürnberger Hausbuch* (N), ff 82r–82v. *MMMM* Ms. 25.

[333] Israel and Jaser 2016, p 223.

people, and not to participate in public dances, public fights (offnen vechtschulen) or other prohibited hobbies.[334]

The Fechtschule, a specificity from the German speaking area, is the most famous sports game featuring fencers. It wasn't necessarily the only one. The knights, of course, stage courteous duels in front of an audience, and possibly scripted in the *pas d'armes*. But fencing could be taught outside the knightly field and practiced without a standard framework. It was moreover part of a positive vision of the body in movement in the Germanic area in the 15th century.[335] The Fechtschulen are especially well documented during the second half of the 16th century. They are organized in the form of fencing assaults, and fulfilled several roles, including examination for a future fencing master, refereed assaults for fencers, source of income for the fencing master, but above all a popular spectacle.[336]

The very first mentions of Fechtschulen, i.e. public sessions of fencing assaults, date back from the early 15th century, and are known to us mainly at that time by laconic mentions without much detail that would allow us to know whether this label clearly designated the same type of event in the 15th century as what took place a century later. The oldest reference that I know of is in an interpretation of a religious text by Ulrich von Pottenstein, written before 1416 (the date of his death), but found in a later copy from the mid-15th century. The excerpt is short, but interesting because it is presented to the reader with an inversion of values, depicting the Fechtschule (here named in an archaic form, Schirmschule, common during the 15th century) as an attribute specific to the nobility, and contrasts it with the teaching that clerics can receive:

> wann da die pfaffen in der schirmschul studirten vnd die fursten in der münich schul[337]

> "If the priests study in the fencing school and the princes in the monk school…"

The second oldest reference is more detailed. It is reported by a 19th century historian who summarizes in modern German an episode from a chronicle written at the beginning of the 16th century. The episode itself would take place in 1444 in Rothenburg-an-der-Tauber, with a fencing master who suspended sword and crown to recruit students by organizing a Fechtschule. Without access to the original text, it is not possible to know whether the historian has modernized the term used in the chronicle. This mention indicates that at this time already, the word Fechtschule designated

---

[334] KRAUS 1950.
[335] DUPUIS 2020.

[336] TLUSTY 2011, pp 217–21.
[337] RANKE 1951, p 186.

Fig. 4: Throwing a fencer into a sack (LM, f 92r)

a particular event bringing fencers together, and not a school in the current meaning of the term.[338]

In 1463, the city of Lucerne seemed annoyed with the organization of a 𝕱𝖊𝖈𝖍𝖙𝖘𝖈𝖍𝖚𝖑𝖊 or 𝕾𝖈𝖍𝖎𝖗𝖒𝖘𝖈𝖍𝖚𝖑𝖊 in front of the town hall and required fencers to ask for permission beforehand.[339]

A few years later, in Nuremberg, a fencing master petitions for organizing such an event during Carnival. The archives of this city then show a handful of such requests from fencing masters: about a dozen are recorded in the archives between 1477 and 1495. The denomination varied between 𝕾𝖈𝖍𝖎𝖗𝖒𝖘𝖈𝖍𝖚𝖑𝖊 and 𝕱𝖊𝖈𝖍𝖙𝖘𝖈𝖍𝖚𝖑𝖊 until the end of the 15th century. One should note in particular one of them where the archive document avoids the prefix 𝖘𝖈𝖍𝖎𝖗𝖒- and only retains the name of 𝕾𝖈𝖍𝖚𝖑 for the event:

> einem schirmaister ist vergönnt uf morgen schul zu halten (1478)

This interest is motivated by the fact that this contracted form can be found in Lecküchner's treatise on two occasions:

> So pestell heymlich etlich dy eynen sack verporgen pey in haben auff der schull dy hintter dem volck stenn[340]

> "So secretly appoint some men who have a hidden bag with them at the swordsmanship competitions and stand behind the people."

> das stuck ist lecherlich zu treyben auff der schull[341]

> "It is a ridiculous element to perform during a swordsmanship competition"

The first sentence occurs in a technique where the opponent is thrown into a bag. This bag here would be kept hidden by accomplices standing behind the public. This audience is gathered there in a standardized framework, the 𝕾𝖈𝖍𝖚𝖑. The second extract is understood as a

[338] KLEINAU 2012.

[339] VON ELGGER 1873, p 259.

[340] LM, f 92r. Translation by ŻABIŃSKI, MITCHELL, FRITZ.

[341] LM, f 183v. Translation by ŻABIŃSKI, MITCHELL, FRITZ.

technique adapted in the context of a leisure or spectacle practice, which means that there is no intention to seriously injure one's opponent, and respecting some of the rules in place.

A third mention exists, albeit only in the first version of Lecküchner's work, where he explicitly mentions a fencing event:

> Haw ym zw dem halz wiltü yn nicht hart wünden so schlag yn auff den armen ist aber daß auff eyner vechtschullz so schlag yn yn den pawch auff seyner lincken seytten grob und pewrisch...[342]

> "Strike him to the neck. If you do not want to hurt him hard, strike to his arm. If it occurs at a Fechtschule tournament, strike to his stomach to his left side, roughly"

Fig. 5: A ridiculous element (LM, f 183v)

It seems clear here that Lecküchner indicates the necessity of adapting the sequence to the context. Here, instead of performing a lethal strike, a blow to the back of the neck, he offers instead a strike on the flank (with a touch of brutality).

Is Lecküchner the only German author of his time to mention the Fechtschulen? The *Nürnberger Hausbuch*, written prior to Lecküchner's manuscript, compiles several texts on different topics, some of them dealing with sword fencing. Here it is possible to find three mentions of the word Schulfechten (ff 14v, 44r, and 52v). Its interpretation is not obvious, but it does not seem to have any relation with the Fechtschulen: according to Eric BURKART, "it would be translated literally as school-fighting, therefore first of all referring to a non-lethal fighting exercise." This implies a more general meaning than a ritualized public contest.[343]

Lecküchner's text is therefore the only one, to my knowledge, among the very large number of fencing treatises to explicitly mention these competitions and public shows. He does so at a time when these shows and physical competition spaces are starting to spread and appear more regularly in administrative sources, which could give the impression that he is reacting to a recent fad. This impression is undoubtedly false, in particular when one notes that on the one hand, it makes only three references to such events, and only two in the second manuscript. On the other hand, the

---

[342] LH, f 15r. Translation by ŻABIŃSKI, MITCHELL, FRITZ.

[343] BURKART 2016, p 468.

associated techniques bring specific adaptations or enrichments to these events, at the end of complex sequences: it is difficult to imagine these contributions without these fencing shows or competitions being already well known and disseminated.

Conclusion

Lecküchner's masterful work is a successful response to the invitation to apply Liechtenauer's principles of fencing to the long knife. He took up his quill at a time when this subject had only been very weakly mentioned by other fencing specialists, contenting themselves with proposing defensive solutions in the event of a direct attack when one only had a sword or a cutting weapon of the same size. It is true that in the 15th century, the wearing of the Messer became commonplace in German-speaking areas, carried on a daily basis by a large part of the population from the peasant to the priest, not to mention the men-at-arms.

Lecküchner has significantly developed Messer fencing both with a view toward combat efficiency when life is at stake, and also by explicitly proposing techniques suitable for fencing games. Though this remains quite rare, in addition to using the words **ernst** and **schimpf** (or their adverbial form) he is able to refer directly to the play space by mentioning the use of these technique in a school (**Schul**) at a time when the practice of sport or leisure fencing is still poorly documented. His work therefore constitutes an interesting secondary witness.

To conclude, I propose to slightly divert the following remarks of Lecküchner, which for me seems to address well to the art of Messer fighting in general: "you have various elements and breakings, for amusement and serious combat. If you do not like one, take another one."[344]

---

[344] "*du hast manigerlay stuck und pruch Schimpfflich und ernestlich gefellt dir eynes nicht so nym eyn anders*". LM, f 171r. Translation by ŻABIŃSKI, MITCHELL, FRITZ.

CASPER J. VAN DIJK **and** OSKAR TER MORS

"The Art that Hans Lecküchner Made and Devised Himself": The Martial Arts Tradition of a Fifteenth Century Bavarian Priest

If you cannot do this, then get instruction in it from someone who can. Even if you know all the masters' arts, if you don't know when, how, or where, and against which parries, cuts, plays, counters, or slices to execute the one and not the other, then it is no use to you.

~ Johannes Lecküchner, 1482

kanstu des nicht so laß dich eß vntterrichten eynen der es kann vnd kunstu aller meyster kunst vnd west nich wenne wye oder wo vnd gegen welchen versetzen hewen stucken pruchen schnytten dytz oder geneß du das treyben solt so ist dir das keyn nütz

he priest Hans Lecküchner of Nürnberg (ca. 1440–1482)[345] composed the largest collection of teachings on the weapon called the langes Messer (literally translated: "long knife") known today. The Messer is a single-edged sword-like weapon with a blade of approximately 60 cm long and a relatively long handle (c. 20 cm), intended to be primarily used one-handed. Lecküchner's teachings on this weapon are recorded in two versions of his fight book *Zedel ym Messer* ("Notes on the art of the knife"): the more extensive and illustrated *Hans Lecküchner-München* (LM), written in 1482 and composed of 432 pages (216 folia), and *Hans Lecküchner-Heidelberg* (LH), which was a preparatory version of LM, written in 1478 and composed of 244 pages (122 folia).

Compared with other, earlier fight books, Lecküchner's work is the first fight book fully dedicated to the Messer and dwarfs the other books in terms of length and number of presented techniques or plays. As a way of illustration, in the *Baumann fight book* (B, 1420s-70s) only 8 pages are dedicated to Messer fighting; in *Hans Talhoffer-München* (HTM, 1467) again just 8 pages are addressed to Messer; and in *Paul Kal-München* (PKM, ca. 1470) only seven pages are spent on discussing the Messer. This difference in size raises the question of whether Lecküchner was part of a Messer martial arts tradition which was not codified earlier, if he dramatically expanded on this tradition, or if (as implied by the first line of LM[346]) he invented a completely new style of Messer combat?

---

[345] FORGENG 2015, p XI.

[346] *"Das is herren hannsen Leckuchner von Nurenberg künst und zedel ym Messer dy er selbs gemacht und get\ icht hatt".* LM, f 1r.

In order to get a grasp on these questions, Lecküchner's fight book can be juxtaposed with earlier works to look for similarities and examine if he was possibly influenced by other authors of fight books and whether these can be considered of the same tradition or style.

Traditionally, the teachings in fight books have been compared using textual and/or iconographic analysis, to establish a possible epistemic lineage between fight books and their authors.[347] For example, using this method, an influence of the Liechtenauer long sword tradition on Lecküchner's teachings can be established. It is possible to trace some parts of LH back to the *Lew fight book* (L, 1450s). A few highly specific plays from Lew that do not appear in any other long sword gloss can be found in Lecküchner as well (the **gehűlz stoss** or "hilt shove", on folio 17v and 25v respectively, is a good example). There is a further intriguing clue to be found in the opening explanation of Lew's **Krumphau** ("crooked cut"), where the text explains that this cut works against "the guard called Ox, as well as **Eber** and **Unterhaw** ("strikes from below")".[348] **Eber** is in this case a misspelling of **Ober** ("above"), as the gloss clearly goes on to describe the **Krumphau** countering cuts from above.[349] Lecküchner follows this scribal error to the letter, which makes it rather more confusing, as the **Eber** ("Boar") is also one of Lecküchner's four **Leger** ("encampments") or guards.

When running a full, side-by-side comparison of the two manuscripts, it appears that Lecküchner has copied the structure of existing long sword glosses. This creates the idea that the manuscript has been largely based on those glosses, but this does not hold up to closer scrutiny. About 59 of the 432 pages in Lecküchner's fight book contain material that was copied from Lew.[350] In other words, 13.6% of Lecküchner's treatise has a direct, textual link with the mainline Liechtenauer tradition. Additionally, counters, continuations and variations on these plays will bring this percentage to around 30%; however, the association with other fight books, in particular those concerning Messer fencing, still remains to be discovered.

As demonstrated, textual analysis can establish previously unknown connections between fight books. As a research method, however, it fails to find the influence authors had on each other when no textual relationality can be established within fight books. This is because text analysis commonly equates the text of fight books with the teachings themselves and thereby overlooks the fact that fight books are a description,

---

347 E.g. HAGEDORN 2016, pp 247–79; HAGEDORN 2017, pp 361–414; FORGENG 2017, pp 8–10.

348 *"Wiss das krump haw ist der vier versetzen eins wider die vier hut Wann damit kriegt mand den ochsen vnd auch den eber, Vnd den untterhaw".* L, f 13v

349 L, f 14r.

350 LM, ff 1v, 2r, 3r, 3v, 9v, 10v–12r, 14v, 17r–18r, 20r, 25r, 25v, 26v–27v, 28v, 29r, 33r–34v, 37r–41r, 46r, 62r–64v, 67r, 68r, 68v, 72v, 73v, 75v–77v, 102v–105r, 183r, 198r.

and thus a derivative, of the physical reality in which a martial art is enacted. Consequently, in order to establish an epistemic connection between fight books, not just the text and illustration style of two or more fight books should be correlated, but also the movements they describe should be analyzed and juxtaposed. Such an approach can then recognize similarities between textually unrelated treatises or identify an earlier non-disclosed martial arts style or tradition.

Therefore, within this article the techniques and plays described by Lecküchner will be compared to techniques and plays in other contemporary Messer treatises from Southern German lands written before Lecküchner created LM, to examine the influence earlier fight books had on Lecküchner. When a connection can be established, it is investigated whether these books can be considered to be of the same martial arts tradition. As discussed, this comparison will not be made through textual similarities but by resemblance of movements and their intentions. LM has been selected because it contains illustrations which make interpretations of plays and techniques easier.

When attempting to establish any relationship between the LM and other sources, it is important to note that this version of Lecküchner's fight book does not appear to have influenced any later fight books. All later copies of *Kunst und Zedel ym Messer* have been based on LH.[351] For the purpose of this article, it is therefore most relevant to compare the LM only to Messer manuscripts that predate it. This has the benefit that it gives us a manageable dataset to use our method on. Examining the relationships between copies of Lecküchner's treatise and later examples of fencing traditions inspired by it would make for interesting research but would quickly outgrow the available space and time.[352]

To understand the relationship between the text in fight books and the movements they describe, the limitation in expressing the knowledge necessary to perform motor-skills is discussed first. Based on this discussion, it is argued that physically training techniques is a requisite in examining them. Then the concept of martial art is discussed and an approach for examining martial art tradition in medieval fight books is presented. After the theory part of this article, movements and gestures presented in HTM, PKM, and B are compared to Lecküchner's teachings using the proposed methodology, and the article is ended with a discussion of this comparison, including a conclusion.

---

[351] "Johannes Lecküchner", *Wiktenauer*. Ed. Michael Chidester http://wiktenauer.com/wiki/Johannes_Lecküchner

[352] Fight books from Southern German lands include *Hans Talhoffer-München* (HTM, 1467); *Baumann fight book* (B, ca. 1470); *Paul Kal-München* (PKM, 1470); *Peter Falkner* (PK, 1495), *Albrecht Dürer* (AD; 1512).

Embodying disembodied texts

Jaquet *et al* defined fight books as:

> "The *terminus technicus* used to indicate a vast and heterogeneous collection of manuscripts and printed books, destined to transmit on paper in a systematized way a highly complex system of gestures or bodily actions, often, but not always, involving the use of weapons of different sorts".[353]

There is, however, a conundrum in using lingual description, and to a certain degree imagery, in transmitting a "complex system of gestures or bodily actions". Humans do not only possess conscious cognitive processes, which the expression of language is part of, but also have unconscious faculties. Motor skills, including combat, are part of this unconscious dimension and are difficult or impossible to communicate verbally.[354] This type of knowledge is dependent on an academic discipline known by a host of different terms or concepts, but will be referred to here as 'embodied-knowledge'.

Embodied-knowledge has a peculiarity: even if it is expressed in language, this does not mean one can learn it by merely reading or hearing it. The philosopher Michael Polanyi, for example, described the rules for riding a bicycle as follows:

> "[F]or a given angle of unbalance the curvature of each winding is inversely proportional to the square of the speed at which the cyclist is proceeding".[355]

Solely by reading the rules and consciously understanding the various terms used, it is still not possible to hop on a bicycle and cycle off, i.e. have the embodied-know-ledge to cycle. A cyclist must practice cycling, to embed it in his/her muscles, central nervous system, and neural pathways in

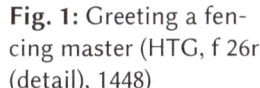

Fig. 1: Greeting a fencing master (HTG, f 26r (detail), 1448)

[353] Jaquet, Verelst, and Dawson 2016, p 9.
[354] E.g. Reed 2010, Bengson and Moffett 2011.

[355] Polanyi 1958, p 50.

the brain, before (s)he can ride off.[356] Due to the inadequacies of language to express embodied-knowledge, teaching physical skills has traditionally been done by a master through example.

This necessity for learning embodied-knowledge from a master and through physical exercise, was of course recognized by authors of medieval fight books. For example, the author of the *Nürnberger Hausbuch* (N) states:

> wisse das man nicht gar eygentlich und bedewtlich von dem fechten mag sagen und schreiben aber aus legen / als man is wol mag czeigen und weisen mit der hant

> "[K]now that one cannot truly and meaningfully speak or write about fighting, just expound on it. Nevertheless, man can show and demonstrate it by hand."[357]

A similar emphasis on learning martial arts through bodily exercise is presented by Joachim Meyer in his *Gründtliche Beschreibung der... Kunst des Fechtens* (1570), where he mentions:

> Die ander ursach ist dise, Nemlich das sich solche Ritterliche Fechkunst / schwerlich last in Bücher schreiben / oder mit Büchstaben verfassen / als die allein durch übung des gantzen leibs ins werck müß gerichtet werden

> "[T]his knightly art of combat hardly allows itself to be written in books, or composed in writing, since it must be executed through the practice of the entire body."[358]

However, if the written or spoken word are not especially suitable for learning embodied-knowledge and medieval authors recognized this themselves, then why were martial arts techniques recorded in books? In order to answer this question, fight books should not be seen as a replacement for learning embodied-knowledge through exercise, but should be interpreted differently, as explained by Fiore de'i Liberi (ca. 1404):

> sença libri non sara çamaii nissuno bon Magistro ne scolaro in quest' arte... quest'arte e si longa che lo non e al modo homo de si granda memoria che podesse tenere a'mente senca libri la quarta parte di quest'arte

> "Without books you cannot be either a good teacher or a good student of this art. ...[T]his art is so vast that there is no one in the world with a memory large enough to be able to retain even a quarter of it".[359]

---

[356] MATSUZAKA 2007.
[357] N, f 15r.

[358] FORGENG 2006, p 41.
[359] LEONI 2012, p 25.

Fig. 2: Training with Dussack under the supervision of a master (Meyer, p [2]-3r, 1570)

In accord with retaining learned teachings Meyer emphasizes:

> vom lernenden vil besser, wann sie ihme neben guter anweisung, in rich-
> tiger ordnund zusamen gesetzt, fürgeschrieben, und für augen gestelt, ins
> gedechtnus eingebildet, ...so kan sich hierauß dei auffwachsent jugent,
> nach dem sie von einem rechten Meister gelernt, und aber denselbigen
> nicht alzeit bey sich hat, erinnern

> "Students can conceive it [i.e. the art of combat], in their memory
> much better when it is assembled, written out, and placed before their
> eyes in a proper pedagogical order. ...[T]he students can refresh their
> memories with it once they have been instructed by a proper master
> whom they cannot always have with them."[360]

Thus, fight books were not devices intended to learn unmastered skill from and they did not replace fencing masters. On the contrary, the oldest fight books were devices meant to help remember previously learned embodied-knowledge and to help structure learned techniques. This use of fight books within the medieval martial pedagogical system is beautifully illustrated by an illustration in the fight book *Jörg Wilhalm-München* (JWM1, 1523).[361] In the depiction, a book is laying on the ground while two individuals practice a movement. The illustration portrays

---

[360] Forgeng 2006, p 41.   [361] JWM1, f 19v.

that fight books stored embodied-know-ledge which only can be un-locked through bodily practice and not by merely reading it.

Knowing that fight books store embodied-knowledge, which was intended to be understood through physical exercise, has implication for examining the origin of combat techniques. First, by comparing verses using text analysis, you do not compare embodied-know-ledge concerning combat techniques between fight books but the

Fig. 3: Fencing with a book nearby (JWM1, f 19v, 1523)

conscious conceptualization of this knowledge by a certain epistemic community. While this comparison of text provides insight into the mental or scholarly frameworks used by different authors or communities, it does not, nevertheless, impart understanding of similarities and transformation of embodied-knowledge across time. For instance, the same embodied-know-ledge can be described by different verses in different fight books. The de-scribed combat techniques in two different fight books can be the same, however, the words used to describe them can be different. Thus, merely comparing the text does not demonstrate their relationship.

Second, when comparing techniques between fight books, the language used to describe them needs to be translated into embodied-knowledge. This can only be done through physical practice and, consequently, it is neces-sary to physically train the techniques described in order to examine them. Here a conundrum exists: without physically practicing the techniques, they cannot be understood; however, the masters needed to unlock the tech-niques are no longer with us. To solve this enigma, the teachings in fight books must be reconstructed as faithfully possible, even if the true nature of these teachings can never be reached due to absence of a master. Through this physical interpretation of the teachings, it is possible to gain insights which are otherwise unattainable.[362] The interpretation of the technique should be photographed, or preferably filmed, to aid in its understanding by other scholars, so they can provide feedback and recreate it.[363] Interpreta-tions of the combat techniques discussed further in the article are available on the Instagram or YouTube pages of the authors.[364]

---

[362] Cf. CLEMENTS 2016, p 190; HANSEN 2008, pp 69–80.
[363] VAN DIJK 2020, pp 30–32.
[364] See: http://www.youtube.com/c/Virtual Fechtschule/featured

and
http://youtube.com/playlist?list=
PLk-YUsqPrVP3AIrkuFxe8WI2zAMIwwFGX

Recognizing different arts

If Lecküchner is to be placed in or possibly to be considered the founder of a martial arts tradition, the concept of martial arts is to be defined first before a method for recognizing it in medieval fight books can be discussed. On first sight this appears to be an easy task; however, the precise definition of martial arts has been debated among scholars and no academic, universally-agreed-upon definition has emerged. The main problem is that whatever definition is taken, either activities that are colloquially known as martial arts are excluded or endeavors that are commonly not considered as martial arts are included.[365]

Based on the difference between object- and meta-language, WETZLER provided a solution to the martial arts definition problem.[366] The object-language describes the world in factual terms, while the metalanguage describes the language that is describing. For examining martial arts, WETZLER argues, researchers should not ask 'how a strike is performed' but should rather ask 'how do historical individuals *believe* a strike is performed'. WETZLER basically defines anything as a martial art, as long as it is considered by an individual to be a martial art. He thus uses self-definition as the primary criteria for a martial art. To examine these self-defined martial arts, they can be grouped into various styles, which are "individual tradition, imagined as a coherent entity from the inside, and more or less clearly distinguishable from the outside".[367] Each style can be described according to how much it fulfills various dimensions of meaning. WETZLER described five dimensions, but more could be added. These dimensions are:

1) Preparation for Violent Conflict
2) Play and Competitive Sports
3) Performance
4) Transcendent Goals
5) Health Care

This metalanguage definition method has been criticized by JUDKINS, however.[368] He argues that some boxing or kickboxing students state they do not practice martial arts but, for instance, a combat sport. In a martial arts studies conference, on the other hand, papers on these sports are commonly included, even if the practitioners themselves do not consider themselves to be martial artists.[369] JUDKINS' answer to the definition problem is to alter the definition to the question asked. The

---

[365] For a discussion see BOWMAN 2020, Introduction.
[366] WETZLER 2015, pp 20–33.
[367] Ibid, p 25.

[368] JUDKINS 2016, p 6.
[369] Ibid, p 9.

academic definition of martial art is then dependent on the question asked.

Based on JUDKINS' work, the definition of martial arts in this article should be adapted to the question asked here, which is: how much overlap is there between various fight books, and is this overlap enough to classify it as a martial arts tradition? Whatever martial arts are, the majority of people would minimally (intuitively) define it as involving a form of 'bodily action', even if this action cannot be defined strictly. Moreover, it is bodily action that fight books describe, as described by JAQUET et al.[370] Consequently, bodily action, and thus embodied-knowledge, should be part of martial arts. Besides bodily action, martial arts have another fundamental part, as captured in WETZLER's dimensions of meaning. That is their cultural, political, and/or societal function. Thus, Lecküchner and other authors of fight books would be part of the same martial arts style or tradition when: their teachings overlap in both (1) their combat movements and gestures, i.e. embodied-knowledge, and (2) the intended meaning of these movements and gestures.

In order to examine the overlap in embodied-knowledge and meaning across fight books, the teachings in the book will be divided into three categories.[371] These categories are a modern, not medieval, classification; they are a heuristic tool intended to better understand the intention of medieval fight books. Moreover, these categories should not be seen as starkly defined groups, but as guidelines to group teachings together. The categories are:

- *Techniques*, which are a conglomeration of bodily movements or gestures that were classified as independent entities in fight books, denoted by their own terms or specific description. This implies that a similar group of movements can be named by a different term in different fight books. Techniques have no inherent meaning because they merely describe a motion without a particular objective, for example, strike someone in the head.

- *Principles*, which describe the application of techniques according to external physical parameters, such as timing, speed, strength, or distance to/of an opponent. They are the rules for applying techniques.

- *Plays*, a set of movements composed of: (1) a beginning situation, commonly a problem to be solved, for instance an attack or a guard to be broken; (2) one or several techniques, each one countering another; and (3) an end-result, the objective or intention of the play. Plays present a concrete example for a single or multiple techniques

---

[370] JAQUET, VERELST, and DAWSON 2016, p 9.   [371] Cf. VAN DIJK 2019, pp 41–42.

and principles. By analyzing the objective of a play, it is possible to understand the intention and possible meaning of the teaching.

Armed with these concepts, it is possible to do a close reading of textually unrelated plays and establish a possible relationship between them. Relationships between martial arts styles can exist on the level of technique, where movements are the same or very similar, even though the method of transmission in the fight book is different. They can also exist on the level of principles, where the tactical concepts in the plays can be compared. Finally, we can use the dimensions of meaning to compare the meaning of different plays.

Plays will be classified either in the dimension of '(Preparation for) Violent Conflict' or 'Play and Competitive Sports', based on the level of the physical trauma that is inflicted in them. For example, in Heidelberg 430, Lecküchner provides several intentions and thus meanings for the same technique. He states, you can either "strike him in the neck; if you do not want to hurt him then strike him on the arm; however, if it is during a 𝔉𝔢𝔠𝔥𝔱𝔰𝔠𝔥𝔲𝔩𝔢 (fight school) then strike him on the belly".[372] Striking an opponent in the neck would classify as the dimension of meaning of 'preparation for violence', as it would inflict a fatal wound, while striking someone on the belly fulfils the dimension of 'play and competitive sports'. By interpreting plays in fight books, we can therefore compare the techniques, principles, and the context of those plays. Hence, for establishing a martial arts tradition in fight books, first the continuation of techniques, as bodily actions, and principles across fight books, should be confirmed. Then, the application of these techniques in plays should be examined, to apprehend if a similar meaning was intended for these techniques by the authors of the fight books.

"Particular plays and counters"

Having previously established that Lecküchner based a sizeable portion of his fight book on Lew's gloss, we will now examine where at least part of the rest might have come from. Interestingly, Lecküchner seems to have taken the seventeen 𝔥𝔞𝔲𝔭𝔱𝔰𝔱𝔲̈𝔠𝔨𝔢 ("principal plays") of the Liechtenauer tradition and added another six to them. These six other 𝔥𝔞𝔲𝔭𝔱𝔰𝔱𝔲̈𝔠𝔨𝔢 are of particular interest when it comes to looking for other influences.[373]

There are four sources on fencing with the Messer that were clearly written before Lecküchner's fight books: the *Nürnberger Hausbuch* (N,

---

[372] "*haw ym zw dem halz wiltü yn nicht hart wünden so schlag yn auff den armen ist aber daß auff eyner vechtschullz so schlag yn yn den pawch*". LH, ff 14v–15r.

[373] For more on Lecküchner's *Haupstücke*, see Jessica Finley's chapter in this book.

1390–1494), *Hans Talhoffer-München* (HTM, 1467); *Paul Kal-Bologna* (PKB, 1460–80) and *Paul Kal-München* (PKM, ca. 1470) (the latter, being more complete, will be the version we will reference); and the *Baumann fight book* (B, ca. 1420s–70s). Of these, we must discard N, as it does not have any plays and only describes very general principles of fencing with the Messer instead. This leaves us three sources each associated with different traditions: PKM is associated with the 𝕲𝖊𝖘𝖊𝖑𝖑𝖘𝖈𝖍𝖆𝖋𝖙 𝕷𝖎𝖊𝖈𝖍𝖙𝖊𝖓𝖆𝖚𝖊𝖗𝖘,[374] HTM is affiliated with the 𝕸𝖆𝖗𝖝𝖇𝖗ü𝖉𝖊𝖗 and other masters such as Martin Syber,[375] and B with the Nuremberg tradition of fencing.[376]

Using our method to compare the plays from these manuscripts with similar ones from Lecküchner might allow us to ascertain a possible relationship between these manuscripts. We will go through the plays in a roughly chronological order, starting with HTM and ending with LM, and group these plays together based on nine archetypical techniques that they all have in common.

Fig. 4: Two guards of the Messer (HTM, f 113r, 1467)

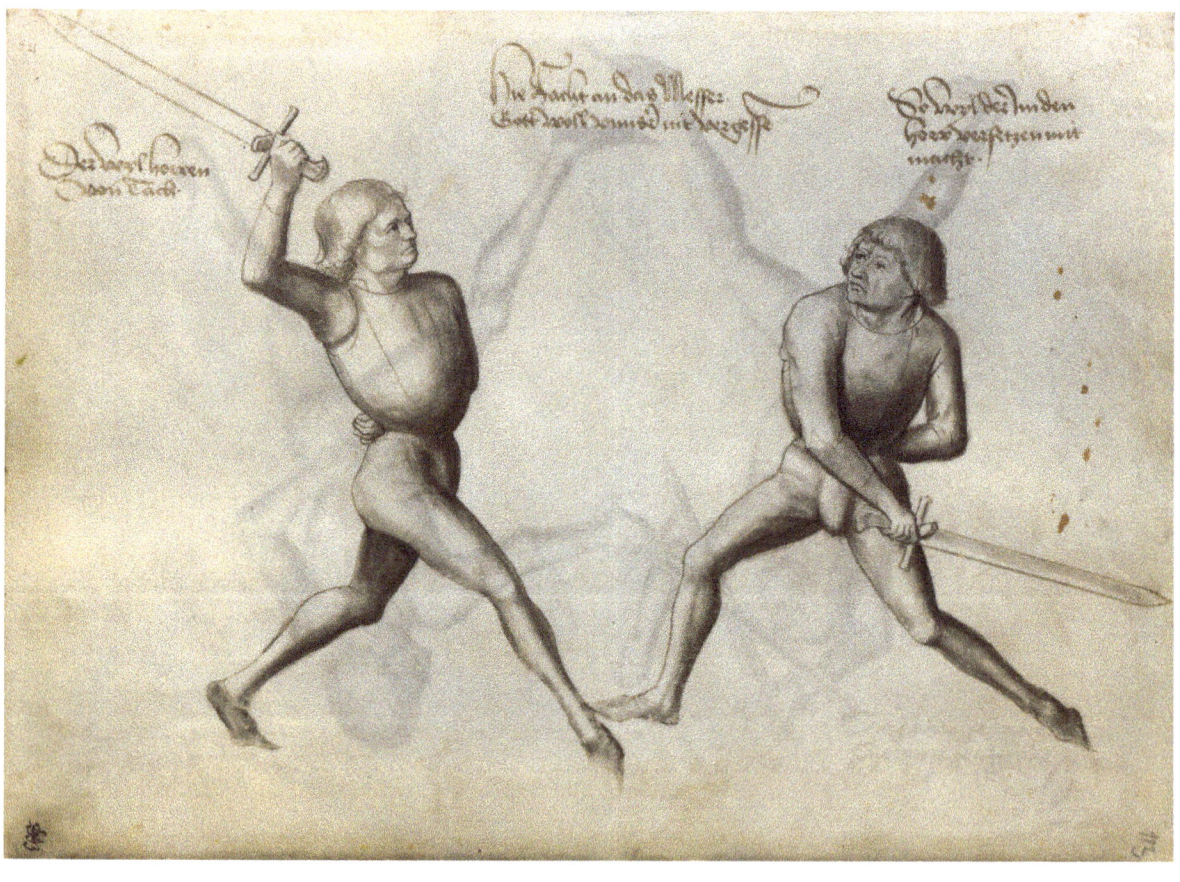

---

[374] Jens P. KLEINAU. "Paulus Kal, a Schirrmeister". *Hans Talhoffer.* https://talhoffer.wordpress.com/2011/07/03/paulus-kal-a-schirrmeister/

[375] "Hans Talhoffer". *Wiktenauer.* Ed. Michael CHIDESTER. http://wiktenauer.com/wiki/Hans_Talhoffer

[376] "Nuremberg Group". *Wiktenauer.* Ed. Michael CHIDESTER. http://wiktenauer.com/wiki.Nuremberg_Group

Grabbing over the arm

Fig. 6 (op.): PKM, ff 72v, 74r, & 74v, 1470

Fig. 5: HTM, ff 113v, 114r & 116v, 1467

HTM 113r–114r and 116v show a play with two different outcomes. The plays of this archetype hinge on a hanging parry with the left hand reaching underneath to gain control of the patient (the actor to 'lose' the play)'s weapon-arm by locking it under the armpit. HTM always shows the agent (the actor to 'win' the play) starting with the Messer at the hip, long edge toward the patient and with the right leg forward. For the principles governing the plays shown in HTM, we find that the patient seems to always initiate and that the agent performs several actions simultaneously with that attack. The play may be completed with a cut to the patient's neck or by pulling back the arm and making a thrust to the belly. Both possible outcomes would lead to serious injuries that would require explaining to the authorities, and as such they seem to fit in the dimension of violent conflict.

PKM 72v/74r comes from the starting situation on 72v, where agent and patient have both cut into a bind, creating the opportunity to grab over and make a hip throw or strike the patient (unspecified whether with the edge or the pommel). Although the technique is very similar to HTM in terms of timing, with regards to principle the starting situation in a bind is clearly different. With regards to the outcome of the play, there is the option to throw and strike, meaning that this play may not be exclusively meant for violent conflict, but also for play and competitive sport. The outcome of

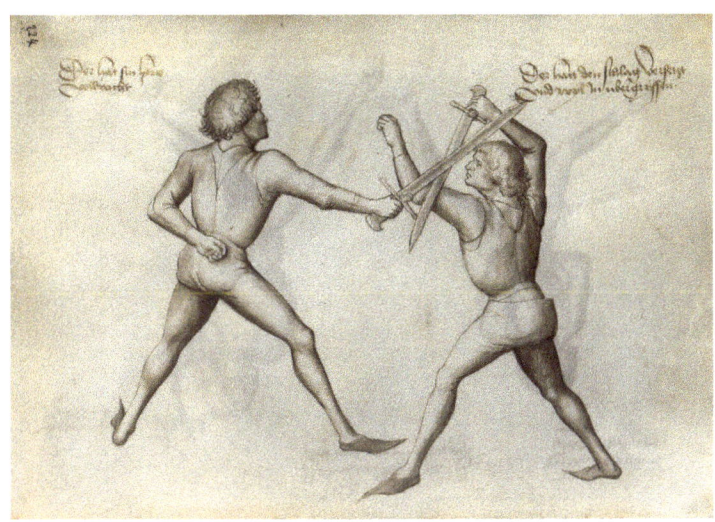

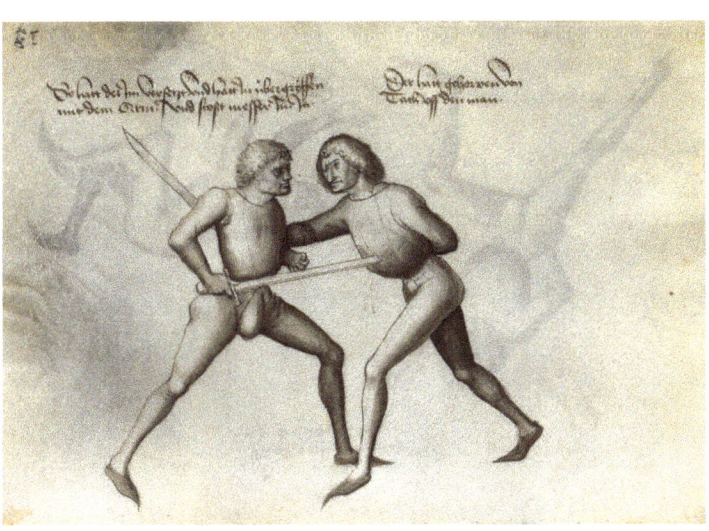

the counter to this play on 74v makes this somewhat unlikely though, as it implies a slice across the throat.

LM 156r/156v shows a play with the same archetypical technique, along with a whole host of counters, continuations, and variations. The agent begins with the left leg forward in this case and will jump forward with the same leg, using the same timing as the other fight books examined but a different starting position. Where LM differs most from the previous two plays is in the outcome (set the point on the throat and two different throws). He does not seem to provide an end to the play that seems explicitly geared to violent conflict and leaves the choice to deal trauma to the reader, making it viable for both play and violent conflict.

There are several counters to this technique: in PKM 74v, the patient is in the process of grabbing the agent's right arm. To counter this, the agent pushes his arm through the patient's armpit, allowing him to fold his blade over the patient's shoulder and in front of the throat. When we look at principles, we find that this particular play is meant as a counter and is once again dependent on simultaneous timing. The play (and the patient) is ended by slicing the throat, which would put this squarely in the dimension of violent conflict.

LM 158r includes a counter that seems almost identical to the one described in PKM. Technically, the movement to go over the shoulder in front of the patient's throat is the same, as is the step forward with the agent's left leg and the principles did not change much either. The only discernible difference is that Lecküchner tells readers to press the patient away from them, rather than slice their throat. This would leave the door open to use this play not only as preparation for violent conflict, but also for play and sports.

Wrenching the hand with the hilt

Fig. 7: HTM ff 113r, 114, & 115v, 1467

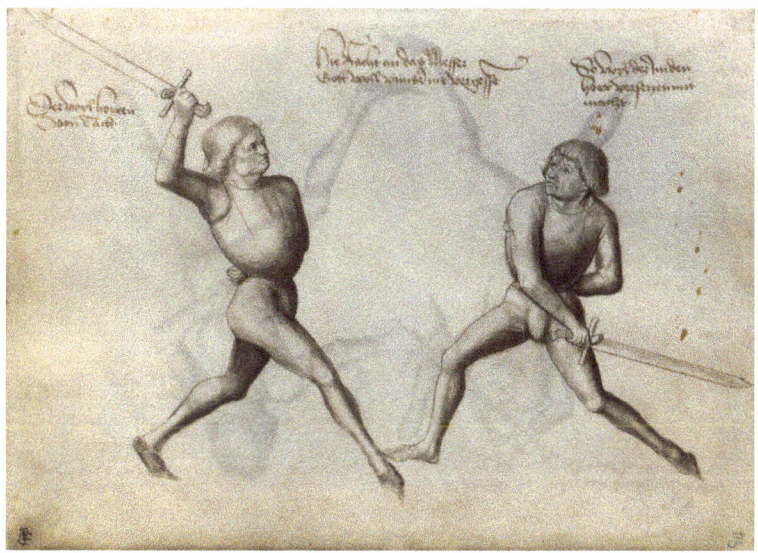

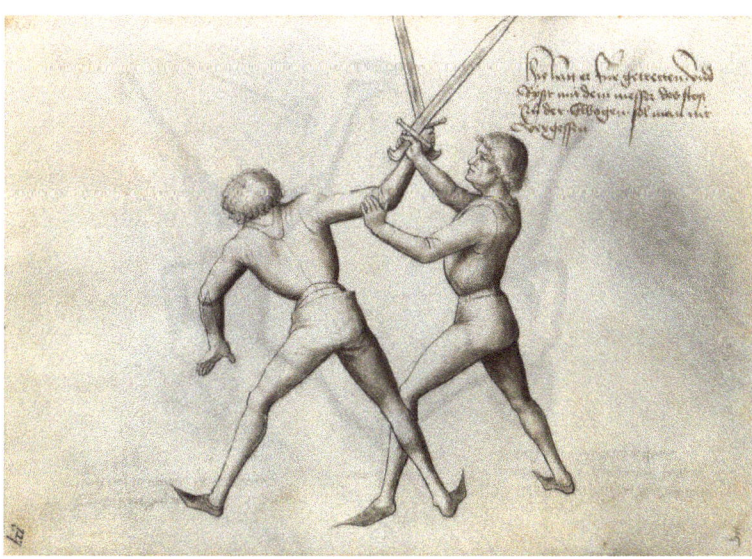

<u>HTM 113r and 114v–115r</u> revolves around catching a cut from above on the close to the cross of the Messer and then winding the handle over the patient's wrist. HTM starts this play from the same position as all his others, with the left leg in front and the knife at the hip with the long edge facing forward. When we examine principles to see how the agent interacts with the patient, we find that the patient once again initiates with a high cut from the right and that the agent first parries and then performs the technique in question. This is the only play in HTM that does not clearly state the intent to injure the patient grievously. The final position, pushing the elbow away, or performing an armbar, makes this a very viable option though. As such, we may assume that this play fits preparation for violent conflict.

<u>B 29r</u> does not provide a starting position, but we can assume that the agent starts with the right leg in front. As with HTM, the parry is done with the inverted hand and on the flat of the blade. The interaction with the patient is the same as well, but the play ends when the agent winds the hilt over the patient's wrist on the outside while

stepping with the left leg behind the patient's leg, cutting them in the head in the process. The notion of striking someone in the head with this particular action also seems to point to a preparation for violent conflict, as this cut is likely quite hard.[377]

LM 84r is more similar to HTM's version, with the noticeable difference that LM recommends letting go of the knife to be able to grab the patient's wrist (although in a later variation he offers the option of holding on to the knife as well). The main difference is that the inverted hand is swapped out for an 𝕰𝖓𝖙𝖗𝖚̆𝖘𝖙𝖍𝖆𝖜 ("beleaguering cut"). Now for the technique itself, the difference is minimal, with the thumb going to the flat of the blade and the short edge being oriented to make a cut. This makes a real difference for the principles involved though, because the play can now be done both offensively and defensively. The play ends

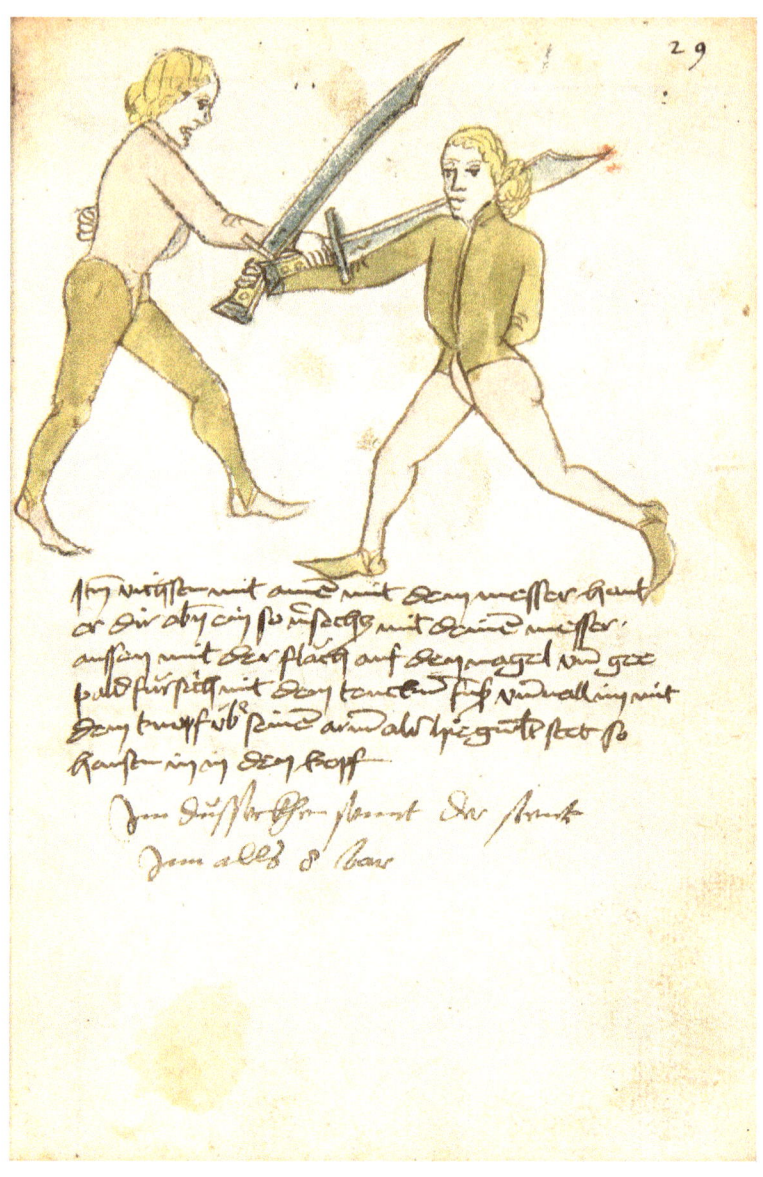

Fig. 8: B, f 29r, ca. 1470

with an armbar, which can point to the meaning being violent conflict and playful fencing. However, the relatively higher risk (in that one has to let go of the Messer to perform it) would be an indication that that meaning was more playful.

LM 171v is rather similar to the plays previously described. As with 84r, it provides the option to perform this play offensively by using the 𝕰𝖓𝖙𝖗𝖚̆𝖘𝖙𝖍𝖆𝖜. Another striking difference is that LM presents the reader with a choice of what injury to inflict at the end: either a full cut through the face, or a slice across it. Although both can be considered part of violent conflict, the latter one could be considered part of playful fencing as well, given the (relatively) low trauma it would cause.

---

[377] As experienced by one of the authors.

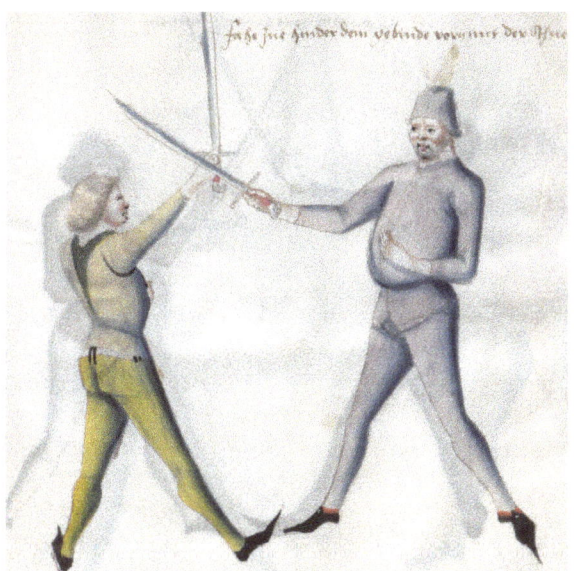

Free Hand-strikes

PKM 71v–72r targets the right hand of the patient, coming from a low guard with the Messer on the hip with an upward cut. The footwork in HTM, PKM, and LM is difficult to interpret. However, when comparing the possibilities in relation to what the patient does, it seems that the agent either steps away from the attack a little or strikes simultaneously with the patient. A play like this is generally quite firmly rooted in the dimension of 'preparation for violent conflict', since deliberately targeting the hand causes high trauma at relatively little risk to the agent.[378]

HTM 115v–116r contains the same archetypical technique, with potentially a footwork variation. The principles and the outcome of this play are not significantly different from PKM. Given the grievous nature of the wounds inflicted on the patient, this play fits preparation for violent conflict to the exclusion of other dimensions.

While Lecküchner has a great many variations on the theme of hand-strikes, LM 185r in particular seems very similar in its technique and principles to the ones described previously. In this case the target is the patient's right hand or forearm to the outside. Given that in the similar Lemstück ("laming plays"), LM provides the option to target the hand or forearm, and that those are indicated for both schimpff (playful combat) and ernste (violent conflict) we may conclude that for Lecküchner, the hand-strikes fit both of these dimensions of meaning.

---

378 VAN DIJK 2019, pp 69–70.

Hand-strikes from a bind

<u>PKM 72v–73r</u> comes from the starting situation described on 72v, where agent and patient have simultaneously cut, entering in an equal bind. This archetype relies on striking down from a bind. In this case, the patient seems to not displace strongly, meaning the cut lands on the inside of the wrist. As with the other plays of this sort, a hit to the hand is a strong argument to assume this play is meant as a preparation for violent conflict.

<u>LM 187v</u> is interesting, as it is similar in technique, but becomes very different when done in relation to an opponent. Lecküchner seems to assume a hard bind that pushes the blade to the side, ensuring the cut lands on the outside of the wrist. Interestingly, the illustration again shows a hit to the arm rather than the hand, indicating that Lecküchner considers this play feasible for play and sport, in addition to violent conflict.

"The left hand is the enemy of the right"

While the name for the archetype we see in <u>PKM 72v and 73v</u> comes from N, we see plenty of examples of this technique in action in the fight books we are examining. They all revolve around using the left hand to wrench the opponent's hand down. PKM shows this play coming from the equal bind on 72v and finishing it on the inside of the patient's wrist. Having done this, the agent is in a position to make a cut to the head, meaning it can be used as preparation for violent conflict, with low risk and potentially high trauma. The fact that control can be established without necessarily inflicting that trauma indicates it can potentially be used in playful combat as well.

Comparing <u>LM 179v</u> to the one in PKM is interesting, as the technique seems to be the same but it functions completely differently in relation to the patient. Where PKM starts the play with an equal bind, LM uses the ꝓogen ("bow") to parry the patient's right high cut, letting the blade hang down to the left, whilst grabbing the wrist from the inside with the left hand and putting in the left leg as well. This means that this play is to be used

Fig. 10: PKM, ff 72v-73r, 1470

Fig. 9 (op.): PKM, ff 71v-72r, 1470; HTM, f 116r, 1467

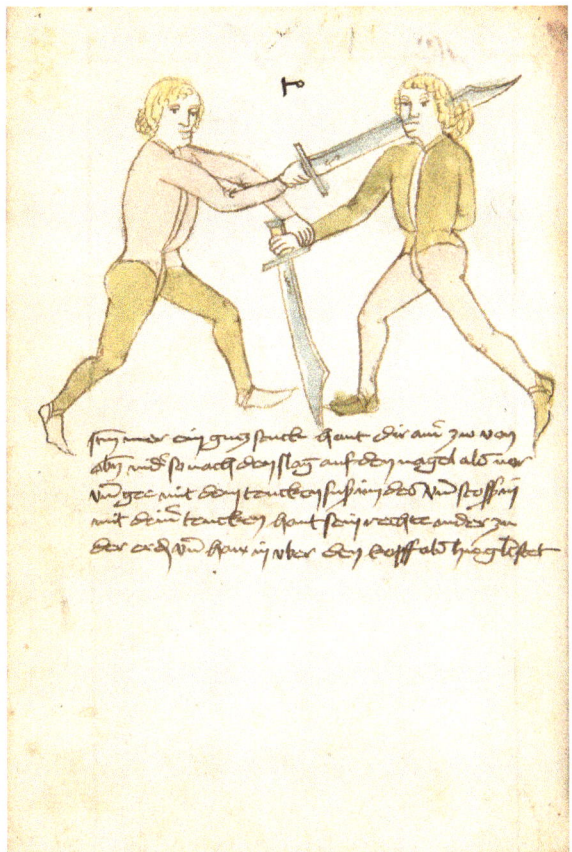

Fig. 11: B, f 29v, 1470

Fig. 12: B, f 30r, 1470

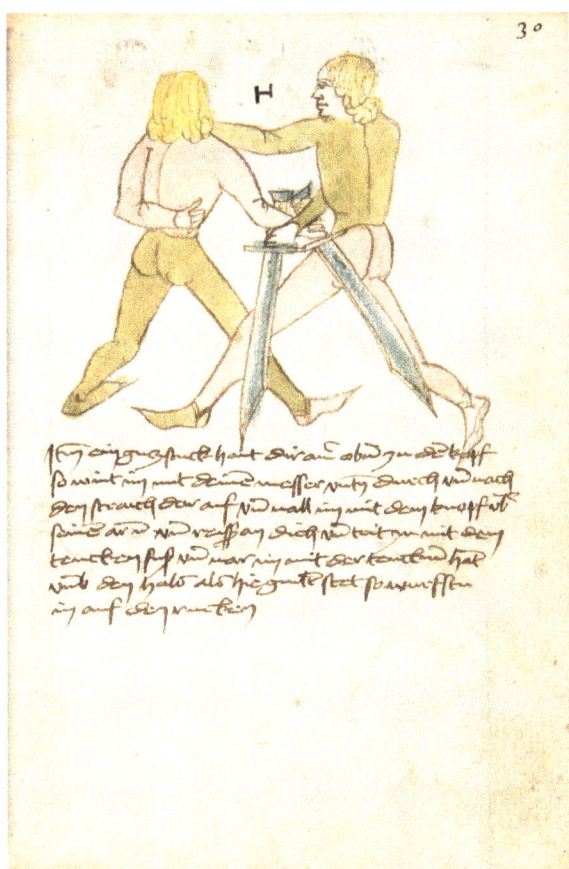

defensively, as the patient needs to initiate. In spite of this interesting difference in principles applied, the contextual meaning of this play is similarly flexible, giving the agent control and options.

This play works on the outside of the patient's arm as well. Coming from a parry on the flat with inverted hand, the agent in B 29v puts in the left leg and pushes the patient's right hand down with the left. The patient initiates this play, which is therefore a defensive one. The play concludes with a cut to the head, that can definitely be used in violent conflict, but the control of the wrist means that theoretically it could be used in playful contexts as well.

While the play in LM 79r is similar to B, it does not appear to describe the inverted hand for the parry. Interestingly, it also describes the play as being viable for both offensive and defensive use, starting from an equal bind after both have made a right high cut. When it comes to contextual meaning, this play once again provides multiple options.

Back-lever throw

B 30r has no technical equivalent in LM, nor in any of the other sources. Given the lack of comparison materials, it is hard to explain why this might be the case. One possibility is that other masters did not consider this technique very useful. Parrying a cut from above, then pulling the knife into one's side in order to perform a back-lever throw may have been considered too risky by Talhoffer and Kal, and Lecküchner might have disliked its lack of versatility. On the other hand, it might be perfectly possible that the author of B added this technique to add their own touch to the fight book. Without any other data though, nothing definitive can be said on this subject.

Thrusting through

In B 30v, the agent winds the knife upward from below and from there goes over the patient's knife with the left arm, pulling it into their own armpit. This means that they can thrust their knife through, first underneath to the outside of the patient's arm, then between their arm and their throat. Ultimately, as with all plays in B, the patient's role is to initiate by striking a high cut from their right and awaiting whatever comes next.

There seems to be no mention of trauma, and given the complicated nature of this play, it is more likely to be used for play and amusement than in any sort of violent conflict (although its context remains unspecified).

When it comes to technique, the only major difference between LM 90r and B is the fact that LM expands on the hold from 30v and adds the possibility to throw the patient. The interaction with the patient is not really described here, meaning that we can scarcely say anything about principles, except that this play seems most viable if the opponent initiates in the same way as in 30v. Likewise, the lack of any contextualization makes it hard to give any meaning to this play, but the complicated and non-lethal outcome seems to indicate play, rather than violent conflict.

Slices

The central technical action in B 31r is that the agent conducts a slicing motion from below. On the level of principles, when the interaction with an opponent comes into play, we find that the agent parries a high cut from the right and performs this slice to the arm when the patient is beginning a second cut. Given that the slice appears to be done in a drawing motion all the way from the strong, i.e. bottom half, to the weak, i.e. top half of the blade, the resulting injury would be quite severe, meaning this play is likely meant as preparation for violent conflict.

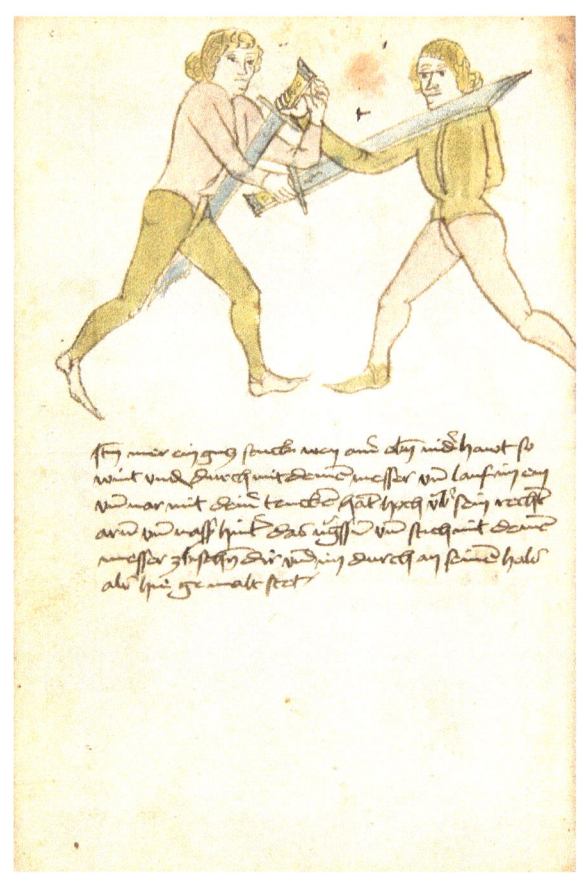

Fig. 13: B, f 30v, 1470

Fig. 14: B, f 31r, 1470

Fig. 15: B, f 31v, 1470

Fig. 16: B, f 32r, 1470

With <u>LM 103v</u>, once again, the technical element in this play is a slicing motion from below. The patient binds with the agent when their left side is in front and then follows up by coming forward with a continuing action. Against this, the agent performs the slicing motion from below, but uses it to push the patient's wrist up using the strong, making this a controlling action. As such, this play could be both a playful exchange as well as a preparation for violent conflict.

The 𝕲𝖊𝖜𝖆𝖕𝖕𝖓𝖊𝖙 𝕳𝖆𝖓𝖉 ("armed hand")

As can be seen in <u>B 31v</u>, the defining feature of this archetype is the way the agent holds the Messer with the left hand on the blade. As in almost all previous plays, the patient initiates and then does nothing. The agent catches this cut on the flat of the Messer, steps behind on the outside and puts the Messer at the patient's throat. The context of this play is interesting though, because a hold like this is suitable for both play and violent conflict. In this case, we should take special note that an accurate interpretation of this play means that the sharp edge would be pressed against the throat.

When it comes to technique, <u>LM 46v–47v</u> seems similar to B, but the main difference is that in LM it comes from an 𝕰𝖓𝖙𝖗ü𝖘𝖙𝖍𝖆𝖜. This results in different principles for timing and initiative, since the 𝕰𝖓𝖙𝖗ü𝖘𝖙𝖍𝖆𝖜 allows for both a defensive and offensive use of this play, although in this case the agent specifically initiates. The long edge is also oriented differently, which leads to different possible outcomes: 46v is a thrust to the face, 47r is a 𝕸𝖔𝖗𝖙𝖙𝖘𝖈𝖍𝖑𝖆𝖌 ("murder stroke"), where the patient gets bludgeoned with the grip and 47v is a throw. The first and last can be used both in a playful or competitive context. The murder stroke, as the name implies, causes unacceptable trauma for that context. All of them seem suitable for

use in violent conflict. Furthermore, the thrust and murder stroke, along with some counters on later folia, make them especially well-suited for armored combat, seeing as the balance of risk and trauma for these plays fits such a context really well.

As a slight variation, in <u>B 32r</u> the agent again catches a cut in the **gewappnet hand**. From there, the left leg comes forward and the agent hooks the grip over the patient's wrist on the inside, pulling it to the right to perform a disarm or armbar. It will surprise nobody that this play has the patient initiate. The context is, like 31v, open to interpretation, but both play and violent conflict are viable. Interestingly and uniquely, <u>LM 138v</u> is identical to the one from B described above in every respect.

As a variation on 31v, <u>B 32v</u> again goes for the patient's throat, but then with the grip to the other side. When it comes to principles, there is nothing new happening here. The outcome is by definition a throw, meaning that a playful context is as likely as a violent one.

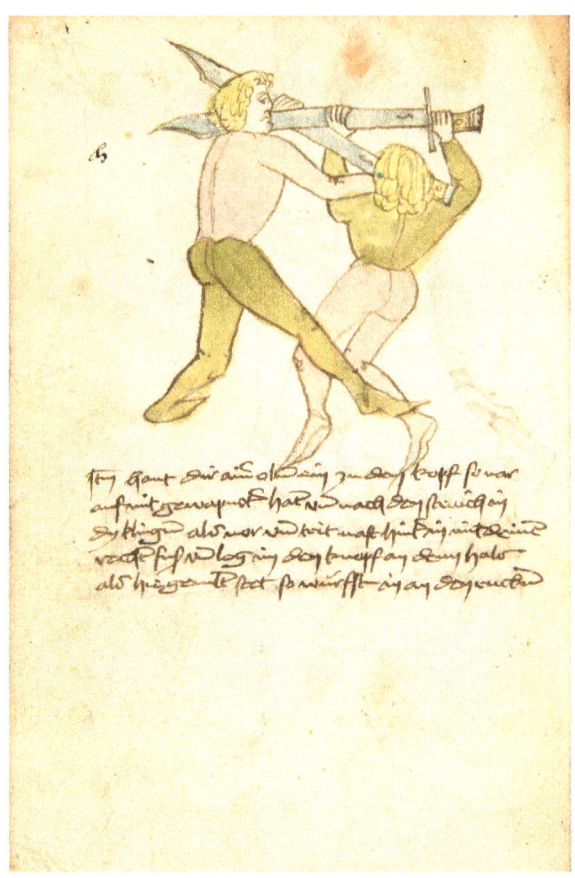

Fig. 17: B, f 32v, 1470

<u>LM 143v</u> follows roughly the same technical movement pattern, as B 32v except that the agent initiates with an **Entrüsthaw** from the left. When considering principles, this means that the agent starts this play. Additionally, the patient seems to be oriented with the right side toward the agent, rather than the left as with 32v. Although from interpretation, it appears that the patient is as likely to step back, making the throw again similar to the one in B. The outcome is no different, meaning that the contextualization is also the same.

Lecküchner's own style

What to think of this analysis of the plays? First, we need to take a small step back to get an overview of the manuscripts that were compared, especially when it comes to the physical aspect of fight books: the techniques and principles. The most important observation here would be that there seems to be no relationship between any of these fight books that can be based on textual similarities, while the plays they describe largely overlap. It would be easy, of course, to assume that this points to some sort of oral tradition being codified in these fight books. This would need some more proof though, and that's where the interpretation of the plays comes in.

When sorting these plays from B, PKM, and HTM based on the technique that is central to them, we are left with nine techniques that can be conceptualized as archetypical, of which eight can be found in LM as well. For those plays that rely on gripping over the arm, there seems to be a remarkable consistency in how this technique works, despite varying starting positions. Both with the free hand-strikes and the hand strikes from a bind, there is some variation with the footwork and the angles at which the knife hits the hand. Although overall, we find very similar techniques here. For wrenching the hand down, there is a remarkable consistency between B and LM, while HTM has some additional options and slightly different foot placement. When using the left hand to control the right, there seem to be different variations in B and PKM, but LM includes both. The back lever throw, on the other hand, has no equivalent in LM or any of the other fight books. It seems possible that this was an exclusive touch added by the master who codified his fencing art in B. An interesting relationship exists between B's and LM's versions of thrusting through, where they are technically the same but the latter offers a continuation while the former does not; LM thus expands the technique. The comparison between the slices is more problematic, since at first glance they vary greatly. However, this is mostly the case for the principles applied, as the techniques themselves are very similar. Finally, the armed hand plays can be found in both B and LM. When examining them at a technical level the similarities are again striking. As such, due to the high consistency of movement patterns across the studied fight books, it seems that in Southern Germany during the later Middle Ages there are a number of archetypical and very common techniques in Messer combat, and except for one technique, they all can be found in LM.

It gets more interesting when the principles come into play. When interacting with a patient, the agent who performs these techniques does so from different starting situations and solves tactical challenges differently. B, PKM, and HTM have a lot in common in this respect: most of the time, the patient starts by throwing a high cut from the right. PKM's equal bind is the exception that proves this rule. LM sometimes follows the same principles, as with his take on over gripping and the hand-strikes. In other cases, Lecküchner made a couple of interesting changes to bring these techniques in line with the principles of the Liechtenauer tradition. The most striking is of course the slice, that is performed with the weak of the blade in B and with the strong of the blade in LM. The principle that governs this technique in B's fight book seems to be a draw cut to the wrist, while with Lecküchner's Liechtenaurian slice, the objective is to control the opponent's arm with the strong by pushing it upward. Far more interesting, however, is the substitution of the turned

hand for the **Entrüsthaw**. From a technical perspective, this change is too minimal to pay attention to; after all, it is a simple matter of using the thumb grip to flip the blade 90 degrees. The tactical implication is far reaching though, as the agent now has the option to perform a technique both from a defensive and an offensive starting position. In some of the cases examined above, LM explicitly uses the **Entrüsthaw** offensively where B uses the turned hand defensively. By connecting these plays with a part of his manuscript that fits into a coherent Liechtenauer-based system, Lecküchner integrates these commonplace techniques into that system.

In conclusion, we can quite confidently say that the lack of textual similarities between all these fight books does not mean they are not related. Unlike with Lew, whose gloss of Liechtenauer's **Zettel** was copied by Lecküchner, the examined books have no textual link, yet describe an almost identical corpus of embodied-knowledge. Although the principles sometimes create a different interaction between fencers, especially when comparing Lecküchner with the rest, no master seems to be describing a different style of fencing. This suggests that a tradition of embodied-knowledge on Messer fighting existed in Southern German Lands around Nürnberg and Augsburg, which each of the studied masters partly codified.

When changing the focus from the physical, i.e. embodied-knowledge, to the mental, i.e. the meaning of each play, the similarities between the compared fight books are more diffused. Juxtaposing Lecküchner's teachings with those of the other discussed fight books, his teachings are more ambiguous regarding the severity of trauma inflicted and most of Lecküchner's plays can be placed in a lower violence spectrum than the earlier fight books (see table 1). None of the plays by Lecküchner discussed here can be placed purely in the dimension of 'violent conflict', while this is the case for most plays in the older fight books. This overlap in meaning of the same archetypical technique across several older masters can be another argument which points to an already existing tradition in Messer fighting prior to Lecküchner's work. Lecküchner then takes these techniques and in many cases offers a less violent ending for his plays. Of this reduction, the most obvious example is the archetypical technique of free hand-hits. PKM and HTM often state to strike the hand or wrist, to maim an opponent by breaking or dismembering the hand. LM, in contrast, provides these same options but moreover describes that it is possible to target the forearm, which is less prone to breaking or being cut off.

This is particularly the case when weaker cuts are made against it, making his versions also suitable for the dimension of 'Play and Competitive Sports'. A similar reduction in trauma by Lecküchner can be observed in all other archetypical techniques, except for The **Gewappnet Hand** and 'thrusting through'. Lecküchner provides other options besides thrusting an opponent in the abdomen or striking him on the head or neck, as is advised in

Archetypical techniques	Folio	End results of each play	'Violent Conflict' or 'Play and Competitive Sports'
Free Hand-hits			
HTM	115v/116r	Cut or strike to the wrist	Violent Conflict
PKM	72r	Cut or strike to the hand	Violent Conflict
LM	185r	Strike to hand or cut to the forearm	Both
Slicing			
B	31r	Slice off the opponent's wrist	Violent conflict
LM	103v	Control the opponent's wrist with the blade	Both
Hand-hits from a bind			
PKM	73r	Cut or strike to the hand	Violent Conflict
Lecküchner	187v	Strike to hand or cut to the forearm	Both
The left is the enemy of the right			
PKM	72v/73v	Multiple options, possible cut to the head	Both
B	29v	Cut to the head	Violent Conflict
LM	179r/179v	Multiple options, while maintaining control over the opponent's Messer	Both
Wrenching the hand down			
HTM	113r/114v/115r	Undisclosed, a possible armbar	Unknown
B	29r	Cut to the head	Violent Conflict
LM	84r/ 171v	Armbar/ Cut or slice face	Both
Arm grabbing			
HTM	113r–114r/116v	Cut to the neck/ Thrust to the abdomen	Violent Conflict
PKM	72v, 74r–74v	Hip throw or strike to the head	Both
LM	156r–156v/158r	Set point on the throat/ Hip throw	Play and Competitive Sports
Thrusting through			
B	30v	Undisclosed, possible throw of an opponent	Unknown
LM	90r	Throw an opponent	Both
The Gewappnet Hand (armed hand)			
B	31v–32v	Press edge into the neck/ disarm or armbar	Both
LM	46v–47v/138v/143v	Thrust into the face/ *Morttschlag*/ Throw/ Disarm or Armbar	Both

Table 1: The result of each play and its dimension of meaning

the earlier fight books. This offering of more options in terms of trauma severity and changing the dimension of meaning of several (archetypical) techniques not part of the Liechtenauer 𝕳𝖆𝖚𝖕𝖙𝖘𝖙ü𝖈𝖐𝖊, is in line with the earlier discussed Liechtenaurisation of these techniques. The trauma in Liechtenauer's system was intended "principally for entertainment, while still adhering to certain basic principles that could be applied to serious combat".[379] Similarly, how Lecküchner added the Liechtenauer

---

[379] Kellet 2015, p 149.

principles of timing and binding to an already existing (oral) tradition, he moreover added the Liechtenauer meaning-system to these techniques as well, while retaining their original violent capabilities.

Conclusion

The book *Notes on the Art of the Knife* by Hans Lecküchner signifies a break with earlier fight books, by presenting a far more complex and larger martial arts system than its predecessors. This increase in size and complexity brings about a question concerning the originality of the works, and if Lecküchner invented his own martial arts style or based his system on earlier fight books. This question is answered here by comparing earlier fight books on the use of the Messer with those of Lecküchner. This comparison is not done by juxtaposing textual similarities, such as terminology or sentence construction, but by correlating interpretations of the physical movements that these texts and illustrations represent. When trained these interpretations become embodied-knowledge. A type of knowledge operating on the unconscious cognitive dimension.

To define if Lecküchner operated within the same or a different martial arts tradition than previous fencing masters, they have been compared on two levels: (1) on the embodied-knowledge which the books represent, including principles defining time and distance to execute it, and (2) if the described movements have an analogous meaning. This meaning has been determined by examining the severity of the trauma inflicted by movements and place the outcome in either the dimension of meaning of 'Violent Conflict' or 'Play and Competitive Sports'. Lethal or grave trauma has been interpreted as being for the dimension of 'violent conflict', while other forms of trauma as being part of 'Play and Competitive Sports'. Plays from Lecküchner, selected from six 𝔥𝔞𝔲𝔭𝔱𝔰𝔱𝔲̈𝔠𝔨𝔢 not present within the Liechtenauer martial arts system, have been compared with the plays from three earlier masters; Talhoffer, Paulus Kal, and Von Baumann.

While there are no textual similarities between any of the examined fight books, using the earlier described method similarities could, nevertheless, be recognized. All techniques described within the earlier fight books, with the exception of one, could be identified within Lecküchner's teachings and eight archetypical techniques could be defined. Moreover, there was not only overlap in techniques between Lecküchner and the other books, but four archetypical techniques were presented in multiple older fight books. A differentiation occurs when principles are taken into account. Where older fight books present their plays as a reactionary, defensive features, Lecküchner's plays commonly can both be deployed offensively and defensively. A similar difference is visible when juxtaposing the meaning of plays. The earlier fight books primarily operate in the dimension of 'Violent

Conflict' with few exceptions, whereas the meaning of the examined Lecküchner plays is more diffused and they can almost all be placed both in the meaning of 'Violent Conflict' and 'Play and Competitive Sports'.

Because Talhoffer, Paulus Kal, and the Von Baumann book present equivalent embodied-knowledge, including principles, operate in a similar dimension of meaning, and roughly originate from the same geographical area in Southern Germany (around Nürnberg and Augsburg), it appears an oral tradition in Messer combat existed in that area, which these masters partly codified or based their teachings on. Lecküchner probably then took this already existing body of knowledge and married it with the principles of the long sword system of Liechtenauer to create a hybrid style. His knowledge of the Liechtenauer system could have been based on his access to L, wherefrom he copied several plays in his book. Lecküchner's style did not only adapt Liechtenauer principles, but moreover changed the meaning or intention of several techniques by offering different physical outcomes. His style of Messer combat, however, still retains the possibility to inflict serious harm. Therefore, both in the physical realm, i.e. how to execute techniques, and in the mental domain, i.e. the meaning of these techniques, Lecküchner took an existing method method of Messer fighting, infused it with Liechtenauerian qualities, and made it his own.

This conclusion is, of course, based on a limited case-study and more techniques need to be assayed to test the hypothesis postulated here. In particular, techniques that have no equivalent in Messer plays of other fight books need to be examined, as these provide more insight into the unique flavor Lecküchner added to Messer or one-handed weapon combat in general. Moreover, the meaning of more plays in Lecküchner need to be studied, to generate a better overview of the meaning of Lecküchner's teachings and if they overlap with Liechtenauer's teachings. With these caveats in mind, the proposed methodology for integrating embodied-knowledge into examining fight books still proves to be a successful strategy in understanding them. Physically practicing movements and gestures described in fight books, and the accompanying embodied-knowledge, can provide otherwise unattainable insights.

Art and Symbolism in the Genre of Fechtbücher

Since [the Dusack] is, after the sword, not only the weapon most used by us Germans, but also an origin and basis of all weapons that are used with one hand, I will here present it and then discuss and explain it in orderly fashion with all its particulars and techniques.

~ Joachim Meyer, 1570

...als der bei uns Teutschen nach dem Schwerdt nicht allein am breuchlichsten, Sonder auch als ein anfange und grund aller Wehr, so zu einer hand gebraucht werden, hieher setzen, unnd volgends mit allen umbstenden, und zugehörenden stucken, der ordnunge nach handlen und erklären.

commonly repeated axiom in the Historical European Martial Arts community is that the Dusack, one of a wide variety of commonly used single-hand swords in the mid- to late-sixteenth century, was nothing more than a training tool. Specifically, its wooden or leather waster version, those shown often in Fechtbücher, was the extent of the Dusack's existence and that any real fighting would have been done with the Messer. Dr. Jeffrey Forgeng, in the introduction to his translation of Joachim Meyer's *Grundtliche Beschreibung der... Kunst des Fechtens*, put it succinctly: "a weapon form that only comes to prominence in the latter part of the fifteenth century is the falchion (lange Messer), a one-handed sword with a broad curved blade, of which Meyer's Dusack is a version for practice".[380]

This assertion has not been given a specific analytical treatment. Most of the research focus on the Dusack has centered either on proving that it was, in fact, a weapon made of steel with a sharp edge,[381] or tracing its use into the sporting culture of the 18th century and detailing the subtle differences in the design of its training forms.[382]

The Dusack in Meyer's treatise is a foiled version of the sword, likely made of wood in various shapes and sizes. But this alone is not enough to justify the assertion that Dusacks were categorically training foils; Meyer's treatise shows training versions of the sword, rapier, and dagger, and those were undisputedly carried by citizens and meant to be used in earnest fights at the time. Why the dismissal of the Dusack as a mere foil?

The first mention of the Dusack in German-language fencing treatises is arguably Andre Paurenfeyndt's 1516 *Ergundering Ritterlicher Kunst des Fechterey*. In the second chapter, he teaches the use of the Messer, an "increasingly useful" weapon, "a predecessor and source of the other weapons

---

[380] Forgeng 2006, p 16.
[381] Norling A.

[382] Amberger 2003

Fig. 1: Messer or Dusack fencing (Paurenfeyndt, p 53, 1516)

that are used in one hand, such as the Dusack or the tucke, the Katzbalger or the hand 𝕯𝖊𝖌𝖊𝖓, and others".[383] In Paurenfeyndt, the Messer is specifically introduced as a weapon meant to train the Dusack—alternately spelled 𝕿𝖊𝖘𝖘𝖆𝖈𝖐—not the other way around.

It is clear that both the Messer and the Dusack are specifically related to the art of fencing itself, and that the art of fencing with a sword in one hand was understood in a general sense: training any specific weapon allowed one to use the learned skills and techniques to whatever weapon was available for any particular fencer. Why is it that out of the huge number of swords wielded in one hand within the Holy Roman Empire in the mid-sixteenth century, the Dusack and the Messer are the ones most often mentioned in fencing treatises? And why did the Dusack come to be understood as a training sword, and not one of many similar options?

Origins of the Dusack

The Dusack remained in use in 𝕱𝖊𝖈𝖍𝖙𝖘𝖈𝖍𝖚𝖑𝖊𝖓 and as part of the culture of fencing instruction and competition well into the 19th century. In 𝕱𝖊𝖈𝖍𝖙𝖘𝖈𝖍𝖚𝖑𝖊 records, the use of the Dusack for bouts sometimes outnumbered the bouts with all other weapons combined.[384] Its familiarity to the

---

[383] Translation by author, based on transcription by Michael CHIDESTER. I have decided to render *Spatel* as "Katzbalger" owing to the Katzbalger's chisel-like flat point, and *Tollich* as "tuck" instead of its more common rendering as "dagger".

[384] AMBERGER 2003; WASSMANNSDORFF 1870.

modern HEMA community is likely due to the popularity of Meyer, and it is through Meyer that the common image of the Dusack as a training or sportive weapon gained prominence.

But even Meyer's version of the weapon is in no way definitive, it was simply a reflection of the local term in use by Meyer and his audience. The origin of 𝔗𝔢𝔰𝔰𝔞𝔠𝔨 and its Germanic derivatives is likely Czech, from their word for fang or tusk.[385] The use of "tusk" relates closely with the posture named 𝔈𝔟𝔢𝔯, or "boar", used by both Meyer and Lecküchner.

Relationships between the Holy Roman Empire and eastern Europe—specifically Poland, Hungary, and Bohemia—were well advanced by the turn of the sixteenth century. Friedrich III, Holy Roman Emperor until his death in 1493, was embroiled in a long succession dispute with Matthias Corvinus of Hungary until the latter's death in 1490, a conflict which involved invasions across both sides of the border and humiliating defeats for Friedrich. Bohemia's politics were tied to the Hapsburg court and a Bohemian army even supported the rebellion of Ruprecht and Elisabeth of the Palatinate in the Landshut War of Succession in 1504. Elisabeth herself was the granddaughter of King Casimir IV of Poland. Maximilian, the son of Emperor Friedrich and the co-ruler of the Empire while his father lived, was nearly killed by Bohemian troops at the Battle of Wenzenbach, the famous "Bohemian Fight" of *Weisskunig*.[386]

Fig. 2: Dusack fencing (Meyer, p [2]-22v, 1570)

---

[385] Norling A.

[386] Seton-Watson 1901, p 1.138; Terjanian 2019, p 47.

Fig. 3: Bohemian Battle from *Der Weisskunig* (1505-16)

While it has been suggested that the Ottoman influence was a major factor in the increasing visibility of Dusacks and other weapons with curved blades, it is clear from art of the first two decades of the century that curved blades were used by Bohemians and Hungarians, as well, and it is possible that the German affinity for that shape of blade was a result of these earlier conflicts.

Typology

The increasing use of the Dusack in 𝔉𝔢𝔠𝔥𝔱𝔟ü𝔠𝔥𝔢𝔯 is consistent with the general trend in the sixteenth century toward increasingly complex hilts. Ewart OAKESHOTT traces the development of the complex single-handed hilt to the first decades of the sixteenth century with what he termed the "Landsknecht hilt". Typified by a thick haft and a flattened round or square pommel, the first examples also used quillons in a figure of eight but changed to include a thumb ring and knuckle and hand guards by the 1530s and 40s. This "actually produced what was almost the first type of basket hilt; there can be little doubt that the forms of basket hilt which began to emerge around about 1600 in England, Scandinavia and southern Germany owed... much of their development to these hilts of the 1540s".[387]

The Landsknecht-hilted swords shared their early development with similar simple swords in one hand, in particular the Swiss 𝔥𝔞𝔲𝔰𝔴𝔢𝔥𝔯, which are believed to have originated from the scramasax of sixth century France.[388] This type is often referred to today as a Swiss dagger. Both Swiss 𝔥𝔞𝔲𝔰𝔴𝔢𝔥𝔯 and early Landsknecht swords were used contemporaneously with the Messer.

By the 1540s there emerged a distinct elaboration of the Landsknecht hilt, which OAKESHOTT simply calls the German saber hilt. The inclusion of a thumb ring, a thumb guard, and a clamshell or wire basket that fully enclosed the hand completed the type.[389] These guard types are difficult to distinguish from later 'Sinclair saber' hilts, and OAKESHOTT traces their development from southern German saber hilts.[390]

---

[387] OAKESHOTT 1980, p 142.
[388] OAKESHOTT 1980, p 142.

[389] OAKESHOTT 1980, pp 155–6.
[390] OAKESHOTT 1980, p 172.

Surviving Dusacks have characteristics of many of these different styles, and also include both curved and straight blades.[391] This is all to say that swords of the type that can be described as Dusacks were in use as far more than wooden training forms as early as the 1520s.

What is more important is the necessity to grapple with the sheer variety of single-hand swords in use in the early to mid-sixteenth century. Apart from the Messer and Dusack—variously named—there were arming swords, Katzbalgers or Landsknecht swords, Swiss 𝕳𝖆𝖚𝖘𝖜𝖊𝖍𝖗𝖊𝖗, falchions, hunting swords, the 𝕯𝖊𝖌𝖊𝖓 in various varieties, 𝕽𝖎𝖙𝖙𝖊𝖗𝖘𝖈𝖍𝖜𝖊𝖗𝖙𝖊𝖗 or riding swords, rapiers, and sabers, of which there are distinct varieties of northern and southern German, Polish, Bohemian, Hungarian, and Ottoman. There is no easy way to differentiate many of these swords, and adding to the confusion is the use of the same word to mean very different weapons. As an example, one of the swords mentioned by Paurenfeyndt is the 𝕿𝖔𝖑𝖑𝖎𝖈𝖍 or "tucke", an alternate version of the word *estoc.* In the modern HEMA community, the *estoc* is generally assumed to be a two-handed sword used short in armored fencing, but both Sydney ANGLO and Ewart OAKESHOTT define the *estoc* or tucke as a "rapier or little sword", which perhaps explains why Paurenfeyndt would use the word in his section on swords in one hand.[392]

Fig. 4: Dussck fencing (Heussler, p 229, 1615)

[391] COTTER-REILLY.

[392] OAKESHOTT 1980, p 172; ANGLO 2001, p 101. See also STONE 1999.

Fig. 5: Dussck, 1560-90 (Nordiska museet, NF. 2011-4340A)

With the understanding of the bewildering complexity of sword types in use for military and civilian purposes, the question now becomes why did the Messer and the Dusack, in particular, come to be so consistently used in fencing literature?

Sechtbücher and genre

Paurenfeyndt's inclusion of both the Messer and the Dusack is interesting, not only because of its reversal of FORGENG's assertion that the Dusack is the training version of the Messer. The Messer and the Dusack have an indisputably tight relationship to Sechtbucher, and one or the other term is almost universally used in Sechtbucher to train fencing with one hand. One of the few exceptions is the late 15th century treatise of *Hugo Wittenwiler* (HW) which, apart from typical entries like two-handed sword, staff, and mounted fencing, taught three short weapons in one hand: the baselard, the Tegen, and the short knife (kurzen Messer).[393]

Sechtbücher can be understood as individual entries in a distinct literary genre. Beyond their shared content and similar structures, they have common conventions, and often reference or reflect earlier texts in the same traditions and linguistic categories as a means of claiming authority and asserting expertise. These overlaps often take the form of shared vocabulary, such as Meyer's use of many of Lecküchner's unique terms, or by claiming association with or knowledge of the secrets of older masters, such as the repeated refrain in 15th century German-language manuscripts that claim a direct association with the teachings of Johannes Liechtenauer. Both Joachim Meyer and Paulus Hector Mair wrote elaborate forewords that traced the lineage of German fencing from antiquity to the present day, laced with references to myth and to heroic folklore.

With this established, it should be no surprise that the genre itself has conventions that are purely literary, that knowingly use signifiers of the distinct social subgroup of fencers. In other words, the emphasis on the Messer and the Dusack may have little basis in the commonality of the weapons as carried by citizens, mercenaries, or knights, but in the shared conception of the Messer and eventually the Dusack as recognized symbols of the art of fencing. Thus, the depiction of certain weapons in Sechtbücher reflect the genre's particular imagery.

---

[393] Kevin MAURER's translation on *Wiktenauer* uses "dagger" for *Tegen,* but I believe that it is a variant spelling of *Degen.*

Art and symbolism

The use of visual metaphor in medieval and early modern art has been heavily studied. Visual art employed a complex system of reference, using characters and scenes from historical or biblical stories and rendering them recognizable to the contemporary audience by the use of contemporary clothing and costume. Thus, Albrecht Altdorfer's *Battle of Issus* shows an Alexander of Macedon in gilded sixteenth century plate armor.[394] Individual figures could be made to look out of place in a class sense through their dress, reinforced by local sumptuary standards, or in a temporal sense, through the use of dated fashions, such as a chaperon hat on a hopeless suitor.[395]

This 'symbolic machinery' rested on the use of particular marriages of idea and image to impart meaning to the viewer, whether that meaning was inspiration, allegory, or even farce.[396] "The work of art's relation to the world it represents would thus no longer depend on its resemblance to that world but would depend on its capacity to denote it."[397] Denoting the world meant the use of commonly understood symbolic signifiers, especially as rendered by costume and weaponry.

Woodcuts, long considered a 'low' art and overlooked in the history of fine art, used the same techniques used by paintings and statuary to create meaning.[398] Just like fine art, woodcuts "taught, reminded, inspired, and protected".[399] The use of imagery as a means of instituting or reinforcing social, political, and religious ideals was, in fact, completely deliberate, stemming from a decree from the Second Council of Nicea in 787, which stated that images should be both an aid to devotion, and serve as a means of instruction.[400] Woodcuts were especially useful as reinforcement mechanisms of the symbolic machine because they were inexpensive and widely consumed, and the popular mode of ever-lower-quality copies of popular works functioned as a kind of social amplification, allowing common symbols to penetrate the complex social strata of the early modern period.

The symbolic elements of common woodcuts can thus be understood as intentionally symbolic, containing visual elements that would be understood by the culture in which they were framed. Art fulfilled a particular cultural function more than simply representing aesthetic beauty. Certain items and costumes contain visual tags that indicate and typify recognizable tropes.

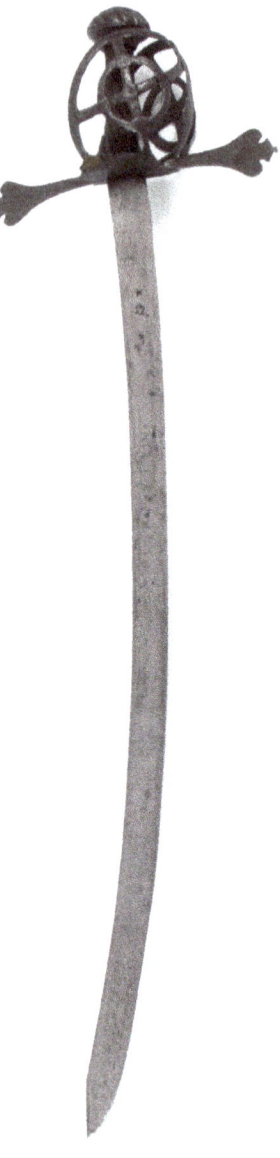

Fig. 6: Dusack, 1560-70 (Nordiska museet, NF. 1924-0996)

---

[394] Albrecht Altdorfer. "Battle of Issus". *WikiArt.* Accessed 20.03.2021.
http://www.wikiart.org/en/albrecht-altdorfer/ the-battle-of-issus-1529
[395] Di Furia 2018, p 10.13.

[396] Nagel and Wood 2005, p 403.
[397] Moxey 1989, p 6.
[398] Moxey 1989, p 3.
[399] Gilbert 1995, p 5.
[400] Moxey 1989, p 3.

Swords and social identity

Swords were ubiquitous elements of the symbolic machine, able in a single item to pinpoint a great degree of meaning to the figure bearing it, and recognizable types of sword used in art should be understood as identifiers of class, community, and social category. Mercenaries, virtuous citizens, knights, peasants, and fencers all variously wore a wide variety of swords, but their signifiers in art remained remarkably consistent.

Social status in Free Imperial Cities was visibly prescribed through laws and peer group privileges. Sumptuary laws, though haphazardly enforced and widely variable, allowed the wealthy to identify themselves in public through conspicuous dress. Class, sex, marital status, property ownership status, and profession were all demarcated by law, which allowed for an instant visual categorization of any person out in public.[401] Peer groups, many of these subject to class and income restrictions, were controlled and amplified from within as men of similar classes organized drinking and other social clubs whose membership was heavily policed. Social clubs often bore symbolic standards or wore particular clothing as a marker of membership.[402]

These visual identifiers were used in art, as well. A particularly useful example comes from the works of Jörg Wilhalm. As a fencing treatise, the depiction of swords is particular and important, and throughout the illustrations the swords used by pairs of fencers are recognizable as Feders, with distinctly rendered Schilts just above the crossguard. However, at the start of Wilhalm's short sword section, he relates an allegory relating to three types of men: a rich man, a strong man, and a weak man. The rich man is obvious from his fur-trimmed cloak and richly dyed clothes; the weak from his cowering demeanor. In the middle is the strong man, prominently bearing at his waist a Landsknecht-styled sword with a figure-of-eight crossguard.

The artistic denotation of 'strong men' or mercenaries with Katzbalgers is so common that it was used to distinguish German mercenaries from other types of armed men. Mercenaries can be distinguished from knights, who, apart from being obvious in armor and on horseback, carry swords and daggers; and

Fig. 7: A rich man, a weak man, and a strong man (JWA3, f Ir, 1522)

---

[401] See CHRISMAN 1995, Chapter 1: The Social Order.   [402] See TLUSTY 2001.

peasants, in common costume with Messers at their waists.[403] In Hans Schauflein's 1525 print of the Battle of Pavia, the German mercenaries are even contrasted by their Katzbalgers to the French mercenaries, who wear rapiers.[404]

Foreigners were often represented by their exotic dress and their association with sabers. In *Weisskunig*, delegations of Hungarians appear in front of Maximilian with heavy-bladed sabers (fig. 7), and many woodcuts showing the "Bohemian Battle" of Wenzenbach put curved sabers in the hands of the Bohemians. Ottomans in later sixteenth century prints almost universally wear sabers. Even 'wild men' in masquerades wore sabers, as can be seen in another image from *Weisskunig*, a clear denotation of an *other* or a *foreigner*.

Civilians, the urban 𝔅urghers of the Imperial cities, also have distinct symbolic markers. Indeed, their responsibility to bear arms was an increasingly important element of the Free Cities' political self-determination.[405] Representations of their skill in arms through art and public display was an important symbol of citizenship. In Matthäus Schwarz's 𝔗rachtenbuch, his personal book of costume, he shows himself wearing dozens of different outfits for different occasions, and he wears a sword in nearly every image. With remarkable consistency, the swords he wears are rapiers. Occasionally, especially when riding, he wears a two-handed sword. He departs from this pattern in a few places, most notably when he shows himself learning fencing, and when he shows himself armed and armored for the

Fig. 8: Hungarian delegation from *Der Weisskunig* (1505-16)

Fig. 9: Masquerade from *Der Weisskunig* (1505-16)

---

[403] See especially MCNEALY and GEISBERG 2019.

[404] MCNEALY and GEISBERG 2019, p 44. Detail of French formation, p 55.

[405] TLUSTY 2016, p 549.

Fig. 10: Matthäus Schwarz at a fencing lesson (Herzog Anton Ulrich Museum, Hs. 27 N. 67a, p 38, 1518)

militia or the city watch. In the former case, he holds a typical Feder; in the latter, he wears shorter swords with wider blades, the hilts appearing in the exact form described by Oakeshott as "Landsknecht hilts".[406]

Genre was never light on meaning, though it depended on an interplay between artist and audience and a shared understanding of aesthetic and symbolic images. Reception, in other words, was as important as intent.[407] If Fechtbücher were a distinct genre with distinct conventions that drew from the broader artistic practices around them, then it is valuable to look carefully at the art as a conscious construction meant to impart particular meaning.

Fencers as a particular cultural subgroup

Fencers as a recognized cultural subgroup can be traced to at least 1100, where the first use of a term for 'fencing' came into use in the French language region. The word *escremie* meant "to defend" or "to protect oneself".[408] The word itself appears to derive in part from the use of the shield, creating a tight relationship between fencing as an art and practice and the combination of the sword or mace with a shield or buckler. The word continued to be used in a Germanic form for a long while, and Matthäus Schwarz's image mentioned above uses the Germanic form schirmen to mean fencing. The use of the word was part of a process that made fencing a distinct profession or vocation.[409]

Clearly, the relationship between the art of fencing and the weapons employed for the purpose had changed in the meantime. By the early sixteenth century the Feder appeared alongside the sword and buckler as a marker for fencers. In Maximilian's triumphal procession illustrations by Burgkmair, several groups of men depicted as fencers are shown with a variety of weapons. Included is a group bearing swords and bucklers as well as one bearing Feders, both with laurel wreaths on their heads.[410] Numerous other images show what are clearly Feders in the hands of men identified either as fencing masters, fencers, or ceremonial bearers in Fechtschulen.

The Dusack came to serve a similar purpose as a genre image. Depictions of fencers and athletes in the 'Children of the Sun' images of the Planetenkinder show foiled Dusacks, and crossed Dusacks are shown

---

[406] Rublack and Hayward 2015, pp 170–1.

[407] De Furia 2018, p 10.3.

[408] Dupuis 2015, p 38.

[409] Dupuis 2015, p 57.

[410] Applebaum 1964, p 7. Images on pp 37, 40.

at the feet of fencing pairs in representations of 𝔉𝔢𝔠𝔥𝔱𝔰𝔠𝔥𝔲𝔩𝔢𝔫 and other folk festivals.[411] Dusacks are shown often in association with the Brotherhood of St. Mark and the Furrier's Guild, whose Imperial privilege of bestowing the rank of Master of the Long Sword made them the first recognized fencing brotherhood in the Holy Roman Empire. An anonymous woodcut depicting a furrier's shop in 1550 shows two Dusacks hanging from the entrance to their shop, an obvious reference to their place in the culture as fenc-

Fig. 11: "Five men with swords", detail from the *Triumphal Procession of Maximilian* (1516-19)

ers, as well as tradesmen.[412] The rival fencing brotherhood, the 𝔉𝔯𝔢𝔦𝔣𝔢𝔠𝔥𝔱𝔢𝔯, also regularly depict Dusacks, either in use or crossed on the ground.

This association reached even beyond imagery; the connection between literature and oral tradition has been well studied, but the analysis typically focuses on literature's influence on fencing rather than the other way around.[413] In Grimmelshausen's *Simplicius Simplicissimus,* a 17th century 𝔅𝔦𝔩𝔡𝔲𝔫𝔤𝔰𝔯𝔬𝔪𝔞𝔫 novel set during the 30 Years' War, the main character, Melchior Sternfels von Fuchshaim, expresses his value as a fencer by boasting of his skill with a Dusack.[414] Fencing is repeatedly emphasized in Fuchshaim's travels, and associations of fencing with furriers—the Brotherhood of St. Mark—is also made explicit.[415]

Fig. 12: A furrier's shop with Dusacks hanging on the wall

The association of particular groups of fencers with particular weapons is no mistake. Feders were instantly recognizable as tools to teach and practice

[411] NORLING B. See also WALDBURG WOLFEGG 1998.

[412] "Geschichte des Kürschnerhandwerks in Württemberg," *Wikimedia Commons.* Accessed March 20, 2021. https://commons.wikimedia.org/wiki/File:Geschichte_des_K%C3%BCrschnerhandwerks_in_W%C3%BCrttemberg,_1967.jpg

[413] See KELLET 2016, pp 62–87.

[414] AMBERGER 2003 noted this relationship, though the common English translations of Grimmelshausen translate Dusack as "cutlass".

[415] Grimmelshausen, ch 18: "I have never been fond of dancing, instead I showed off my slim body when fencing with my furrier."

Fig. 13: Rapier and crossed Dusacks inside a crown (Meyer, title page (detail), 1570)

the art of fencing, and Dusacks came to perform a similar function by the mid-sixteenth century. This trend in representation involved far more than imagery in 𝕱𝖊𝖈𝖍𝖙𝖇𝖚̈𝖈𝖍𝖊𝖗, but was common in popular woodcuts and literature as well.

Conclusions

Paurenfeyndt's inversion of the relationship between the Dusack and Messer should be enough to complicate the place of the Dusack in the HEMA community. The bewildering array of commonly used weapons by the middle of the century, from short-bladed 𝕳𝖆𝖚𝖘𝖜𝖊𝖍𝖗 and blunt-tipped Katzbalger all the way to long rapiers with complex hilts, add to the necessity to rethink artistic representation of fencing culture with its reality. 𝕱𝖊𝖈𝖍𝖙𝖇𝖚̈𝖈𝖍𝖊𝖗 are literature, they exist within popular genre, and make use of elements of that genre as a symbolic machine to communicate with their audience.

The Messer and the Dusack were both meant to be generic stand-ins to train any of these weapons, as is explicitly stated in multiple 𝕱𝖊𝖈𝖍𝖙𝖇𝖚̈𝖈𝖍𝖊𝖗. The increasing visibility of the Dusack, with its typological origin in the 'Landsknecht' hilted short sword, follows a general trend of sixteenth century weaponry toward increasingly complex hilts on common sword design. The importation of the terminology was due in part to the close political and military connection between the Holy Roman Empire and their Hungarian and Bohemian neighbors.

The existence of dozens of different types of weapons, differentiated by hilt design and other aesthetic features, as well as regional vocabulary used to describe them, makes the specific use of the Dusack and the Feder in 𝕱𝖊𝖈𝖍𝖙𝖇𝖚̈𝖈𝖍𝖊𝖗 and in broader popular culture even more important. These weapons were part of a visual code that identified their wielders as *fencers*, that is, devotees of the art of fencing, a cultural subgroup distinct from mercenaries or men-at-arms, knights, peasants, or urban 𝕭𝖚𝖗𝖌𝖍𝖊𝖗𝖘. Fencers had long been considered a particular social group, distinct in their social framing and separate from related armed classes, and had traditionally been represented by their association with the sword and buckler. By the sixteenth century, the visual symbolism had shifted to the Feder and the Messer, until the latter was replaced by the Dusack. The Dusack remained a prominent symbol of German fencing brotherhoods and German fencers, in art and literature, until the 18th century.

Bibliography

Primary sources

For a list of manuscript primary sources, see Manuscript Abbreviations *at the beginning of this book.*

Heussler

    Heussler, Sebastian. *Neu Kunstlich Fechtbuch Darinnen 500 stuck im ainfachen Rapier, wie auch ettliche im Rapier und Dolch deß weitberümbten Fecht. Zum andern mal in truck geben und mit schön Stucken verbessert.* Nürnberg: Ludwig Lochner, 1615. <VD17 23:267689Q>

Meyer

    Meyer, Joachim. *Gründtliche Beschreibung, der freyen Ritterlichen unnd Adelichen kunst des Fechtens, in allerley gebreuchlichen Wehren, mit vil schönen und nützlichen Figuren gezieret und fürgestellet.* Strassbourg: Thiebolt Burger, 1570. <VD16 M 5087>

Pauli

    Pauli, Johannes. *Schimpf und Ernst.* Strasbourg: Johann Grüninger, 1522. <VD16 P 937>

Paurenfeyndt

    Paurenfeyndt, Andre. *Ergrundung Ritterlicher kunst der Fechterey durch Andre paurenfeindt Freyfechter czu Vienna in Osterreich, nach klerlicher begreiffung und kurczlicher verstendnusz.* Vienna: Hieronymus Vietor, 1516. <VD16 P 1071>

Sainct-Didier

    Sainct-Didier, Henri. *Traicté contenant les secrets du premier livre sur l'espée seule.* Paris: Jean Mettayer et Matthurin Challenge, 1573.

Secondary sources

ALCÁNTARA 2017

    ALCÁNTARA, Viridiana, et al. "Legacy of medieval ridge and furrow cultivation on soil organic carbon distribution and stocks in forests". *CATENA* **154**(1): 85–94. 2017. Doi: 10.1016/j.catena.2017.02.013.

ANDRESEN 2017

    ANDRESEN, Suse. "In fürstlichem Auftrag. Die gelehrten Räte der Kurfürsten von Brandenburg aus dem Hause Hohenzollern im 15. Jahrhundert". *Schriftenreihe der Historischen Kommission bei der Bayerischen Akademie der Wissenschaften* **97**. Göttingen, 2017.

ANGLO 2001

ANGLO, Sydney. *The Martial Arts of Renaissance Europe*. Yale University Press: 2000.

APPLEBAUM 1964

APPLEBAUM, Stanley. *The Triumph of Maximilian I: 137 Woodcuts by Hans Burgkmair and Others*. Dover Publications, 1964.

BAADER 1851

Nürnberger Polizeiordnungen aus dem XIII. bis XV. Jahrhundert (Bibliothek des litterarischen Vereins in Stuttgart LXIII). Ed. Joseph BAADER. Stuttgart, 1851.

BACKES 1992

BACKES, Martina. *Das literarische Leben am kurpfälzischen Hof zu Heidelberg im 15. Jahrhundert. Ein Beitrag zur Gönnerforschung des Spätmittelalters (Hermaea 68)*. Tübingen, 1992.

BARTSCH 1883

BARTSCH, Karl. "Lecküchner, Hans". *Allgemeine Deutsche Biographie (ADB)* **18**: 108. Leipzig, 1883.

BARTSCH 1887

———. *Die altdeutschen Handschriften der Universitäts-Bib-liothek in Heidelberg (Katalog der Handschriften der Universitäts-Bibliothek in Heidelberg I)*. Heidelberg, 1887.

BÄTZ 2002

BÄTZ, Klaus. "Aus der 1000-jährigen Geschichte Herzogenaurachs". *Stadtbuch Herzogenaurach 1002–2002*. Herzogenaurach, 2002.

BAUFELD 1996

BAUFELD, Christa. *Kleines frühneuhochdeutsches Wörterbuch*. Tübingen: Max Niemeyer Verlag, 1996.

BECKER 2020

BECKER, Paul. "Hans Talhoffer: Insights into the life of a fencing master in the 15th century". *Alte Armatur und Ringkunst: The Royal Danish Library Ms. Thott 290 2°*. Ed. Michael CHIDESTER. Somerville, MA: HEMA Bookshelf, 2020. pp 47–68.

BENGSON AND MOFFETT 2011

Knowing How: Essays on Knowledge, Mind, and Action. Ed. John BENGSON and Marc A. MOFFETT. Oxford: Oxford University Press, 2011.

BERM 1997

BERM, Heinrich. "Erlebnisraum Insel Schütt: hier standen früher Wildbad und Fechthaus". *Nürnberger Altstadtberichte* **22**: 37–62. 1997.

BLANCHARD 1986

BLANCHARD, Ian. "The Continental European Cattle Trades, 1400–1600". *The Economic History Review* **39**(3): 427–460. 1986. Doi: 10.2307/2596351

BOWMAN 2020

BOWMAN, Paul. *Deconstructing Martial Arts*. Cardif: Cardiff University Press, 2020.

BURKART 2016

BURKART, Eric. "The Autograph of an Erudite Martial Artist: A Close Reading of Nuremberg, Germanisches Nationalmuseum, Hs. 3227a". *Late Medieval and Early Modern Fight Books*. Brill, 2016.

BURGER 1961

Nürnberger Totengeläutbücher I: St. Sebald (1439–1517) (Freie Schriftenfolge der Gesellschaft für Familienforschung in Franken 13). Ed. Helene BURGER. Neustadt/Aisch, 1961.

CAPPELLI 1973

CAPPELLI, Adriano. *Dizionario di Abbreviature latine ed italiane, Sesta edizione*. Milano: Ulrico Hoepli Editore, 1973.

CARRUTHERS AND ZIOLKOWSKI 2004

The Medieval Craft of Memory. Ed. Mary CARRUTHERS and Jan M. ZIOLKOWSKI. Philadelphia: University of Pennsylvania Press, 2004.

CHIDESTER 2016

The Flower of Battle of Master Fiore Friulano de'i Liberi. Ed. Michael CHIDESTER. Wiktenauer, 2016.

CHRISMAN 1995

CHRISMAN, Miriam Usher. *Conflicting Visions of Reform: German Lay Propaganda Pamphlets, 1519–1530*. Brill, 1995.

CLASSEN 2007

CLASSEN, Albrecht. *Erotic Tales of Medieval Germany*. Tempe, Arizona: Arizona Center for Medieval and Renaissance Studies, 2007.

CLEMENTS 2016

CLEMENTS, John. "Problems of Interpretation and Application in Fight Book Studies". *Late Medieval and Early Modern Fight Books*: 188–215. Ed. Daniel JAQUET, Karen VERELST, and Timothy DAWSON. Leiden: Brill, 2016.

DALBY 1965

DALBY, David. *Lexicon of the German Medieval Hunt: A Lexicon of Middle High German terms (1050–1500), associated with the Chase, Hunting with Bows, Falconry, Trapping and Fowling*. Berlin: Walter De Gruyter & Co, 1965.

VAN DIJK 2019

DIJK, Casper J. van. "'For Pleasure, or Until Death': Late Medieval Fight Books Contextualized (1350–1550)". Leiden University, 2019.

VAN DIJK 2020

———. "Feeling a Bronze Spear: A Phenomenological Approach for Understanding the Combative Use of Bronze Spears (1800–600 BC) during Formalised Fights" (Unpublished Master's Thesis). Leiden, 2020.

DÖRNHÖFFER 1910

DÖRNHÖFFER, Friedrich. *Albrecht Dürers Fechtbuch*. Wien and Lepzig: F. Tempsky and G. Freytag, 1910.

DUPUIS 2013

DUPUIS, Olivier. "Des couteaux à clous ou pourquoi l'épée seule est si peu représentée dans les jeux d'épées et livres de combat au Moyen Âge". *L'art chevaleresque du combat*. Ed. Daniel JACQUET. Neuchâtel: éditions ALPHIL-Presses universitaires suisses, 2013.

DUPUIS 2015

———. "The Roots of Fencing from the Twelfth to the Fourteenth Centuries in the French Language Area". *Acta Periodica Duellatorum* **3**(38): 37–62. 2015.

EBERT-WOLF 1963/64

EBERT-WOLF, Marianne. *Geschichte des Nürnberger Lebkuchens vom Handwerk zur Industrie, in: Mitteilungen des Vereins für Geschichte der Stadt Nürnberg* **52**: 491–531. 1963/64. pp.

VON ELGGER 1873

VON ELGGER, Carl. *Kriegswesen und Kriegskunst der schweizerischen Eidgenossen im XIV., XV. und XVI Jahrhundert*. Lucerne: Militärisches Verlagsbureau, 1873.

ERLER 1895

ERLER, Georg. *Die Matrikel der Universität Leipzig, Bd. 1: Die Immatrikulationen von 1409–1559 (Codex Diplomaticus Saxoniae Regiae, II. Hauptteil Bd. XVI)*. Leipzig, 1895.

ERLER 1897

———. *Die Matrikel der Universität Leipzig, Bd. 2: Promo-tionen (Codex Diplomaticus Saxoniae Regiae, II. Hauptteil Bd. XVII)*. Leipzig, 1897.

EYRE 2021

EYRE, S. R. "The curving Plough-strip and its Historical Implications". *The Agricultural History Review* **3**(2): 80–94. 1955. *British Agricultural Historical Society*. http://www.bahs.org.uk/AGHR/ARTICLES/03n2a2.pdf. Accessed 2 March 2021.

FLEISCHMANN 2008

FLEISCHMANN, Peter. *Rat und Patriziat in Nürnberg Die Herrschaft der Ratsgeschlechter in der Reichsstadt Nürnberg vom 13. bis zum 18. Jahrhundert. Neustadt an der Aisch. Nürnberger Forschungen* **31**. 2008.

FLEISCHMANN 2012

———. *Nürnberg im 15. Jahrhundert (Das bayerische Jahrtausend)*. München, 2012.

FORGENG 2006

FORGENG, Jeffrey L. *The Art of Combat. A German Martial Arts Treatise of 1570.* London: Greenhill books, 2006.

FORGENG 2015

———. *The Art of Swordsmanship by Hans Leckuchner.* Woodbridge: Boydell & Brewer, 2015.

FORGENG 2017

———. *The Art of Sword Combat. A 1568 German Treatise on Swordsmanship.* London: Frontline Books, 2017.

FORGENG 2018

———. *The Medieval Art of Swordsmanship.* Leeds, 2018.

FORGENG 2021

———. *Das Tower-Fechtbuch. Ein Meisterwerk der mit-telalterlichen Kampfkunst.* Darmstadt, 2021.

FROMMBERGER-WEBER 1973

FROMMBERGER-WEBER, Ulrike. "Spätgotische Buchmalerei in den Städten Speyer, Worms und Heidelberg (1440–1510). Ein Beitrag zur Malerei des nördlichen Oberrheingebietes im ausgehenden Mittelalter". *Zeitschrift für Geschichte des Oberrheins ZGO* **121**: 35–145. 1973.

DI FURIA 2018

DI FURIA, Arthur J. "Genre: Audience, Origins, and Definitions". *Genre Imagery in Early Modern Northern Europe.* Routledge: Abingdon-on-Thames, 2018.

GEORGES 1909

GEORGES, Karl Ernst. *Ausführliches Lateinisches Wörterbuch, unveränderter Nachdruck der 8. bessserten und vermehrten Auflage von Heinrich Georges.* Hannover and Leipzig: Wissenschaftliche Buchgesellschaft Darmstadt, 1909.

GILBERT 1995

GILBERT, Bennett. *The Art of the Woodcut in the Italian Renaissance Book.* New York: Grolier Club, 1995.

GRIEB 2007

Nürnberger Künstlerlexikon. Bildende Künstler, Kunsthandwerker, Gelehrte, Sammler, Kulturschaffende und Mäzene vom 12. bis zur Mitte des 20. Jahrhunderts: 580. Ed. Manfred H. GRIEB. München, 2007.

GRIMM and GRIMM 2004

GRIMM, Jacob and Wilhelm GRIMM. *Deutsches Wörterbuch, Elektronische Ausgabe, 1. Auflage.* 2004.

HAGEDORN 2016

HAGEDORN, Dierk. "German Fechtbücher from the Middle Ages to the Renaissance". *Late Medieval and Early Modern Fight Books*: 188–215. Ed. Daniel JAQUET, Karen VERELST, and Timothy DAWSON. Leiden: Brill, 2016.

HAGEDORN 2017

———. *Jude Lew Das Fechtbuch*. VS-Books Torsten Verhuelsdonk, 2017.

HANSEN 2008

HANSEN, Cordula. "Experiment and Experience – Practice in a Collaborative Environment". *Experiencing Archaeology by Experiment. Proceedings of the Experimental Archaeology Conference, Exeter.* 69–80. Ed. Penny CUNNINGHAM, Julia HEEB, and Roeland PAARDEKOPEN. Oxford: Oxbow Book, 2008.

HATT 1929

HATT, Jean-Jacques. *Une ville du XVe siècle, Strasbourg.* Strasbourg: 1929.

HEGEL 1872

"Chronik des Heinrich Deichsler". *Die Chroniken der fränkischen Städte. Nürnberg Bd. 4 (Die Chroniken der deutschen Städte vom 14. bis ins 16. Jahrhundert Bd. 10)* (2nd edition). Ed. G. HEGEL. Göttingen, 1872 (ND 1961).

HENNIG 2001

HENNIG, Beate. *Kleines Mittelhochdeutsches Wörterbuch, 4., verbesserte Auflage.* Tübingen: Max Niemeyer Verlag, 2001.

HENNIG 2014

———. *Kleines Mittelhochdeutsches Woerterbuch.* Walter De Gruyer GmbH, 2014.

HILS 1985A

HILS, Hans-Peter. *Meister Johann Liechtenauers Kunst des langen Schwertes.* Frankfurt am Main: Peter Lang, 1985.

HILS 1985B

———. "Lecküchner, Hans". *Verfasserlexikon* **5**: col. 641-644. 1985.

ISRAEL AND JASER 2016

ISRAEL, Uwe and Christian JASER. *Agon und Distinktion: Soziale Räume des Zweikampfs zwischen Mittelalter und Neuzeit.* Lit Verlag, 2016.

JAQUET, WERELST, and DAWSON 2016

JAQUET, Daniel, Karen VERELST, and Timothy DAWSON. *Late Medieval and Early Modern Fight Books: Transmission and Tradition of Martial Arts in Europe.* Leiden: Brill, 2016.

JUDKINS 2016

JUDKINS, Benjamin N. "The Seven Forms of Lightsaber Combat: Hyper-Reality and the Invention of the Martial Arts". *Martial Arts Studies* **1**(2): 6. 2016.

KELLET 2015

KELLET, Rachel E. "'...Vnnd Schüß Im Vnder Dem Schwert Den Ort Lang Ein Zů Der Brust': The Placement and Consequences of Sword-

Blows in Sigmund Ringeck's Fifteenth- Century Fencing Manual".
Wounds and Wound Repair in Medieval Culture: 128–49. 2015.

KELLET 2016

———. "Only a Flesh-Wound? The Literary Background to Medieval
German Fight Books". *Late Medieval and Early Modern Fight Books*:
62–87. Ed. Daniel JAQUET, Karen VERELST, and Timothy DAWSON.
Leiden: Brill, 2016.

KIEFHABER 1793

KIEFHABER, Johann Carl Sigmund. *Der Beylagen der Materialien zur
Nürnbergischen Geschichte erste Sammlung*. Nürnberg(?), 1793.

KIST 1965

KIST, Johannes. *Die Matrikel der Geistlichkeit des Bistums Bamberg
1400–1556 (Veröffentlichungen der Gesellschaft für fränkische Geschichte IV/7)*. Würzburg, 1965.

KIST 1936

———. "Die Ordinanden des Bistums Bamberg von 1436 bis 1470". *Archiv für Sippenforschung* **13**: 101–111, 136–142, 177–178, 208–209, 243–
249, 277–280, 313–317, 341–345, 368–370. 1936.

KRAUS 1950

KRAUS, J.A., "Unbekannte Klerikerstatuten". *Freiburger Diözesan-Ar-
chiv* **70**: 154–78. 1950.

KNEFELKAMP 1989

KNEFELKAMP, Ulrich. *Stiftungen und Haushaltsführung im Heilig-
Geist-Spital in Nürnberg, 14.–17. Jahrhundert*. Selbstverlag, 1989.

LACHMAN 1841

Ulrich von Lichtenstein. Ed. Karl LACHMAN. Berlin, 1841.

LANDOIS 2014

LANDOIS, Antonia. *Gelehrtentum und Patrizierstand: Wirkungs-kreise
des Nürnberger Humanisten Sixtus Tucher (1449–1507) (Spätmittelalter, Humanismus, Reformation 77)*. Tübingen, 2014.

LEISKE 2018

LEISKE, Patrick. "Höfisches Spiel und tödlicher Ernst. Das Bloßfechten
mit dem langen Schwert in den deutschsprachigen Fechtbüchern des
späten Mittelalters und der frühen Neuzeit Ostfildern". *Mediaevistik*
31(1). 2018.

LENG 2008

LENG, Rainer. *Katalog der deutschsprachigen illustrierten Handschriften des Mittelalters (KdiH), begonnen von Hella Frühmorgen-Voss,
fortgeführt von Norbert H. Ott zusammen mit Ulrike Bodemann und
Christine Stöllinger-Löser, Bd. 4/2 (Lfg. 1/2) (Fecht- und Ringbücher)*.
München, 2008.

LENG 2017

———. "Text und Bild in deutschsprachigen illustrierten

Fechthandschriften des Mittelalters". *Die Kunst des Fechtens (Interdisziplinäre Beiträge zu Mittelalter und Früher Neuzeit 7)*. Ed. Elisabeth VARVRA, Matthias Johannes BAUER. Heidelberg, 2017. pp 211–234.

LEONI 2012

LEONI, Tom. *Fiore de' Liberi Fior Di Battaglia*. Freelance Academy Press, 2012.

LEXER 1885

LEXER, Matthias. *Mittelhoch-deutsches Taschenwörterbuch, 3. Auflage*. Stuttgart: Hirzel Verlag, 1885.

LEXER 1992

———. *Mittelhochdeutsches Taschenwoerterbuch*. Stuttgart: S. Hirzel Verlag, 1992.

LINDNER 1940

LINDNER, Kurt. *Geschichte des deutschen Weidwerks: Die Jagd im frühen Mittelalter*. Berlin: Walter de Gruyter & Co., 1940.

LOCHNER 1860

Dr. LOCHNER: "Zur Geschichte der Fechtschulen in Nürnberg". *Beilage zum Anzeiger für Kunde der deutschen Vorzeit NF 7*. col. 407–408. 1860.

MASSMANN 1844

MASSMANN. "Ueber historische Fechtbücher". *Serapeum: Zeitschrift für Bibliothekwissenschaft, Handschriftenkunde und ältere Litteratur* **5**: 52. 1844.

DE MACHAUT 1994

DE MACHAUT, Guillaume. *The Tale of the Alerion*. Trans. Minette GAUDET and Constance B. HIEATT. Toronto: University of Toronto Press, 1994.

MATSUZAKA 2007

MATSUZAKA, Yoshiya, Nathalie PICARD, and Peter L. STRICK. "Skill Representation in the Primary Motor Cortex after Long-Term Practice". *Journal of Neurophysiology* **97**(2): 1819–32. 2007

MCNEALY and GEISBERG 2019

MCNEALY, Marion and Max GEISBERG. *Landsknechts on Campaign*. Kennewisk, WA: Nadel and Faden Press, 2019.

MEYER 1949

MEYER, Otto. "Uraha Sacra. Vom Geist der Frömmigkeit im spätmittel-alterlichen Herzogenaurach". *Herzogenaurach. Ein Heimatbuch*. Ed. Valentin FRÖHLICH. Herzogenaurach, 1949.

MEYER 1981

———. "Eine Herzogenauracher Pfarrhausordnung". *Varia Franconiae historica: Aufsätze, Studien, Vorträge zur Geschichte Frankens*

(Mainfränkische Studien) **24**(2): 532–558. Ed. Dieter WEBER, Gerd ZIMMERMANN. Würzburg, 1981.

MILLER and ZIMMERMANN 2007

MILLER, Matthias and Karin ZIMMERMANN. *Die Codices Palatini germanici in der Universitätsbibliothek Heidelberg (Cod. Pal. germ. 304–495) (Kataloge der Universitätsbibliothek Heidelberg VIII).* Wiesbaden, 2007.

MITTLER and WERNER 1986

MITTLER, Elmar and Wilfried WERNER. *Mit der Zeit. Die Kurfürsten von der Pfalz und die Heidelberger Handschriften der Bibliotheca Palatina.* Wiesbaden, 1986.

MOXEY 1989

MOXEY, Keith. *Peasants, Warriors, and Wives: Popular Imagery in the Reformation.* Chicago: University of Chicago Press Books, 1989.

MÜLLER 1992

MÜLLER, Jan-Dirk. "Bild – Verse – Proskommentar am Beispiel con Fechtbüchern. Probleme der Verschriftlichung einer schriftlosen Praxis." *Pragmatische Schriftlichkeit im Mitterlalter: Erscheinungsformen und Entwicklungsstufen* **65**: 251–282. 1992.

MÜLLER 1994

———. "Hans Lecküchners Messerfechtlehre und die Tradition. Schriftliche Anweisungen für eine praktische Disziplin". *Wissen für den Hof. Der spätmittelalterliche Verschriftlichungsprozeß am Beispiel Heidelberg im 15. Jahrhundert (Münstersche Mittelalter-Schriften 67)*: 355–384. Ed. Jan-Dirk MÜLLER. München, 1994.

MÜLLNER 1972

MÜLLNER, Johannes. *Die Annalen der Reichsstadt Nürnberg von 1623. Teil II: Von 1351–1469.* Nürnberg, 1972.

NAGEL and WOOD 2005

NAGEL, Alexander and Christopher S. WOOD. "Interventions: Toward a New Model of Renaissance Anachronism". *Art Bulletin* **87**(3). 2005.

OAKESHOTT 1980

OAKESHOTT, Ewart. *European Weapons and Armor: From the Renaissance to the Industrial Revolution.* London: Lutterworth Press, 1980.

OMAN 1885

OMAN, Charles W.C. *The Art of War in the Middle Ages A.D. 378–1515.* London: B.H. Blackwell, 1885.

POLANYI 1958

POLANYI, Michael. *Personal Knowledge. Towards a Post-Critical Philosophy.* London: Routledge, 1958.

RANKE 1951

RANKE, Friedrich. "Zum Wortschatz der österreichischen

Umgangsprache um 1400". *Beiträge zur Sprachwissenschaft und Volkskunde: Festschrift für Ernst Ochs zum 60 Geburtstag*. Ed. Karl Friedrich MÜLLER. Lahr: Moritz Schauenburg Verlagsbuchhandlung, 1951. pp. 180–89.

RASPE 1905
RASPE, Theodor. *Die Nurnberger Miniaturmalerei bis 1515*. Strassburg: J. H. E. Heitz, 1905.

REED 2010
REED, Nick, Peter MCLEOD, and Zoltan DIENES. "Implicit Knowledge and Motor Skill: What People Who Know How to Catch Don't Know". *Consciousness and Cognition* **19**(1): 63–76. 2010.

RUBLACK and HAYWARD 2015
RUBLACK, Ulinka, and Maria HAYWARD. *The First Book of Fashion: The Books of Clothes of Matthäus & Veit Konrad Schwarz of Augsburg*. Bloomsbury, 2015.

SANDER 1902
SANDER, Paul. *Die reichsstädtische Haushaltung Nürnbergs. Dargestellt auf Grund ihres Zustandes von 1431 bis 1440* (2 vols.). Leipzig, 1902.

SAYCE 1967
SAYCE, Olive. *Poets of the Minnesang*. Oxford University Press, 1967.

SCHEFFLER-ERHARD 1959
SCHEFFLER-ERHARD, Charlotte. *Alt-Nürnberger Namenbuch (Nürnberger Forschungen 5)*. Nürnberg, 1959.

SCHMELLER 1837
SCHMELLER, Johann Andreas. *Bayerisches Wörterbuch*. 1837.

SCHMITT 1985
Prêcher d'exemples, récits de prédicateurs du Moyen Âge. Ed. Jean-Claude SCHMITT. Paris: Stock, 1985.

SCHNEIDER 1978
SCHNEIDER, Karin. *Die deutschen Handschriften der Bayerischen Staatsbibliothek München. Cgm 501–690 (Catalogus codicum manu scriptorum Bibliothecae Monacensis 4)*. Wiesbaden: Otto Harrassowitz, 1978.

SETON-WATSON 1901
SETON-WATSON, R. W. *Maximilian I: Holy Roman Emperor*. Stanhope Historical Essay, 1901.

SIEBENKEES 1794
SIEBENKEES, Johann Christian: "Von den ehemahligen Fechtschulen in Nürnberg". *Materialien zur Nürnbergischen Geschichte* **3**(14): 65–75. Nürnberg, 1794.

SODER VON GÜLDENSTUBBE 1978
SODER VON GÜLDENSTUBBE, Erik. "Die katholischen Pfarrer von

Herzo-genaurach". *Heimatbuch der Stadt Herzogenaurach*: 117–154. Bamberg, 1978.

STONE 1999

STONE, George Cameron. *A Glossary of the Construction, Decoration and Use of Arms and Armor: in All Times*. Dover Publications, 1999.

TERJANIAN 2019

The Last Knight: The Art, Armor, and Ambition of Maximilian I. Ed. Pierre TERJANIAN. New York: Metropolitan Museum of Art, 2019.

TLUSTY 2001

TLUSTY, B. Ann. *Bacchus and Civic Order: The Culture of Drink in Early Modern Germany*. University of Virginia Press, 2001.

TLUSTY 2011

———. *The Martial Ethic in Early Modern Germany. Civic Duty and the Right of Arms*. New York: Palgrave Macmillan, 2011.

TLUSTY 2016

———. "Martial Identity and the Culture of the Sword in Early Modern Germany". *Late Medieval and Early Modern Fight Books*: 547–570. Ed. Daniel JAQUET, Karen VERELST, and Timothy DAWSON. Leiden: Brill, 2016.

THIEBAUX 1974

THIEBAUX, Marcelle. *The Stag of Love: The Chase in Medieval Literature*. Cornell University Press, 1974.

TOBLER 2010

TOBLER, Christian H. *In St. George's Name*. Freelance Academy Press, 2010.

TOEPKE 1884

TOEPKE, Gustav. *Die Matrikel der Universität Heidelberg von 1386–1662, Bd. 1*. Heidelberg, 1884.

VOGEL 1998

VOGEL, Thomas. *Fehderecht und Fehdepraxis im Spätmittelalter am Beispiel der Reichsstadt Nürnberg (1404–1438) (Freiburger Beiträge zur mittelalterlichen Geschichte 11)*. Frankfurt am Main, u.a. 1998.

WALDBURG WOLFEGG 1998

WALDBURG WOLFEGG, Christoph Graf zu. *Venus and Mars: The World of the Medieval Hausbuch*. Prestel Publishing, 1998.

WASSMANNSDORFF 1870

WASSMANNSDORFF, Karl. *Sechs Fechtschulen*. Heidelberg: Karl Groos, 1870.

WASSMANNSDORFF 1888

———. "Nachtrag über das sog. Lebkhommer'sche Fechtbuch und dessen Verhältnis zu der Breslauer Dürer Handschrift von 1512". *Monatsschrift für das Turnwesen, mit besonderer Berück-sichtigung des Schulturnens und der Gesundheitspflege* **7**: 138–145. Berlin, 1888.

WEISS 1980

WEISS, Hildegard. *Lebenshaltung und Vermögensbildung des mittleren Bürgertums.* München: Beck Verlag, 1980.

WELLE 2014

WELLE, Rainer. *"...vnd mit der rechten fauſt ein mordſtuck".* Baumanns Fecht- und Ringkampfhandschrift, 2 vols.* München, 2014.

WELLE 2021

———. *Albrecht Dürer und seine Kunſt des Zweikampfes. Auf den Spuren der Handschrift 26232 in der Albertina Wien.* Kumberg, Austria: Sublilium Schaffer, 2021.

WETZLER 2015

WETZLER, Sixt. "Martial Arts Studies as Kulturwissenschaft. A Possible Theoretical Framework". *Martial Arts Studies* **1**: 20–33. 2015.

WIERSCHIN 1965

WIERSCHIN, Martin. *Meiſter Johann Liechtenauers Kunſt des Fechtens.* München: C. H. Beck, 1965.

WILL 1805

WILL, Georg Andreas. *Nürnbergisches Gelehrten-Lexicon, fortgeſetzt von Chriſtian Conrad Nopitsch 6 (2 Supplementband).* Altdorf, 1805.

WINKLER 1956

WINKLER, Johann. *Der Güterbeſitz der Nürnberger Kirchenſtiftungen unter der Verwaltung des Landalmosenamtes im 16. Jahrhundert, in: Mitteilungen des Vereins für Geschichte der Stadt Nürnberg* **47**: 160–296. 1956.

VON ZINGERLE 1909

Mittelalterliche Inventare aus Tirol und Vorarlberg. Mit Sacherklärungen. Ed. Oswald VON ZINGERLE. Innsbruck, 1909.

ZMORA 1997

ZMORA, Hillay. *State and nobility in early modern Germany. The knightly feud in Franconia 1440–1567.* Cambridge, 1997 (2nd edition 2002).

Websites

ACUTT 2014

ACUTT, Jamie. "German Commonplace or Cuſtomary fencing (Critical edition)". *Academia. edu.* 2014. Accessed 4th July 2021.
http://www.academia.edu/12496348/German_Commonplace_or_Customary_fencing_Critical_edition

AMBERGER 2003

AMBERGER, Chriſtoph. "Behind the woodshed: Little-known aſpeĉts of Dussack play through the ages". *Journal of Weſtern Martial Art.*

2003. Accessed 4th July 2021.

http://ejmas.com/jwma/articles/2003/jwmaart_amberger_0503.htm

Cᴏᴛᴛᴇʀ-Rᴇɪʟʟʏ 2018

COTTER-REILLY, Keith. "The Tessack of Norway". *Atlanta Historical Fencing Academy.* Accessed 4th July 2021.

http://fencingatl.com/blog/2018/8/9/the-tessak-of-norway

Dᴜᴘᴜɪs 2020

DUPUIS, Olivier. "When Fencers and Wrestlers were the Children of the Sun". *Martial Culture in Medieval Town.* Accessed 4th July 2021.

http://martcult.hypotheses.org/780

Kʟᴇɪɴᴀᴜ 2011

KLEINAU, Jens-Peter. "Paulus Kal, a Schirrmeister". *Hans Talhoffer. A Historical Martial Arts blog by Jens P. Kleinau.* Accessed 4th July 2021.

http://talhoffer.wordpress.com/2011/07/03/paulus-kal-a-schirrmeister/

Kʟᴇɪɴᴀᴜ 2012

———. "1444 Two fencing masters in Rothenburg". *Hans Talhoffer. A Historical Martial Arts blog by Jens P. Kleinau.* Accessed 4th July 2021.

http://talhoffer.wordpress.com/2012/12/03/1444-two-fencing-masters-in-rothenburg/

Lᴏʀʙᴇᴇʀ 2006

LORBEER, Carston, et al. "Das ist herr hannsen Lecküchner von Nurenberg künst vnd zedel ym messer". *Gesellschaft für pragmatische Schriftlichkeit.* 2006. Accessed 4th July 2021.

http://www.pragmatische-schriftlichkeit.de/cgm582.html.

ᴛᴇʀ Mᴏʀs 2021

TER MORS, Oskar. "Longsword and Messer Terminology". *YouTube.com.* 2021. Accessed 4th July 2021.

http://youtu.be/8ikzswtMm3I

Nᴏʀʟɪɴɢ A

NORLING, Roger. "The Dussack, A Weapon of War". *HROARR.* Accessed 4th July 2021.

http://hroarr.com/article/the-dussack/

Nᴏʀʟɪɴɢ B

———. "The Rose and the Pentagram". *HROARR.* Accessed 4th July 2021.

http://hroarr.com/article/the-rose-and-the-pentagram/

Tᴏʙʟᴇʀ 2011

TOBLER, Christian H. "An Anonymous Dueling Shield Treatise." *Freelance Academy Press.* July 2011. Accessed 4th July 2021.

http://www.freelanceacademypress.com/duelingshield.aspx

Yale 2006

Yale University. "On Visual Exegesis". *The Speculum Theologiae in*

Beinecke MS 416. 2006. Accessed 4th July 2021.

http://brbl-archive.library.yale.edu/exhibitions/speculum/index.html

ŻABIŃSKI 2012

ŻABIŃSKI, Grzegorz, Russell A. MITCHELL, Falko FRITZ. "A Falchion / Langes Messer Fencing Treatise by Johannes Lecküchner (1482)". *Hammaborg–Historischer Schwertkampf.* Accessed 4th July 2021.

http://www.hammaborg.de/wp-content/uploads/2020/01/zabiski_mitchell_fritz_leckuchner.pdf

"Hans Talhoffer". *Wiktenauer.* Ed. Michael CHIDESTER. Accessed 4th July 2021.

http://wiktenauer.com/wiki/Hans_Talhoffer

"Johannes Lecküchner". *Wiktenauer.* Ed. Michael CHIDESTER. Accessed 4th July 2021.

http://wiktenauer.com/wiki/Johannes_Leck%C3%BCchner

"Nuremberg Group". *Wiktenauer.* Ed. Michael CHIDESTER. Accessed 4th July 2021.

http://wiktenauer.com/wiki/Nuremberg_Group

Contributors

DANIEL BURGER was born in Weißenburg in Bavaria in 1971. He studied history and German language and literature at the Catholic University of Eichstätt and received his doctorate (Dr.phil.) in 1999. After a scientific traineeship at the Germanisches Nationalmuseum in Nuremberg, he graduated from the Bavarian School of Archives in Munich in 2003-2005 and became an archivist; since 2009, he has worked at the Nuremberg State Archives. His academic focus is Bavarian and Franconian regional history, architectural history (especially on castles and fortresses), and archival history. For several years he has been a member of the HEMA club Schwertbund Nurmberg e.V., training in long sword, Messer, dagger, and ḫarnіѕϛϳ ϝeϛϳten.

MICHAEL CHIDESTER is the Editor-in-Chief of Wiktenauer and owner of HEMA Bookshelf. He is a Research Scholar of the Meyer Freifechter Guild, a founding member of the Society for Historical European Martial Arts Studies (SHEMAS), a member of the Western Martial Arts Coalition (WMAC), and a Lifetime Member of the HEMA Alliance. He has lectured on historical martial arts across North America and Europe and authored several books, including *The Illustrated Meyer*, *The Long Sword Gloss of GNM Manuscript 3227a*, and *The Flower of Battle: MS M 383*.

CASPER J. VAN DIJK studied medieval archaeology and history with a specialization in weapon use at Leiden University, the Netherlands. Currently, he works as a junior curator of late medieval material culture at the Fries Museum in Leeuwarden, the Netherlands. Moreover, he is the chairman of the Dutch national Historical European Martial Arts Federation. Previously he published a new halberd typology as well as on medieval cannons, and acted as an advisor for archaeological experiments with bronze age weapons.

OLIVIER DUPUIS is a teacher in HEMA since 2001 and an independent researcher. Beside working on critical editions of historical treatises, he studies the practice of fencing and the organizations of fencers during the Medieval and Renaissance times in France and Germany.

JESSICA FINLEY has had a sword in her hands for the past twenty years. Her initial interest was in stage combat, but not too long after beginning that pursuit began to ask "...but how did they really fight?", and from that question branched out to German Medieval Martial Arts. She is the head instructor at Ritterkunst Turhalle in Lawrence, Kansas. She has taught and competed internationally at events including the Western Martial Arts

Workshop, Paddy Crean International Workshop, Swordfish, and Long-point, and has taught intensives at various events and schools as well as weekend private intensives at her home. She also has a background study-ing Judo under the tutelage of Arden Cowherd of Topeka Judo Club. Her previous titles include *Medieval Wrestling: Modern Practice of a Fifteenth Century Art*, which is an interpretation of Master Ott's wrestling treatise of German wrestling techniques. She also researches medieval clothing con-struction and fabric armor and has presented her findings at the Interna-tional Congress on Medieval Studies at Kalamazoo and published an article in *Medieval Clothing and Textiles* (Boydell and Brewer) on her study of a 15th century quilted armor. Jessica currently lives in Lawrence, Kansas with her family. In between "kid-wrangling" her children and remodeling her house, she writes books and does her assigned physical therapy exercises.

ADAM FRANTI is a Fechter and Research Scholar in the Meyer Freifechter Guild, and earned his master's degree in history in 2018. He is a founder of the Midwest Historical Fencing league as well as the Lansing Longsword Guild in Lansing, Michigan. He has taught or lectured at Longpoint, the Meyer Symposium, the Western Martial Arts Workshop, and Icebreaker, with his research focusing on the cultural background of the mid-16th cen-tury Holy Roman Empire. He currently works as an Adjunct Professor of History at Grand Rapids Community College.

FALKO FRITZ (*1976) lives in Hamburg, Germany and started historical fencing in 2007. He became a member of Hammaborg–Historischer Schwertkampf in 2008 and was promoted to a trainer and member of the steering committee soon after. In addition to the long sword his main inter-est is the langes Messer and polearms. In 2010, he published the first partial translation of Lecküchner's treatise on Hammaborg's website and became co-author of Grzegorz ŻABIŃ-SKI's full translation in 2012. From 2014 to 2018 he was vice-president of the International Federation of Historical Eu-ropean Martial Arts (IFHEMA) and he frequently visits national and inter-national HEMA events as a participant or workshop instructor. In addition to the sports aspect of HEMA, Falko has a strong interest in the social role of martial arts in the late medieval society and regularly incorporates the findings in his training.

DIERK HAGEDORN works for the IT subsidiary of a large German airline for numerous creative issues such as user experience, interface, and corpo-rate design. He is an instructor for the long sword at Hammaborg–Histor-ischer Schwertkampf, a club for historical fencing in Hamburg; he is also a passionate wearer of armor. He has transcribed 30 of the approximately 80 surviving fight books in the German language and made them available on

the website of his club or published them in book form. Among his numerous publications are editions about the manuscripts by Peter von Danzig, Lew the Jew, Hans Talhoffer, and Jörg Wilhalm.

OSKAR TER MORS is a medieval historian from Leiden, the Netherlands, specialized in dueling and urban culture. Currently working as a history teacher, he still occasionally finds some time off from work and HEMA-practice to do fencing related research. Although Oskar practices a large number of weapons and does so by working from a multitude of fight books, most of his work tends to focus on Lecküchner's *Kunst und Zedel ym Messer,* such as the full interpretation of the manuscript that was published on YouTube. Additionally, Oskar has worked on several archaeological experiments as an expert user.

I0541854

www.ingramcontent.com/pod-product-compliance
Lightning Source LLC
Chambersburg PA
CBHW040523150626
46551CB00032B/2338

* 9 7 8 1 9 5 3 6 8 3 1 7 5 *